Probability, Choice, and Reason

Probability, Choice, and Reason

Leighton Vaughan Williams

CRC Press
Taylor & Francis Group
Boca Raton London New York

CRC Press is an imprint of the
Taylor & Francis Group, an **informa** business
A CHAPMAN & HALL BOOK

First edition published 2022
by CRC Press
6000 Broken Sound Parkway NW, Suite 300, Boca Raton, FL 33487-2742

and by CRC Press
2 Park Square, Milton Park, Abingdon, Oxon, OX14 4RN

© 2022 Leighton Vaughan Williams

CRC Press is an imprint of Taylor & Francis Group, LLC

Library of Congress Cataloging-in-Publication Data
A catalog record has been requested for this book

ISBN: 978-0-367-53893-4 (hbk)
ISBN: 978-0-367-53891-0 (pbk)
ISBN: 978-1-003-08361-0 (ebk)

DOI: 10.1201/9781003083610

Typeset in Palatino
by Deanta Global Publishing Services, Chennai, India

For Mum and Dad, and my wife, Julie.

Contents

Preface.. xiii
Author Biography... xvii

1. Probability, Evidence, and Reason 1
  1.1   Bayes' Theorem: The Most Powerful Equation in the World 1
       1.1.1   Appendix 7
       1.1.2   Exercise... 9
       1.1.3   Reading and Links.. 11
  1.2   Bayes and the Taxi Problem.. 13
       1.2.1   Appendix 16
       1.2.2   Exercise... 17
       1.2.3   Reading and Links.. 18
  1.3   Bayes and the Beetle 19
       1.3.1   Appendix 20
       1.3.2   Exercise... 20
       1.3.3   Reading and Links.. 20
  1.4   Bayes and the False Positives Problem.................................... 20
       1.4.1   Examples 22
       1.4.2   Appendix 23
             1.4.2.1   Sensitivity and Specificity................... 23
             1.4.2.2   Vaccine Efficacy 24
       1.4.3   Exercise... 25
       1.4.4   Reading and Links.. 26
  1.5   Bayes and the Bobby Smith Problem 27
       1.5.1   Appendix 29
       1.5.2   Exercise... 30
       1.5.3   Reading and Links.. 31
  1.6   Bayes and the Broken Window 31
       1.6.1   Appendix 33
       1.6.2   Exercise... 33
  1.7   The Bayesian Detective Problem.. 34
       1.7.1   Epilogue 35
       1.7.2   Exercise... 35
  1.8   Bayesian Bus Problems.. 36
       1.8.1   Exercise... 37
  1.9   Bayes at the Theatre 37
       1.9.1   Appendix 39
       1.9.2   Exercise... 40
       1.9.3   Reading and Links.. 40

1.10   Bayes in the Courtroom 40
       1.10.1 Exercise...44
       1.10.2 Reading and Links...44

2. **Probability Paradoxes** 47
   2.1   The Bertrand's Box Paradox..47
         2.1.1 Exercise...48
         2.1.2 Reading and Links...49
   2.2   The Monty Hall Problem 49
         2.2.1 Appendix 52
                2.2.1.1   Alternative Derivation...54
         2.2.2 Exercise...55
         2.2.3 Reading and Links...55
   2.3   The Three Prisoners Problem..56
         2.3.1 Exercise...58
         2.3.2 Reading and Links...58
   2.4   The Deadly Doors Problem 59
         2.4.1 Exercise...60
         2.4.2 Reading and Links...60
   2.5   Portia's Challenge...61
         2.5.1 Exercise...62
         2.5.2 Reading and Links...62
   2.6   The Boy–Girl Paradox..62
         2.6.1 Appendix 67
         2.6.2 Exercise...68
         2.6.3 Reading and Links...68
   2.7   The Girl Named Florida Problem 68
         2.7.1 Appendix 71
         2.7.2 Exercise...72
         2.7.3 Reading and Links...72
   2.8   The Two Envelopes Problem 73
         2.8.1 Exercise...75
         2.8.2 Reading and Links...75
   2.9   The Birthday Problem 75
         2.9.1 Exercise...79
         2.9.2 Reading and Links...79
   2.10 The Inspection Paradox...80
         2.10.1 Exercise...82
         2.10.2 Reading and Links...82
   2.11 Berkson's Paradox 82
         2.11.2 Exercise...84
         2.11.3 Reading and Links...85
   2.12 Simpson's Paradox 85
         2.12.1 Exercise...87
         2.12.2 Reading and Links...88

2.13   The Will Rogers Phenomenon..88
     2.13.1   Exercise..89
     2.13.2   Reading and Links...89

3. Probability and Choice..91
  3.1   Newcomb's Paradox...91
     3.1.1   Exercise..93
     3.1.2   Reading and Links..93
  3.2   The Sleeping Beauty Problem 93
     3.2.1   Exercise..96
     3.2.2   Reading and Links..96
  3.3   The God's Coin Toss Problem...97
     3.3.1   Exercise..100
     3.3.2   Reading and Links..100
  3.4   The Doomsday Argument...100
     3.4.1   Exercise..102
     3.4.2   Reading and Links..103
  3.5   When Should You Stop Looking and Start Choosing?...............103
     3.5.1   Exercise..108
     3.5.2   Reading and Links..109
  3.6   Why Do We Always Seem to End Up in the Slower Lane? 110
     3.6.1   Exercise..111
     3.6.2   Reading and Links..111
  3.7   Pascal's Wager...111
     3.7.1   Exercise..113
     3.7.2   Reading and Links..113
  3.8   The Keynesian Beauty Contest 113
     3.8.1   Exercise..115
     3.8.2   Reading and Links..115
  3.9   Benford's Law 115
     3.9.1   Exercise..118
     3.9.2   Reading and Links..118
  3.10 Faking Randomness 119
     3.10.1 Exercise..120
     3.10.2 Reading and Links..120

4. Probability, Games, and Gambling 123
  4.1   The Chevalier's Dice Problem 123
     4.1.1   Exercise..128
     4.1.2   Reading and Links..129
  4.2   The Pascal–Fermat "Problem of Points" 130
     4.2.1   Appendix 131
     4.2.2   Exercise..132
     4.2.3   Reading and Links..132
  4.3   The Newton–Pepys Problem 132

	4.3.1	Exercise	138
	4.3.2	Reading and Links	138
4.4	Staking to Reach a Target Sum		138
	4.4.1	Exercise	139
	4.4.2	Reading and Links	139
4.5	The Favourite-Longshot Bias		140
	4.5.1	Appendix	143
	4.5.2	Exercise	146
	4.5.3	Reading and Links	147
4.6	The Poisson Distribution		148
	4.6.1	Exercise	150
	4.6.2	Reading and Links	150
4.7	Card Counting		150
	4.7.1	Exercise	151
	4.7.2	References and Links	151
4.8	Can the Martingale Betting System Guarantee a Profit?		152
	4.8.1	Appendix	154
	4.8.2	Exercise	155
	4.8.3	Reading and Links	155
4.9	How Much Should We Bet When We Have the Edge?		156
	4.9.1	Exercise	157
	4.9.2	Reading and Links	158
4.10	The Expected Value Paradox		158
	4.10.1	Exercise	161
	4.10.2	Reading and Links	161
4.11	Options, Spreads, and Wagers		162
	4.11.1	Appendix	166
	4.11.2	Exercise	167
		A. Buy Call Option	167
		B. Buy Put Option	168
		C. Sell Call Option	168
		D. Sell Put Option	169
	4.11.3	Reading and Links	170

5. Probability, Truth, and Reason			**171**
5.1	Does Seeing a Blue Tennis Shoe Increase the Likelihood That All Flamingos Are Pink?		171
	5.1.1	Exercise	173
	5.1.2	Reading and Links	173
5.2	The Simulated World Question		174
	5.2.1	Exercise	176
	5.2.2	Reading and Links	176
5.3	Quantum World Thought Experiments		176
	5.3.1	Exercise	179
	5.3.2	Reading and Links	179

5.4   The Fine-Tuned Universe Puzzle.. 179
    5.4.1   Exercise.. 185
    5.4.2   Reading and Links... 185
5.5   Occam's Razor 186
    5.5.1   Exercise.. 190
    5.5.2   Reading and Links... 190

6. Anomalies of Choice and Reason.. 191
6.1   Efficiency and Inefficiency of Markets..................................... 191
    6.1.1   Exercise.. 196
    6.1.2   Reading and Links... 196
6.2   Curious and Classic Market Anomalies 197
    6.2.1   Exercise.. 201
    6.2.2   Reading and Links... 202
6.3   Ketchup Anomalies, Financial Puzzles, and Prospect Theory 203
    6.3.1   Exercise.. 208
    6.3.2   Reading and Links... 208
6.4   The Wisdom of Crowds.. 209
    6.4.1   Exercise.. 213
    6.4.2   Reading and Links... 214
6.5   Superforecasting 215
    6.5.1   Exercise.. 218
    6.5.2   Reading and Links... 218
6.6   Anomalies of Taxation.. 219
    6.6.1   Exercise.. 220
    6.6.2   Reading and Links... 220

7. Game Theory, Probability, and Practice 223
7.1   Game Theory: Nash Equilibrium 223
    7.1.1   Exercise.. 229
    7.1.2   Reading and Links... 229
7.2   Game Theory: Repeated Game Strategies 230
    7.2.1   Exercise.. 231
    7.2.2   Reading and Links... 232
7.3   Game Theory: Mixed Strategies.. 232
    7.3.1   Appendix 234
    7.3.2   Exercise.. 235
    7.3.3   Reading and Links... 236

8. Further Ideas and Exercises.. 237
8.1   The Four Card Problem.. 237
    8.1.1   Exercise.. 238
    8.1.2   Reading and Links... 238
8.2   The Bell Boy Paradox.. 238
    8.2.1   Exercise.. 239

8.3   Can a Number of Infinite Length Be Represented by
a Line of Finite Length?... 239
       8.3.1   Exercise.. 239
8.4   Does the Sum of All Positive Numbers Really Add Up
to a Negative Number?... 239
       8.4.1   Reading and Links.. 241
8.5   Zeno's Paradox... 241
       8.5.1   Exercise.. 243
       8.5.2   Reading and Links.. 243
8.6   Cool Down Exercise... 243
       8.6.1   Exercise.. 243
       8.6.2   Reading and Links.. 243
Reading and References.. 244

Solutions to Exercises.. 253

Index... 289

Preface

This book is designed as an invaluable resource for those studying the sciences, social sciences, and humanities on a formal or informal basis, especially those with an interest in engaging with ideas rooted in chance and probability and in the theory and application of choice and reason. Notably, statistics and probability are topics that students often find difficult to get to grips with, and this book fills and makes accessible a significant gap in this area across a range of disciplines. These include economics, engineering, finance, law, marketing, mathematics, medicine, psychology, and many others. It will also appeal to those taking courses in probability and in statistics.

The target student audience includes university, college, and high school students who wish to expand their reading, as well as teachers and lecturers who want to liven up their courses while retaining academic rigour. More generally, the book is designed with the intelligent and enquiring layperson in mind, including anyone who wants to develop their skills to probe numbers and anyone who is interested in the many statistical and other paradoxes that permeate our lives.

The underpinning of the book is that much of our thinking on a range of subjects is flawed because we base much of our thinking on faulty intuition. The content is primarily about the tools and framework of logical thought that we can use to address and overcome these fundamental cognitive flaws.

By using the framework and tools of probability and statistics, we can overcome these barriers to provide solutions to many real-world problems and paradoxes. We show how to do this and find answers that are frequently very contrary to what we might expect. Along the way, we venture into diverse realms and thought experiments which challenge the way that most of us see the world, and we explore the big questions of choice and reason.

The tools of so-called Bayesian reasoning run through important sections of the book. The ideas extend well beyond a Bayesian framework, however, and include several topics and anomalies that are attractive on their own merits and from which we might learn broader lessons. The reader will also explore ideas, concepts, and applications rooted in game theory.

A recurring theme at the heart of this book, however, is the conflict between intuition and logic.

Imagine, for example, a bus that arrives every 30 minutes, on average, and you arrive at the bus stop at some random time, with no idea when the last bus left. How long can you expect to wait for the next bus to arrive? Half of 30 minutes, i.e. 15 minutes? Intuitively, that sounds right, but you'd be lucky to wait only 15 minutes. It's likely to be somewhat longer, and the laws of probability and statistics show why.

In medical trials, the success rate for a new drug is better than for an old drug on each of the first two days of the trials. The new drug must, therefore, have recorded a higher success rate than the old drug, judged over the entire two days of the trials. Sounds right, but it's not so. After the two days, the old drug turns out to be more successful than the new drug even though it performed worse on each of the first two days. Is this possible? It's like saying that a player performs better than another player in successive seasons but performs worse overall. Can that happen? Yes. It can and does.

How many restaurants should you look at before starting to decide on a place to eat? How many used cars should you pass on before you start looking seriously for one? How many potential partners should you consider before looking for the special one? In each case, we can derive the answer from a simple formula.

A doctor performs a test on all her patients for a virus. The test she gives them is 99% accurate, in the sense that 99% of people who have the virus test positive, and 99% of the healthy people test negative. Now the question is: If the patient tests positive, what is the chance the doctor should give to the patient having the virus? The intuitive answer is 99%, but that is likely to be a gross over-estimate of the true probability.

You meet a man at a sales convention, who mentions his two children, one of whom, you learn, is a boy. You never found out anything about the other child. What should be your best estimate of the probability that the other child is a girl? It's not a half, as you might think intuitively. If you had met the same man in different circumstances, accompanied by his son, now what is the probability that the man's other child is a girl? Isn't it the same as before? In fact, it's quite different. It does matter in estimating the chance of the other child being a girl that you bumped into the young boy instead of being told about him. It matters that his name was Barrington, and it would be slightly different if it were Bob.

You turn up to watch your local team play football. There are 22 players on the pitch, plus the referee. What's the chance that two or more of them share a birthday? Well, there are 365 days in the year and only 23 people on the pitch, so the chance is likely to be slim, you might think. In fact, it's more likely than not that at least two of those on the pitch share the same birthday. But the referee is unlikely to be one of them.

Can we improve our forecasts of football match outcomes by studying the rate of fatalities from horse kicks of Prussian cavalry officers? Yes, we can.

Can we devise a game where you auction a dollar and be pretty much guaranteed to turn a profit on the deal? There is a way to do this.

You have arranged to meet a stranger on a particular day for an important appointment, but you forgot to name the time and place, and neither of you have the contact details of the other. Where and when should you turn up?

You need to double your remaining money to pay off a pressing debt, and you decide to take to the casino tables. If you don't double up tonight, you are doomed to a dusty demise. What staking plan should you adopt to maximise your chances of survival? The answer may be surprising.

It's possible to win at Blackjack by counting cards, memorising what cards have already been dealt. You need a good memory for that, don't you? You don't.

Choose a number between 0 and 100. You win a prize if your number is equal or closest to two thirds of the average number chosen by all other participants. What number should you choose?

Select a newspaper or magazine with a lot of numbers about naturally occurring phenomena, such as the populations of different countries or the heights of mountains. Now circle the numbers. Would you expect a very big difference between the numbers starting with a 1, 2, 3, 4, 5, 6, 7, 8, or 9? Yes, you would. And that fact can help identify fraudsters.

As a prize for winning a competition, you're offered a chance to open a gold, silver or lead casket, in one of which the host has placed a cheque for £10,000. The others are empty. You choose the gold casket and the silver casket is opened. It is empty. You are generously offered a chance to swap to a different casket before the reveal. Should you take the offer? The solution is counter-intuitive.

The penalty-taker must decide which way to shoot. The goalkeeper must decide which way, and whether, to dive. How can they use game theory to maximise their chances of success? The answer involves thinking both inside and outside the box.

Can we profit on the stock market by waiting till Halloween or by investing on a cold, overcast day?

Are professional golfers more successful when putting for par than for birdie?

Does seeing a blue tennis shoe increase the likelihood that all flamingos are pink?

Do we live in a simulation or is this world the real thing?

How long can we expect humanity as we know it to survive?

We ask and seek to resolve these and many more questions involving probability, choice, and reason. Not least, we solve the greatest mystery of them all – why we always seem to end up in the slower lane.

Exercises, references, and links are provided for those wishing to cross-reference or to probe further, and many of the chapters contain a technical appendix. Solutions to the exercises are provided at the end of the book.

Author Biography

Leighton Vaughan Williams, BSc (Econ), PhD, FHEA, is Professor of Economics and Finance at Nottingham Business School, Nottingham Trent University, as well as Director of the Betting Research Unit and of the Political Forecasting Unit. He has researched and published extensively in the areas of probability, risk, and choice under uncertainty and given expert witness evidence before national and international courts of law and select committees of the House of Commons and House of Lords. He has served as a senior adviser to the UK government and teaches undergraduates and postgraduates how to apply Bayesian methods, and the tools of probability and statistics, to real-world problems and paradoxes.

1

Probability, Evidence, and Reason

This chapter introduces Bayes' Theorem, named in honour of the Reverend Thomas Bayes. Bayes' Theorem offers a way to update the probability of a hypothesis being true, given some new evidence, using a simple but very powerful mathematical equation. Bayesian updating is in this way a solution to the problem of how to combine pre-existing (prior) beliefs with new evidence. We also introduce the Bayes Factor, which is the ratio of the likelihood of one hypothesis to the likelihood of another. It is essentially a measure of which hypothesis better explains the world, given the evidence. We examine the Prosecutor's Fallacy and Laplace's Rule of Succession and show some applications of Bayesian reasoning. These include the classic taxi problem, the beetle problem and the false positives problem, the latter taking us into the realms of health and medicine. We also look at the application of Bayesian reasoning in the real-world courtroom. Stylised examples include the Bayesian detective, the Bobby Smith problem, and Bayes at the theatre.

1.1 Bayes' Theorem: The Most Powerful Equation in the World

How should we change our beliefs about the world when we encounter new data or information? A theorem bearing the name of Thomas Bayes, an eighteenth-century clergyman, is central to the way we should answer this question.

The original presentation of the Reverend Thomas Bayes' work, "An Essay toward Solving a Problem in the Doctrine of Chances", was given in 1763, after Bayes' death, to the Royal Society, by Bayes' friend and confidant, Richard Price.

In explaining Bayes' work, Price proposed, as a thought experiment, the example of a person who enters the world and sees the sun rise for the first time. Perhaps he has spent his entire life entombed in a dark cave. As this person has had no previous opportunity to observe dawn, he is not able to decide whether this is a typical or unusual occurrence. It might even be a unique event. Every day that he sees the same thing happen, the degree of confidence he assigns to this being a permanent aspect of nature increases. His estimate of the probability that the sun will rise again tomorrow as it did yesterday and the day before, and so on, gradually approaches but never quite reaches 100%.

The Bayesian viewpoint is just like that, the idea that we learn about the world and everything in it through a process of gradually updating our beliefs. In this way, we edge closer to the truth as we obtain more data, more information, more evidence.

The Bayes Business School, formerly City University of London's business school, explained their choice of name in similar terms: "Bayes' theorem suggests that we get closer to the truth by constantly updating our beliefs in proportion to the weight of new evidence. It is this idea … that is the motivation behind adopting this name" (Significance, June 2021, p. 3).

As such, the perspective of Reverend Bayes differs from that of philosopher David Hume. For Hume, assumptions about the future, such as that the sun will rise again, cannot be rationally justified based simply on the past because no law exists that the future will always resemble the past. Bayes instead sees reason as a practical matter, to which we can apply the laws of probability in a systematic way.

To Bayes, therefore, we step ever nearer to the truth based on new evidence and the proper application of the laws of probability. This is called Bayesian reasoning. According to this approach, we can see probability as a bridge between ignorance and knowledge. Bayes' Theorem is, in this way, concerned with conditional probability. It tells us the probability, or updates the probability, that a theory or hypothesis is correct, given that we observe some new evidence. A particularly good thing about Bayesian reasoning is that the mathematics of it is so straightforward.

At its heart, then, Bayes' Theorem allows us to use all the information available to us. Our beliefs, our judgments, our subjective opinions, what we have already learned from the previous body of knowledge to which we have had access. We can incorporate this in updating our estimate of the probability that a hypothesis is true. As such, we can be explicit and open about the uncertainty in our data and our beliefs. The problem with implicit reasoning, or intuition, is that our intuition is often wrong and subject to systematic biases. Instead, we should be trained to think in a Bayesian way about the world.

Often the conclusions generated by the application of Bayes' Theorem will challenge intuition. This is because the world is, in many ways, a counter-intuitive place. Accepting that fact is the first step towards mastering life's logical maze.

Intuition also often lets us down because our in-built judgment of the weight that we should attach to new evidence tends to be skewed relative to pre-existing evidence.

New evidence also tends to colour our perception of the pre-existing evidence. Moreover, we tend to see evidence that is consistent with something being true as evidence that it is in fact true. Bayes' Theorem is the map that helps guide us through this maze.

Essentially, though, Bayes' Theorem is just an algebraic expression with three known variables and one unknown. Yet this simple formula is the foundation stone of that bridge between ignorance and knowledge, which can lead to critical predictive insights. Bayesian reasoning allows us to use this formula to update the probability that a theory or hypothesis is true when some new evidence comes to light.

There are three things a Bayesian needs to estimate.

1. A Bayesian's first task is to assign a starting point probability to a hypothesis being true before some new evidence arises. This is known as the "prior" probability. Let's assign the letter "a" to this.

2. A Bayesian's second task is to estimate the probability that the new evidence would have arisen if the hypothesis was correct. This is sometimes known as the "likelihood". Let's assign the letter "b" to this.

3. A Bayesian's third task is to estimate the probability that the new evidence would have arisen if the hypothesis was false. Let's assign the letter "c" to this.

Based on these three probability estimates, Bayes' Theorem offers a way to calculate the revised probability of the hypothesis being true, given the new evidence. A notable point is that the equation is true as a matter of logic. The result it produces will be as accurate as the values inputted into the equation. The formula is also so straightforward that it can be jotted down on the back of a hand.

We can represent the formula for Bayes' Theorem as follows:

Updated (posterior) probability given new evidence = ab/ [ab + c (1 − a)].

Bayesian updating is thus a straightforward solution to the problem of how to combine pre-existing (prior) beliefs with new evidence. The solution is essentially to combine the probabilities. To do this properly, we use Bayes' Theorem. It is of particular use when we have a conditional probability of two events, and we are interested in the reverse conditional probability. For example, when we have P (A given B) − the probability of A given B − and want to find P (B given A) − the probability of B given A.

Looked at from another angle, Bayes' Theorem allows us to calculate the probability of certain events occurring conditional on other events that may occur.

The probability that event B happens given that event A has occurred is given by the formula:

$$P (B|A) = P (A|B) . P (B)/P (A)$$

This is the conditional probability of event B given event A, which is calculated by multiplying the conditional probability of event A given event B by the probability of event B, divided by the probability of event A.

This idea can also be applied to beliefs. So, P (BIE) can be understood as the degree of belief, B, given evidence, E. P (B) is our prior degree of belief before we encountered evidence, E. Employing Bayes' Theorem allows us to convert our prior belief into a posterior belief. When new evidence is observed, we can perform the same calculation again, this time our previous posterior belief becoming our next prior belief. And so on. As McGrayne (2011, preface) puts it, "by updating our initial belief about something with objective new information, we get a new and improved belief. To its adherents, this is an elegant statement about learning from experience".

More generally, the probability that a hypothesis is true, P (H), given new evidence, P (E), is written as: P (H I E).

$$Now, P(HIE) = P(EIH).P(H)/P(E).$$

Where P (E) is the probability of the evidence.

$$P(E) = P(EIH) \cdot P(H) + P(EI1-H).P(1-H)/P(E).$$

P (1 − H) can also be written as P (H′).

In words, probability of the evidence = probability of the evidence given that the hypothesis is true *times* probability the hypothesis is true *plus* probability of the evidence given that the hypothesis is not true *times* probability that the hypothesis is not true.

In a, b, c notation, a = P (H), b = P (E I H), c = P (E I 1 − H) or P (E I H′).

The problem with P (E) is that it's often difficult to calculate it in many real-world cases. In such cases, it may sometimes be preferable to use the Bayes Factor, which is a formula for comparing the plausibility of one hypothesis with another.

The Bayes Factor

Bayes' Theorem states that:

$$P(HIE) = P(H).P(EIH)/P(E).$$

But it's not always clear how to measure P (E), the probability of the evidence.

An alternative approach which doesn't require knowledge of P (E) is by using the proportional form of Bayes' Theorem.

The proportional form of Bayes' Theorem sees the posterior probability of a hypothesis, P (H I E), as proportional to the prior probability, P (H), multiplied by the likelihood, P (E I H).

From this we derive a ratio of how well each of our hypotheses explains the evidence we have observed.

The formula is: P (H1) × P (E I H1) / [P (H2) × P (E I H2)], where P (H1) is the probability that hypothesis 1 is true and P (H2) is the probability that hypothesis 2 is true.

If the ratio is five, for example, this means that H1 explains the evidence five times as well as H2, and vice versa if the ratio is 1/5.

Using this formula, assume that P (H1) = P (H2), i.e. our prior belief in each hypothesis is the same. In this case, P (H1) / P (H2) = 1.

This leaves: P (E I H1) / P (E I H2). This ratio is known as the Bayes Factor. Bayes Factor = P (E I H1) / P (E I H2).

The Bayes Factor is thus the ratio of the likelihood of one hypothesis to the likelihood of the other. It is essentially a measure of which hypothesis better explains the world, given the evidence.

This reasoning assumes, however, that the prior probability of each hypothesis is the same. This may not be true. Before observing the evidence, one hypothesis may be considered more likely than another. We should also consider, therefore, the ratio of prior probabilities: P (H1) / P (H2).

When used in conjunction with the Bayes Factor, this ratio is commonly termed the "prior odds", written as O (H1), which is a measure of how likely H1 is relative to the competing hypothesis before we observe the new evidence.

From these ratios, we can calculate the posterior (or updated) odds.

Posterior odds = O (H1) . P (E I H1) / P (E I H2).

The posterior odds is a measure of how many times better our hypothesis explains the evidence compared to a competing hypothesis.

Take as an example a hypothesis that a machine is a perfect Coin Toss Predictor, in that it can unfailingly calculate how a coin will land face up as soon as it is thrown. You toss a fair coin a series of times.

If the Predictor is perfect, it will always calculate correctly, so P (E I H1) = 1, i.e. the probability of calling it correctly (E) given that the hypothesis, H1 (it is a perfect Predictor), is true = 1.

The alternative hypothesis, H2, is that the Predictor is simply guessing. In this case, P (E I H2) = 0.5 from one toss of the coin.

Say you toss the coin five times, and the Predictor calls it correctly five times. In this case, the probability of doing this by chance = 0.5 × 0.5 × 0.5 × 0.5 × 0.5 = $(0.5)^5$ = 1/32. We can multiply the probabilities as each coin toss is an independent event.

Here, Bayes Factor = 1 / $(0.5)^5$ = 32.

If the original (prior) probability we attach to the Predictor being genuine is equal to it being a guessing machine, then:

Posterior odds = O (H I E) = O (H) . P (E I H1) / P (E I H2) = 1 × 32 = 32. This means that the hypothesis that the machine is a perfect Coin Toss Predictor explains the evidence we have witnessed 32 times better than the alternative hypothesis, that it is a guessing machine.

If, on the other hand, the prior probability we assign to the coin being a genuine perfect Predictor compared to a guesser is 1/64, then the posterior odds = 1/64 × 32 = 1/2. Now, we believe that it is twice as likely that the machine is a guessing machine than a perfect Coin Toss Predictor.

How big do the posterior odds have to be to prove convincing? To some extent that depends on what you are using them for. If it's to help resolve a casual disagreement among friends, a small positive number might be enough. If your life depends on getting it right, you might prefer that number to end in quite a few zeros!

Prosecutor's Fallacy

The Prosecutor's Fallacy is to represent P (HIE) as an equivalent to P (EIH). In fact, P (HIE) = P (EIH) P (H) / P (E) ... Bayes' Theorem. Therefore, P (HIE) only equals P (EIH) when P (H) = P (E), i.e. P (H) / P (E) = 1. Bayes' Theorem can be expanded to: P (HIE) = P (EIH) P (H) / [P (EIH) P (H) + P (EIH') P (H')]

Laplace's Rule of Succession and Bayesian "Priors"

Laplace's Rule of Succession, named after Pierre-Simon Laplace, is a rule-of-thumb way of calculating how likely it is that something that has happened before will happen again. The method is to count how many times it has happened in the past plus one (successes, S + 1) and divide that by the number of opportunities for it to have happened, plus two (trials, T + 2). For a person emerging from a dark cave into the world for the first time and watching the sun rise seven times, for example, the estimate that it will rise again the next day is: (S + 1) / (T + 2) = (7 + 1) / (7 + 2) = 8/9 = 88.9%. Every time it rises again makes it even more likely that the pattern will repeat so that by the end of a year, the estimated probability goes up to (365 + 1) / (365 + 2) = 99.7%. And so on. The 1 and 2 in the Laplace equation, (S + 1) / (T + 2), represents a version of the Bayesian "prior". The 1 and 2 can be replaced by any numbers in the same proportion, such as 5 and 10 or 10 and 20, depending upon how anchored we are to our prior beliefs or understanding of the world.

Larger numbers (e.g. S + 10, T + 20) lead to slower updating in response to new evidence. So (S + 10) / (T + 20) after seven days updates to a probability of (7 + 10) / (7 + 20) = 17/27 = 63.0%, compared to 88.9% for (S + 1) / (T + 2).

So, smaller numbers indicate that we are more open to quickly updating our beliefs based on new evidence. In other words, updating takes place more quickly and readily with smaller numbers in the Laplace equation.

In conclusion, the core contributions of Bayesian analysis to our understanding of the world are threefold.

1. Bayes' Theorem makes clear the importance of not just new evidence but also the (prior) probability that the hypothesis was true before the arrival of the new evidence. This prior probability may in common intuition be given too little (or too much) weight relative to the latest evidence. Bayes' Theorem makes the assigned prior probability explicit and shows how much weight to attach to it.

2. Bayes' Theorem allows us a way to update the probability that a hypothesis is true. It does so by combining the prior probability with the probability that the new evidence would arise if the hypothesis is true and the probability that it would arise if the hypothesis is false.

3. Bayes' Theorem shows that the probability that a hypothesis (H) is true given the evidence (E) is not equal to the probability of the evidence arising given that the hypothesis is true, except in limiting circumstances. Specifically, P (H given E) does not equal P (E given H) except when P (H) = P (E).

1.1.1 Appendix

Bayes' Theorem consists of three variables.

- a is the prior probability of the hypothesis being true (the probability we attach before the arrival of new evidence). In traditional notation, this is represented as P (H).
- b is the probability that the new evidence would arise if the hypothesis is true. In traditional notation, we represent this as P (EIH). We use the notation P (AIB) to represent the probability of A given B.
- c is the probability the new evidence would arise if the hypothesis is not true. In traditional notation, we represent this as P (EIH'). H' is the notation for H not being true.

(1 − a) is, therefore, the prior probability that the hypothesis is not true. In traditional notation, we represent this as P (H') or 1 − P (H), i.e. one minus the probability that the hypothesis is true.

Using the a, b, c notation, the probability that a hypothesis is true given some new evidence ("posterior probability") = ab/ [ab + c (1 − a)].

Deriving Bayes' Theorem

If A and B are two independent events, we can write the conditional probability that A occurs, given that B has occurred, as P (AIB), which is the probability of A *given* B.

Now, the formula for Bayes' Theorem states that: P (AIB) = P (BIA) . P (A) / P (B).

This follows from the definition of conditional probability: P (AIB) = P (A and B) / P (B).

To see how Bayes' Theorem is derived, note that P (AIB) = P (A and B) / P (B) and also that P (BIA) = P (B and A) / P (A).

But the event "A and B" is the same as "B and A". Dividing one equation by the other, we have P (AIB) / P (BIA) = P (A) / P (B).

Now, multiply both sides by P (BIA).

$$P(AIB) = P(BIA) \cdot P(A) / P(B) \dots \text{Bayes' Theorem}$$

More Formal Derivation

The conditional probability of H given E is conventionally represented as P (HIE). It can be defined as the probability that the hypothesis, H, occurs, given that the evidence, E, occurs.

Now, the probability that both H and E occur, P (H∩E), is the conditional probability of H occurring given E multiplied by the probability that E occurs.

$$P(H \cap E) = P(HIE) \cdot P(E)$$

Similarly,

$$P(E \cap H) = P(EIH) \cdot P(H)$$

Now,

$$P(H \cap E) = P(E \cap H)$$

So:

$$P(HIE) P(E) = P(EIH) P(H)$$

Dividing both sides by P (H) (which we take to be non-zero), the result follows:

$$P(HIE) P(E) / P(H) = P(EIH)$$

So, P (HIE) = P (EIH) P (H) / P (E) ... Bayes' Theorem.

Expanding the denominator, P (E) = P (EIH) P (H) + P (EIH') P (H'), where P (H') represents the probability that the hypothesis is not true, i.e. P (H') = 1 − P (H).

$$P(HIE) = P(EIH) P(H) / \left[P(EIH) P(H) + P(EIH') P(H') \right] \dots \text{Bayes' Theorem}$$

This can also be stated as:

$$P(HIE) = P(EIH)P(H)/[P(EIH)P(H)+P(EIH')P(1-H)]...\text{Bayes' Theorem}$$

Intuitive Presentation

Bayes' Theorem can be derived from the equation P (HIE) . P (E) = P (H) . P (EIH).

The intuition underlying this equation is that both sides are alternative ways of looking at the same thing. It is the combined probability of observing the evidence relating to a hypothesis and the probability that the hypothesis is true, P (H and E).

$$\text{So, P (HIE) . P (E) = P (H) . P (EIH).}$$

Dividing both sides of the equation by P (E),

$$P(HIE) = P(H) \cdot P(EIH)/P(E)...\text{Bayes' Theorem}$$

$$P(E) = P(EIH) \cdot P(H)+P(EIH') \cdot P(H')$$

$$P(HIE) = P(H) \cdot P(EIH)/[P(H) \cdot P(EIH)+P(EIH') \cdot P(H')]...\text{Bayes' Theorem}$$

This can also be stated as:

$$P(HIE) = P(EIH)P(H)/[P(EIH)P(H)+P(EIH')P(1-H)]...\text{Bayes' Theorem}$$

This is equivalent to the formula:

$$\text{Posterior probability} = ab/[ab+c(1-a)], \text{ where } a = P (H);$$

$$b = P (EIH); c = P (EIH').$$

1.1.2 Exercise

Question a.

Write the Bayesian equation (using a, b, and c) for deriving the posterior (updated) probability of a hypothesis being true after the arrival of new evidence. Explain what a, b, and c represent.

Question b.

If P (H) is the probability that a hypothesis is true before some new evidence (E), what is the updated (or posterior) probability after the new evidence? Use the terms P (H), P (EIH), P (HIE), P (H′), and P (EIH′) to construct the Bayesian equation.

Question c.

How do the terms used in Question b relate to a, b, and c in the Bayesian formula referred to in Question a?

Question d.

1. Is the probability that a hypothesis is true, given the evidence, P (HIE), equal to the probability of the evidence, given that the hypothesis is true, P (EIH)? In other words, does P (HIE) = P (EIH)?

2. Is the probability of feeling warm given that you are out in the sun equal to the probability of being out in the sun given that you are feeling warm?

Question e.

For a person emerging from a dark cave into the world for the first time and watching the sun rise seven times, the estimate that it will rise again is 88.9%, if we use a Bayesian "prior" of 1, 2. Calculate the updated probabilities that the sun will rise again if we use a Bayesian "prior" of 5, 10? What is the significance of using a 5, 10 prior compared to a 1, 2 prior?

Question f.

Uncle Austin and Uncle Idris each present you with a die. One is fair, and one is biased. The fair die (A) lands on all numbers (1–6) with equal probability. The biased die (B) lands on 6 with a 50% chance and each of the other numbers (1–5) with an equal 10% chance each.

Now, choose one of the two dice at random. You can't tell by inspection whether it is the fair or the biased die. You now roll the die, and it lands on 6. What is the probability that the die you rolled is the biased die?

Answer guide: state the hypothesis to be that you chose the biased die.

What is P (H)? What is the probability that the die is biased before the evidence that the die landed on a 6?

What is P (EIH)? Note that the evidence is that the die landed on a 6.

What is P (EIH′), i.e. the probability that you would throw a 6 if the die was not biased?

What is P (HIE)?

Alternatively, you can use the formula: ab / [(ab + c (1 − a)].

Question g.

Auntie Beatrice and Auntie Kit each present you with a coin. One of these is a fair coin, and the other is weighted. The fair coin (Coin 1) lands on heads and tails with equal likelihood, the weighted coin (Coin 2) lands on heads with a 75% chance.

Now, choose one coin. You can't tell by inspection whether it is the fair or the weighted coin. You select a coin and toss it, and it lands on heads. What is the probability that you tossed Coin 2 (the weighted coin)?

Question h.

You own a pair of dice (one blue and one red) and are told that your colleague can always guess how they will land as soon as they leave your hands. They are your own dice so you know there are no tricks. You throw the dice in the air and he calls out: blue will show 5 and red will show 6. That's exactly how they end up.

In terms of the probability that your colleague is a genuine perfect Dice Predictor compared to a guesser, what is the Bayes Factor? What are the posterior odds if you were originally perfectly split between believing in his powers and believing he was a guesser? What are the posterior odds if you originally believed that he was 100 times more likely to be a guesser than a genuine perfect Dice Predictor?

Question i.

Every time you meet to play tennis, your friend, May, tosses a coin to determine who serves first. You know that she prefers to serve first and are a tiny bit, but only a tiny bit, suspicious that the coin tosses are not fair and are designed to land heads, so that she can choose to serve first. You play once a week and over the course of 12 weeks, she always calls heads and the coin lands heads every time. What is the Bayes Factor for the hypothesis that she is cheating compared to a hypothesis that the coin tosses were fair? What if, despite your hypothesis, you are almost (though not totally) certain that your friend is playing fair, so you assign a probability of 1 in 1,000 that she would cheat? What are the posterior odds now for the hypothesis that the coin toss was rigged to always land heads? What do the posterior odds represent here?

1.1.3 Reading and Links

Bayes Theorem. 2016. *A Take Five Primer. An Iterative Quantification of Probability.* Corsair's Publishing. 24 March. http://comprehension360.corsairs.network/bayes-theorem-a-take-five-primer-fc7f7ade7abe

Bayes, T. and Price, R. 1763. An Essay towards solving a Problem in the Doctrine of Chances. By the late Rev. Mr. Bayes, communicated by Mr. Price, in a letter to John Canton, M.A. and F.R.S. *Philosophical Transactions of the Royal Society of London.* 53: 370–418. https://web.archive.org/web/20110410085940/http://www.stat.ucla.edu/history/essay.pdf

BBC Sounds. 2021. Bayes: the clergyman whose maths changed the world. More or Less: Behind the Stats. 2 May. https://www.bbc.co.uk/sounds/play/p09g10xn

Ellerton, P. 2014. Why facts alone don't change minds in our public debates. *The Conversation.* 13 May. https://theconversation.com/why-facts-alone-dont-change-minds-in-our-big-public-debates-25094

Flam, F.D. 2014. The odds, continually updated. *New York Times*. 29 September. https://www.nytimes.com/2014/09/30/science/the-odds-continually-updated.html?referringSource=articleShare

Hooper, M. 2013. Richard Price, Bayes' theorem and god. *Significance*. February, 36–39. https://www.york.ac.uk/depts/maths/histstat/price.pdf

Johnson, E.D. and Tubau, E. 2015. Comprehension and computation in Bayesian problem solving. Frontiers in Psychology, 27 July, 6: 938. https://www.frontiersin.org/articles/10.3389/fpsyg.2015.00938/full

Kurt, W. 2019. *Bayesian Statistics: The Fun Way. Understanding Statistics and Probability with Star Wars, Lego, and Rubber Ducks*. San Francisco, CA: No Starch Press.

Lee, M., and King, B. 2017. Bayes' theorem: The maths tool we probably use every day. But what is it? *The Conversation*. 23 April. https://theconversation.com/bayes-theorem-the-maths-tool-we-probably-use-every-day-but-what-is-it-76140

LessWrong. 2011. A history of Bayes' theorem. *Lukeprog*. 29 August. https://www.nytimes.com/2014/09/30/science/the-odds-continually-updated.html?referringSource=articleShare

Marianne. 2016. Maths in a minute: The prosecutor's fallacy. + plus magazine. 11 October. https://plus.maths.org/content/maths-minute-prosecutor-s-fallacy

McGrayne, S.B. 2011. *The Theory that Would Not Die: How Bayes' Rule Cracked the Enigma Code, Hunted Down Rusian Submarines, and Emerged Triumphant from Two Centuries of Controversy*. New Haven, CT: Yale University Press.

McRaney, D. 2016. YANSS 073 – How to get the most out of realizing you are wrong by using Bayes' theorem to update your beliefs. 8 April. [Podcast]. https://youarenotsosmart.com/2016/04/08/yanss-073-how-to-get-the-most-out-of-realizing-you-are-wrong-by-using-bayes-theorem-to-update-your-beliefs/

Olasov, I. 2016. Fundamentals: Bayes' Theorem. 22 April. https://www.khanacademy.org/partner-content/wi-phi/wiphi-critical-thinking/wiphi-fundamentals/v/bayes-theorem

Puga, J., Krzywinski, N., and Altman, N. 2015. Points of significance: Bayes' theorem. 12, 4, April, 277–278. https://www.nature.com/articles/nmeth.3335.pdf?origin=ppub

Significance. 2021. A school named Bayes. June, 18, 3.

Stylianides, N. and Kontou, E. (2020). Bayes Theorem and Its Recent Applications. MA3517 Mathematics Research Journal, March, 1-7. file:///C:/Users/epa3willilv/Downloads/3488-9410-1-PB.pdf

Taylor, K. (2018). The Prosecutor's Fallacy. Centre for Evidence-Based Medicine, 16 July. https://www.cebm.ox.ac.uk/news/views/the-prosecutors-fallacy

Tijms, H. 2019. Chapter 4: Was the champions league rigged? In *Surprises in Probability – Seven Short Stories*. CRC Press. Taylor & Francis Group, Boca Raton, pp. 23–30.

AMSI. 2020. Bayes' theorem: The past and the future. 19 June. [Podcast]. https://amsi.org.au/2020/06/19/bayes-theorem-the-past-the-future-acems-podcast/

Rationally Speaking Podcast. 2012. RS58 – Intuition. 8 April. [Podcast]. http://rationallyspeakingpodcast.org/show/rs58-intuition.html

SuperDataScience. SDS 096: Bayes theorem. [Podcast]. https://soundcloud.com/superdatascience/sds-096-bayes-theorem

Wiblin, R., and Harris, K. 2018. How much should you change your beliefs based on new evidence? 7 August. [Podcast]. https://80000hours.org/podcast/episodes/spencer-greenberg-bayesian-updating/

A Derivation of Bayes' Rule. Ox educ. 29 July 2014. YouTube. https://youtu.be/_DsO4ZSYpHUA Visual Guide to Bayesian Thinking. Galef, J. 17 July 2015. YouTube. https://youtu.be/BrK7X_XlGB8

Bayes Theorem. 3Blue1Brown. 22 December 2019. YouTube. https://youtu.be/ HZGCoVF3YvM

Bayes: How One Equation Changed the Way I Think. Galef, J. 4 June 2013. YouTube. https://youtu.be/za7RqnT7CM0

Bayes' Theorem – The Simplest Case. Bazett, T. 19 November 2017. YouTube. https:// youtu.be/XQoLVI31ZfQ

Bayes' Theorem/Law, Part 1. patrickJMT.3 January 2013. YouTube. https://youtu.be/ E4rlJ82CUZI

Bayes' Theorem/Law, Part 2. patrickJMT. 3 January 2013. YouTube. https://youtu.be/ zh1E8cGoV7k

FRM: Bayes' Formula. Bionic Turtle. 9 January 2008. YouTube. https://youtu.be/ pPTLK5hFGnQ

Prior Indifference Fallacy. Fuggetta, M. 24 January, 2014. YouTube. https://youtu.be/ PkcUM3Mr_F4

The Bayesian Trap. Veritasium. 2017. 5 April 2017.. YouTube. https://youtu.be/ R13BD8qKeTg

---

1.2 Bayes and the Taxi Problem

To help explain how we can apply Bayes' Theorem in practice, let's start with the classic Bayesian Taxi Problem. It goes something like this. New Amsterdam has 1,000 taxis – 850 are blue and 150 are green. One of these taxis knocks down a pedestrian and then is driven away without stopping. We have no prior reason to believe that the driver of a blue taxi is more likely to have knocked down the pedestrian than of a green taxi, or vice versa. There is one independent witness, however, who did see the event. The witness says the colour of the taxi was green. The witness is given a rigorous observation test, which recreates as carefully as possible the event in question, and her judgment has proved correct 80% of the time.

So what is the probability that the taxi was green?

The intuitive answer is in the region of 80%, as the only evidence is that of the witness, and the test of her powers of observation shows that she is right 80% of the time. That is not the Bayesian approach, however, which is to consider the evidence in the light of the baseline, or prior, probability that the taxi was green before the witness evidence came to light. We can derive the prior probability from an identification of the proportion of taxis in New Amsterdam that are green, which is 15% (of the 1,000 taxis, 150 are green).

Now, the (posterior) probability that a hypothesis is true after obtaining new evidence, according to the a, b, c formula of Bayes' Theorem, is equal to:

$$ab / [ab + c(1 - a)]$$

In this case, the hypothesis is that the taxi that knocked down the pedestrian was green, where:

- a is the prior probability, i.e. the probability that a hypothesis is true before the new evidence arises. This is 0.15 (15%) because 15% of the taxis in New Amsterdam are green.
- b is the probability the new evidence would arise if the hypothesis is true. This is 0.8 (80%). There is an 80% chance that the witness would say the taxi was green if it was indeed green.
- c is the probability the new evidence would arise if the hypothesis is false. This is 0.2 (20%). There is a 20% chance that the witness would be wrong and identify the taxi as green if it was blue.

Inserting these numbers into the formula, ab / [ab + c (1 − a)], gives:

Posterior probability $= 0.15 \times 0.8 / [0.15 \times 0.8 + 0.2(1 - 0.15)] = 0.41 = 41\%$.

In other words, the actual probability that the taxi that knocked down the pedestrian was green is not 80% (despite the witness evidence) but about half of that. The baseline probability is important. A common error is to place too much weight on new evidence about an event (the judgment of the witness) and too little on the general frequency of that event (in this case, represented by the proportion of green cabs in the taxi population).

If new evidence subsequently arises, Bayesians are not content to leave the probabilities alone. Say, for example, that a second witness appears and is also given the observation test, revealing a reliability score of 90%. Again, we have no reason to doubt the integrity of this second witness. A Bayesian now inserts that number (0.9) into Bayes' formula (b = 0.9) so that c (the probability that the witness is mistaken) = 0.1. The new baseline (or prior) probability, a, is no longer 0.15, as it was before the first witness appeared, but 0.41 (the probability incorporating the evidence of the first witness). In this sense, yesterday's posterior probabilities are today's prior probabilities.

Inserting into Bayes' Theorem, the new posterior probability = 0.86 = 86%. This is the new baseline probability underpinning any further new evidence which might arise.

There are three critical illustrative cases of the Bayesian Taxi Problem which bear highlighting. The first is a scenario where the new witness scores 50% on the observation test. Here is a case where intuition and Bayes' formula converge. A witness who is right only half the time is also wrong half the time, and so any evidence they give is worthless. Bayes' Theorem tells us that this is indeed so, as the posterior probability ends up being equal to the prior probability.

The second illustrative case is where a new witness is 100% reliable about the colour of the taxi. In this case, b = 1 and c = 0. Intuition tells us that the evidence of such a witness solves the case. If the infallible witness says the taxi was green, it was green. Bayes' Theorem agrees.

Now for the third illustrative case. If the new witness scores 0% on the observation test, this indicates that they always identify the wrong colour for the taxi. If they say it is green, it is not green. So the chance (posterior probability) that the cab is green if they say so is zero, which accords with Bayes' Theorem.

More generally, information that informs us that a witness is usually wrong is valuable, as it can be reversed to beneficial effect. A witness who always identifies a green taxi as blue and vice versa, and is 100% consistent in doing so, yields us reliable information by merely reversing their designated colour.

So if the witness says the taxi is blue, we can now identify the taxi as definitely being green. This now converges on the second illustrative case.

Similarly, a witness who is, say, right only 25% of the time in identifying the colour of the taxi in the observation test also yields us valuable information. By reversing the defined colour, this produces a 75% reliability score, which can be inserted accordingly into Bayes' Theorem to update the probability that the taxi that knocked down the pedestrian was green. In other words, a witness who is 25% reliable and identifies the cab as green is equivalent to the witness being 75% reliable in determining the taxi as blue, and vice versa.

The only observation evidence that is worthless, therefore, is evidence that could have been produced by the flip of a coin.

The Bayesian Taxi Problem is an instance of what is known as the Base Rate Fallacy. This occurs when we undervalue prior information when making a judgement as to how likely something is. If presented with general (base rate) information and specific information (pertaining only to a particular case), the fallacy arises from a tendency to focus on the latter at the expense of the former. For example, if someone is an avid book enthusiast, we might think it more likely that they work in a bookshop or a library than as, say, a nurse. There are, however, many more nurses than librarians and bookshop assistants. Our mistake is not to take sufficient account of the base rate numbers for each occupation.

And the conclusion to the case? CCTV evidence was later produced in court, which was able to identify the taxi and the driver conclusively. The pedestrian never regained consciousness. The driver of what transpired to be a blue taxi told the jury that the pedestrian unexpectedly stepped out and lightly brushed against the passenger side door. He thought at the time that it was a minor incident and was completely unaware that the victim had slipped and hit his head awkwardly. This account was rejected by the jury, who accepted the prosecution's contention that the driver had acted with premeditation and malicious intent. They based their decision on their view that a driver who was so motivated would indeed have driven off. It was all they needed to reach their unanimous verdict of first-degree murder.

James Parker, a 29-year-old long-time resident of New Amsterdam, of previous good character, with no prior convictions or any known motive for the crime, is currently serving a sentence of life in a maximum-security prison with no possibility of parole. No member of the jury, it turned out, had ever heard of the Reverend Thomas Bayes or the Prosecutor's Fallacy.

1.2.1 Appendix

In the original taxi problem scenario:

a = 0.15 (15% of taxis are green)
b = 0.8 (the witness is correct 80% of the time)
c = 0.2 (the witness is wrong 20% of the time)

Inserting these numbers into the formula gives:

Posterior probability = (0.15 × 0.8) / (0.15 × 0.8 + 0.2 × 0.85) = 0.12 / (0.12 + 0.17) = 41% (rounded to the nearest per cent).

This is the new baseline probability underpinning any new evidence which might arise.

If new evidence subsequently arises, this should be used to update the new baseline probability of 0.41.

Say, for example, that a new witness is correct 90% of the time (wrong 10% of the time). New posterior probability = 0.41 × 0.9 / (0.41 × 0.9 + 0.1 × 0.59) = 0.369 / (0.369 + 0.059) = 86% (rounded to the nearest per cent). This is also the new baseline probability underpinning any further new evidence which might arise.

Solution to the three illustrative cases of the Bayesian Taxi Problem:

1. A scenario where the new witness scores 50% on the observation test. In terms of the equation, b = 0.5 and c = 0.5.

$$\text{Posterior probability} = ab / \left[ab + c(1-a) \right] = 0.5a / \left[0.5a + 0.5(1-a) \right]$$

$$= 0.5a / (0.5 + 0.5a - 0.5a) = 0.5a / 0.5 = a$$

So when b and c both equal 0.5 in regard to new evidence, this evidence has no impact on the probability of the hypothesis being tested being true. The posterior probability equals the prior probability. In this case, the evidence of the witness can be discounted.

2. The second illustrative case is where a new witness is 100% accurate about the colour of the taxi. In this case, b = 1 and c = 0. Intuition tells us that the evidence of such a witness solves the case. If the infallible

witness says the taxi was green, it was green. Bayes' formula agrees. Inserting b = 1 and c = 0 into the formula gives:

$$ab / \left[ab + c(1-a) \right] = a / (a+0) = a / a = 1$$

So the new (posterior) probability that the taxi is green = 1.

3. This leads directly to the third illustrative case. If the new witness scores 0% on the observation test, this indicates that they always identify the wrong colour for the taxi. If they say it is green, it is not green. So the chance (posterior probability) that the taxi is green if they say so is zero. This accords with the formula.

$$ab / \left[ab + c(1-a) \right] = 0 / \left[0 + (1-a) \right],\ \text{assuming a is not equal to } 1 = 0$$

Note that a cannot equal 1 if c equals 1. It would represent a logical contradiction, implying within the context of the illustration that every taxi is green but a witness who is always wrong says that the taxi is green.

1.2.2 Exercise

For the purpose of this exercise, use the a, b, c method to derive the solutions.

Question a.

New Amsterdam has 1,000 taxis, and 800 of them are yellow and 200 are white. The driver of one of these taxis knocks down a pedestrian and drives away. There is no prior reason to believe that the driver of a yellow taxi is more likely to have knocked down the pedestrian than of a white taxi, or vice versa. There is one witness, however, who saw the event and says the colour of the cab was white.

The witness, Reverend Latimer Williams, is given a well-respected observation test and is right 80% of the time.

What is our best estimate now of the probability that the taxi was white?

Question b.

What if a second witness now comes forward?

We determine that the probability that this witness is correct when identifying the colour of the taxi as 70%.

The witness, Mr. Henry Morris, says the colour of the taxi was white.

What is the new posterior (updated) probability that the taxi that knocked down the pedestrian is white?

Question c.

What if a third witness now comes forward?

We determine that the probability that this witness is correct when identifying the colour of the taxi as 50%.

The witness, Mr. Edmund Coss, says the colour of the taxi was white.

What is the new posterior (updated) probability that the taxi that knocked down the pedestrian is white?

Question d.

A witness, Mr. Smith, is correct 50% of the time.

A witness, Mr. Jones, is correct 100% of the time.

A witness, Mr. Evans, gets it wrong 100% of the time.

Which of the three witnesses is the most useful/least useful to investigators?

1.2.3 Reading and Links

Bedwell, M. 2015. Slow thinking and deep learning: Tversky and Kahneman's cabs. *Global Journal of Human-Social Science*, 15, 12. file:///C:/Users/epa3willilv/Downloads/1634-1-1640-1-10-20160225%20(1).pdf

Gesmann, M. 2014. Hit and run. Think Bayes! 29 July. https://magesblog.com/post/2014-07-29-hit-and-run-think-bayes/

plusadmin. 1997. Solution to the taxi problem. + plus magazine. 1 May. https://plus.maths.org/content/solution-taxi-problem

plusadmin. 1997. Solution to the taxi problem revisited. + plus magazine. 1 May. https://plus.maths.org/content/solution-taxi-problem-revisited

Salop, S.C. 1987. Evaluating uncertain evidence with Sir Thomas Bayes: A note for teachers. *Economic Perspectives*, 1, 1, Summer, 155–160. https://pubs.aeaweb.org/doi/pdf/10.1257/jep.1.1.155

Talwalkar, P. 2013. Mind your decisions. The taxi-cab problem. 5 September. https://mindyourdecisions.com/blog/2013/09/05/the-taxi-cab-problem/

Taxi Cab Problem, Exercise in Value and Information. Putt, B. 12 April 2020. YouTube link. https://www.youtube.com/watch?v=ySwwiji0B7o

The Decision Lab. 2021. Why do we rely on event-specific information over statistics? Base Rate Fallacy, explained. https://thedecisionlab.com/biases/base-rate-fallacy/

Tversky, A., and Kahneman, D. 1982. Evidential impact of base rates. In Kahneman, D., Slovic, P. and Tversky, A. (eds.), *Judgment under Uncertainty: Heuristics and Biases*. https://www.cambridge.org/core/books/judgment-under-uncertainty/evidential-impact-of-base-rates/CC35C9E390727085713C4E6D0D1D4633

Woodcock, S. 2017. Base rate fallacy. In: Paradoxes of probability and other statistical strangeness. UTS, 5 April. http://newsroom.uts.edu.au/news/2017/04/paradoxes-probability-and-other-statistical-strangeness

Woolley, R. 2016. Do I call or fold? How Bayes' theorem can help navigate Poker's uncertainty, part 1. 15 February. https://www.pokernews.com/strategy/call-or-fold-bayes-theorem-poker-uncertainty-24077.htm

Woolley, R. (2016). Do I call or fold? How Bayes' Theorem can help navigate Poker's uncertainty, part 2. 22 February. https://www.pokernews.com/strategy/call-or-fold-bayes-theorem-poker-uncertainty-2-24133.htm

Base rate fallacy. Wikipedia. https://en.wikipedia.org/wiki/Base_rate_fallacy

Base Rate Fallacy. Yang, C. 22 March 2017. YouTube. https://youtu.be/Fs8cs0gUjGY

Counting Carefully – The Base Rate Fallacy. Simple Scientist. 20 May 2013. YouTube. https://youtu.be/VeQXXzEJQrg

Know Your Bias: Base Rate Neglect. Deciderata. 25 July 2016. YouTube. https://youtu.be/YuURK_q2NR8

1.3 Bayes and the Beetle

A nature lover spots what might be a rare type of beetle due to the pattern on its back. In the rare category, 98% have the pattern. In the common category, only 5% have the pattern. The rare category accounts for only 0.1% of the population. How likely is the beetle to be rare?

Intuition might suggest that we have come across a rare insect when observing the unusual pattern. This is because there is a very high chance of a rare beetle having this pattern and a low chance of a common beetle having this pattern. Bayes' Theorem tells us something entirely different.

To calculate just how likely the beetle is to be rare given that we see the pattern on its back, we apply Bayes' Theorem.

$$\text{Posterior probability} = ab / \left[ab + c\left(1-a\right) \right]$$

Where a is the prior probability of the hypothesis (beetle is rare) being true. b is the probability we observe the pattern, and the beetle is rare (hypothesis is true). c is the probability we observe the pattern, and the beetle is not rare (hypothesis is false).

In this case, a = 0.001 (0.1%); b = 0.98 (98%); c = 0.05 (5%).

So, updated probability = ab / [ab + c (1 − a)] = 0.0192. So there is just a 1.92% chance that the beetle is rare when the distinctive pattern is spotted on its back.

Why the counter-intuitive result? Few beetles are rare, so it would take a lot more evidence than observing the rare pattern to alter the prior expectation that the beetle is not rare.

So the probability that the beetle is rare (the hypothesis) given that we observe the distinctive pattern (the evidence) is 1.92%. What is the chance, however, that we will observe the distinctive pattern if the beetle is rare? In other words, what is the chance of observing the evidence (the pattern) if the hypothesis (the beetle is rare) is correct? That is 98%.

To believe these two things are the same is a common mistake known as the Inverse (or Prosecutor's) Fallacy. In this instance, it is to believe that the chance of observing the pattern given that the beetle is rare (98%) is the same as the chance that the beetle is rare given the observation of the pattern (the actual probability that the beetle is rare, which is 1.92%).

1.3.1 Appendix

We can also solve the beetle problem using the traditional notation version of Bayes' Theorem.

$$P(HIE) = P(EIH).P(H)/[P(EIH).P(H) + P(EIH').P(H')]$$

In this case, P (H) = 0.001 (0.1%); P (EIH) = 0.98 (98%); P (EIH') = 0.05 (5%).

So, P (HIE) = 0.98 × 0.001 / [0.98 × 0.001 + 0.05 × 0.999)] = 0.00098 / 0.00098 + 0.04995 = 0.00098 / 0.05093 = 0.0192. So there is just a 1.92% chance that the beetle is rare when the entomologist spots the distinctive pattern on its back.

Note also that P (HIE) = 0.0192, while P (EIH) = 0.98.

The Prosecutor's Fallacy is to conflate these two expressions.

1.3.2 Exercise

A nature lover spots what might be a rare category of beetle, due to the pattern on its back. In the rare category, 95% have the pattern. In the common category, only 2% have the pattern. The rare category accounts for only 1% of the population. How likely is the beetle to be rare?

In solving this question, what are a, b, and c?

Solve again, using traditional notation, in the case where 5% (instead of 2%) of those in the common category have the pattern.

1.3.3 Reading and Links

AgileKiwi. 2011. Bayes' theorem demystified. 30 December. http://www.agilekiwi. com/off-topic/bayes-theorem-demystified/

CS201. Bayes' theorem. Excerpts from Wikipedia. https://mathcs.clarku.edu/~jma gee/cs201/slides/BayesTheorem.pdf

Thompson, J. 2011. Bayes' theorem. 20 November. https://www.jeffreythompson.org/ blog/2011/11/20/bayes-theorem/

1.4 Bayes and the False Positives Problem

A patient goes to see the doctor. The doctor performs a test on all her patients for a virus, estimating in advance of the test that 1% of those who visit the surgery have the virus. The test is 99% accurate, in the sense that 99% of people with the virus test positive, and 99% of those who do not have the virus test negative.

Let us say that the first patient tests positive. What is the chance that the patient has the virus?

The intuitive answer is 99%, as the test is 99% accurate. But is that right?

The information we are given relates to the probability of testing positive given that you have the virus. What we want to know, however, is the probability of having the virus given that you test positive. This is a crucial difference.

Common intuition conflates these two probabilities, but they are very different. If the test is 99% accurate, this means that 99% of those with the virus test positive. But this is *not* the same thing as saying that 99% of patients who test positive have the virus. This is another example of the "Inverse Fallacy" or "Prosecutor's Fallacy". In fact, those two probabilities can diverge markedly.

So what is the probability you have the virus if you test positive, given that the test is 99% accurate? To answer this, we can use Bayes' Theorem.

The probability that a hypothesis is true after obtaining new evidence, according to the a, b, c formula of Bayes' Theorem, is equal to: ab / [ab + c (1 − a)], where:

- a is the prior probability, i.e. the probability that a hypothesis is true before you see the new evidence. Before the new evidence (the test), this chance is estimated at 1 in 100 (0.01), as 1% of the people who visit the surgery have the virus. So, a = 0.01.

- b is the probability of the new evidence if the hypothesis is true. The probability of the new evidence (the positive result on the test) if the hypothesis is true (the patient has the virus) is 99% since the test is 99% accurate. So, b = 0.99.

- c is the probability of the new evidence if the hypothesis is false. The probability of the new evidence (the positive result on the test) if the hypothesis is false (the patient does not have the virus) is just 1% because the test is 99% accurate. So, c = 0.01.

Using Bayes' Theorem, the updated (posterior) probability = ab / [ab + c (1 − a)] = 1/2.

So there is a 50% chance that the patient has the virus if testing positive.

It is basically a competition between how rare the virus is and how rarely the test is wrong. In this case, there is a 1 in 100 chance that you have the virus before taking the test, and the test is wrong one time in 100. These two probabilities are equal, so the chance that you have the virus when testing positive is 1 in 2, despite the test being 99% accurate.

But what if the patient is showing symptoms of the virus before being tested?

In this case, we should update the prior probability to something higher than the prevalence rate in the entire tested population. The chance you have the virus when you test positive rises accordingly. To the extent that a doctor only checks for something that there is corroborating support for, the chance that the test result is correct grows. For this reason, any positive test result should be taken very seriously, statistics aside.

1.4.1 Examples

In the original setting with the test results showing positive for a virus, a = 0.01, b = 0.99, c = 0.01. Substituting into Bayes' equation, ab / [ab + c (1 − a)], gives:

$$\text{Posterior probability} = 0.01 \times 0.99 / \left[0.01 \times 0.99 + 0.01 \left(1 - 0.01 \right) \right]$$

$$= 0.01 \times 0.99 / \left[0.01 \times 0.99 + 0.01 \times 0.99 \right] = 1/2$$

Let's take another example.

The probability of a true positive (test comes back positive for the virus and the patient has the virus) is 90%. The chance of a false positive (test comes back positive, yet the patient does not have the virus) is 7%.

The probability that a random patient has the virus based on the prevalence of the virus in the tested population is 0.8%.

Applying Bayes' Theorem:

a = 0.8% (0.008) − this is the prior probability that the hypothesis is true (patient has the virus)

b = 90% (0.9) − probability of a true positive

c = 7% (0.07) − probability of a false positive

So, updated probability that the patient has the virus given the positive test result =

$$ab/[ab+c(1-a)] = 0.008 \times 0.9 / [0.008 \times 0.9 + 0.07 \times (1 - 0.008)]$$

$$= 0.0072 / [0.0072 + 0.06944] = 0.0072 / 0.07664 = 0.0939 = 9.39\%$$

This can be shown using the raw figures to produce the same result. We can choose any number for the total tested, and the result is the same. Let's choose 1 million, say, as the number tested.

So total tested = 1,000,000

Total with virus = 0.008 × 1,000,000 = 8,000

Total without virus = 992,000

True positive = 0.9 × 8,000 = 7,200
False positive = 0.07 × 992,000 = 69,440
Tested positive = 69,440 + 7,200 = 76,640

Updated (posterior) probability that the patient who tests positive has the virus = true positives / tested positive = 7,200 / 76,640 = 0.0939 = 9.39%.

1.4.2 Appendix

1.4.2.1 Sensitivity and Specificity

In terms of false positive analysis, especially in a medical context, the concepts of Sensitivity and Specificity are often used.

Sensitivity (the true positive rate) is the proportion of true positives who have a positive test result. In a medical context, for example, it is the proportion of people with a disease that are correctly identified (test positive) with the disease.

Specificity (the true negative rate) is the proportion of true negatives. In a medical context, for example, it is the proportion of people without a disease that are correctly identified (test negative).

To calculate these, note that TP (true positive) is someone who has a disease and tests positive for it. FN (false negative) is someone who has the disease and tests negative for it. FP (false positive) is someone who does not have the disease but tests positive for it. TN is someone who does not have the disease and tests negative for it (true negative).

Therefore, Sensitivity (True Positive Rate) = TP / (TP + FN). This is the probability of a positive test given that the patient has the disease.

1 – Sensitivity is the probability of testing negative given that the patient has the disease, i.e. P (T– I D+).

Specificity (True Negative Rate) = TN / (TN + FP). This is the probability of a negative test given that the patient is free from the disease.

1 – Specificity is the probability of testing positive given that the patient does not have the disease, i.e. P (T+ I D–).

Put simply, Sensitivity is the proportion of those with the disease who test positive. Stated another way, Sensitivity is a measure of how well the test detects disease when it is really there. A sensitive test has few false negatives. Specificity is the proportion of those without the disease who test negative. It is a measure of how well the test rules out disease when it is really absent. A specific test has few false positives.

Sensitivity is not the same as *Precision (Positive Predictive Value, PPV)*, which is the ratio of true positives to all true and false positives.

$$PPV = TP / (TP + FP)$$

PPV is the probability that you have the disease if you have tested positive for it.

NPV (Negative Predictive Value) = TN / (TN + FN).

NPV is the probability that you don't have the disease if you test negative for it.

The *Likelihood Ratio* is the probability that a test is correct divided by the probability that it is incorrect. In medicine, Likelihood Ratios can be used to determine whether a test result usefully changes the probability that a disease exists.

Two versions of the Likelihood Ratio (Positive LR and Negative LR) exist, one for positive and one for negative test results.

We calculate the *Positive Likelihood Ratio* as:

LR+ = P (test positive and *have* the disease) / P (test positive and *do not* have the disease), i.e.

$$LR+ = P(T+ID+)/P(T+ID-)$$

$$LR+ = Sensitivity / (1 - Specificity)$$

We calculate the *Negative Likelihood Ratio* as:

LR– = P (test negative and *have* the disease) / P (test negative and *do not* have the disease), i.e.

$$LR- = P(T-ID+)/P(T-ID-)$$

$$LR- = (1 - Specificity) / Sensitivity$$

The interpretation of the Likelihood Ratios is intuitive. The larger the Positive Likelihood Ratio, the greater the likelihood of disease when testing positive; the smaller the Negative Likelihood Ratio, the less the likelihood of disease when testing negative.

1.4.2.2 Vaccine Efficacy

Vaccine Efficacy = 1 – risk in vaccinated arm / risk in unvaccinated arm. Take the following numbers, derived from a real Covid-19 trial.

Percentage of infected in vaccinated group = 34/38,995 = 0.00087

Percentage of infected in unvaccinated group = 60/6,583 = 0.00911

Vaccine Efficacy = 1 – 0.00087 / 0.00911 = 1 – 0.0955 = 0.9045 = 90.45%

A Vaccine Efficacy of 90.45% roughly means that you have just over a 90% reduced risk of becoming sick from the disease you are vaccinated against compared to an otherwise similar unvaccinated person.

There are different types of efficacy, however, such as efficacy to prevent infection, efficacy to prevent disease and efficacy to prevent severe disease.

Finally, Vaccine Efficacy refers to estimates from controlled randomised trials, whereas vaccine effectiveness refers to real-world outcomes.

1.4.3 Exercise

1. A patient goes to the surgery to be tested for a virus. We know that 1% of the people who visit the surgery have the virus. The test is 95% accurate, in the sense that 95% of people who have the virus test positive and 95% of those without the virus test negative. What is the probability the patient has the virus if the test is positive?

2. A tennis tournament administers a test for banned substances to all of the tournament entrants. The test is 90% accurate if the person is using the banned substances (90% chance it will show positive if the player is guilty) and 85% accurate if the person is not using them (85% chance it will show negative if the player is not taking the banned substances). And 10% of all tournament entrants are using the banned substances. Now, what is the probability that an entrant is guilty if they test positive?

3. Sixty-six people have the flu and test positive for it. Four people have the flu and test negative for it. Three people don't have the flu but test positive for it. 827 people don't have the flu and test negative for it.

   a. What is the Sensitivity of the test, as a ratio

   b. What is the Specificity of the test, as a ratio?

   c. What is the Positive Predictive Value, as a ratio?

   d. What is the Negative Predictive Value, as a ratio?

   e. What is the Positive Likelihood Ratio, in terms of specificity and sensitivity?

   f. What is the Negative Likelihood Ratio, in terms of specificity and sensitivity?

4. Thousand people are tested for the flu. 100 people have the flu. Of these, 90 test positive and 10 test negative. 900 do not have the flu. 150 of these test positive and 750 test negative.

   a. What is the Sensitivity of the test?

   b. What is the Specificity of the test?

5. a. Explain the meaning of Sensitivity in one sentence.

   b. Explain the meaning of Specificity in one sentence.

   c. In one or two sentences, how would you interpret the size of the Positive and Negative Likelihood Ratios?

6. Vaccine Efficacy calculation = percentage of infected in the control group minus percentage of infected in the vaccinated group, divided by the percentage of infected in the control group.

Take the following numbers derived from a randomised clinical trial.

Number in the vaccinated group = 56,230

Number infected in the vaccinated group = 52

Number in the placebo group = 8,428

Number infected in the placebo group = 88

Calculate the Vaccine Efficacy.

In a sentence, how does vaccine efficacy differ from vaccine effectiveness?

1.4.4 Reading and Links

Attia, J. 2003. Diagnostic tests. Moving beyond sensitivity and specificity: Using likelihood ratios to help interpret diagnostic tests. *Australian Prescriber*, 26(5). https://www.nps.org.au/assets/16fac94ac4fea0f1-5c419b1d2622-9a593de92d40f0579350390894d1066fac2c465761b9f3e953aeb3f399fb.pdf

Dunsson, D.B. 2001. Commentary: Practical Advantages of Bayesian Analysis of Epidemiologic Data. American Journal of Epidemiology, 153, 12, 1222-1226.

Eddy, D.M. 1982. Probabilistic reasoning in clinical medicine: Problems and opportunities. Chapter 18. https://personal.lse.ac.uk/robert49/teaching/mm/articles/Eddy1982_ProbReasoningInClinicalMedicine.pdf

Horgan, J. 2016. Bayes' theorem: What's the big deal? John Horgan. *Scientific American*. 4 January. https://blogs.scientificamerican.com/cross-check/bayes-s-theorem-what-s-the-big-deal/

Lewis, M.A. 2020. Bayes' theorem and Covid-19 testing. *Significance*. 22 April. https://www.significancemagazine.com/science/660-bayes-theorem-and-covid-19-testing

Science Direct. 2013. Sensitivity and specificity. From: Handbook of clinical neurology. https://www.sciencedirect.com/topics/medicine-and-dentistry/sensitivity-and-specificity

StatsDirect. StatisticalHelp. Screening test errors. https://www.statsdirect.com/help/Default.htm#clinical_epidemiology/screening_test.htm

StatsNinja, 2018. How to decipher false positives (and negatives) with Bayes' theorem. 11 October. https://thestatsninja.com/2019/03/03/how-to-decipher-false-positives-and-negatives-with-bayes-theorem/

Woodcock, S. 2017. Paradoxes of probability and other statistical strangeness. *The Conversation*. 4 April. https://theconversation.com/paradoxes-of-probability-and-other-statistical-strangeness-74440

An intuitive (and short) explanation of Bayes' theorem. Better Explained. https://betterexplained.com/articles/an-intuitive-and-short-explanation-of-bayes-theorem/

Are You REALLY Sick? (False Positives). Numberphile. 28 March 2016. YouTube. https://youtu.be/M8xlOm2wPAA

Base Rate Fallacy. Yang, C. 22 March 2017. YouTube. https://youtu.be/Fs8cs0gUjGY

Bayes' Theorem. Fuggetta, M. 21 December 2013.. YouTube. https://youtu.be/49lpesJcpFw

Bayes Theorem Example. Glen, S. 24 December 2014.. YouTube. https://youtu.be/Jht31ML2Hxl

Bayes' Theorem Example. Surprising False Positives. Bazett, T. 6 February, 2018. YouTube. https://youtu.be/HaYbxQC6lpw

Diagnoses, Predictive Values, and Whether You're Sick or Not: NPV and PPV. Healthcare Triage. 1 June 2015. YouTube. https://youtu.be/dHj7ygeqelw

2020. False Positives & Negatives for COVID-19 Tests. Using Bayes' Theorem to Estimate Probabilities. Bazett, T. 21 May 2020. YouTube. https://youtu.be/VuskwslW02M

Sensitivity and Specificity Explained. Physiotutors. 18 March 2017. YouTube. https://youtu.be/UsOv0DcXk6w

Sensitivity and Specificity – Explained in 3 Minutes. Global Health with Greg Martin. 30 April 2014. YouTube. https://youtu.be/FnJ3L-63Cf8

Sickness and Stats (Extra Footage). Numberphile 2. 29 March 2016. YouTube. https://youtu.be/sFNJc6gsap4

Test Characteristics: How Accurate Was that Test? Healthcare Triage. 7 April 2014. YouTube. https://www.youtu.be/UF1T7KzRnrs

The Bayes Theorem: What Are the Odds? Healthcare Triage. 15 April 2014. YouTube. https://www.youtu.be/Ql2jEJ-6e-Y

1.5 Bayes and the Bobby Smith Problem

Bobby Smith is an excellent schoolboy tennis player, but only 1 in 1,000 such children in our stylised world go on to become professional players.

There is a test, however, that can in this world measure potential. This is provided to a thousand of these aspiring tennis players, all of whom want to enter the tennis academy. Only 5% of those tested will gain entry to the academy but then fail to become professional players.

We also learn that graduation from the academy is a condition of entry to the professional tour. So we can rule out anyone who does not go to the academy as a future professional player.

Bobby scores well on the test and is admitted to the academy.

Without any other information, what is the actual chance now that Bobby will become a professional tennis player?

Well, no professional player has ever failed to enter the academy, and only a very small proportion of those tested gain entry to the academy. If the test is this good, therefore, it looks very likely that Bobby will have a bright future as a tennis star.

Is this true, however? Without the test, the chance that a player like Bobby would achieve professional status is admittedly very low. But does that matter? With the test result in hand, it might seem intuitive that there is no need to ask this question.

In fact, this is to confuse the probability of a hypothesis being true (Bobby becomes a professional tennis player) given some evidence (entrance to the academy) – question 1 – with its inverse, i.e. the probability of the evidence arising (entrance to the academy) given the hypothesis is true (Bobby will become a professional player) – question 2. This is a fallacy, known as the "Inverse Fallacy" or "Prosecutor's Fallacy".

In our example, the probability of the evidence arising (entrance to the academy) given that Bobby will make it to professional circles is a sure thing. All future professional players will be graduates of the academy. This is the answer to question 2. What we seek to know, however, is the probability that Bobby will become a professional player given that he enters the academy – question 1. This is a very different question.

So what is the actual chance that Bobby will become a professional tennis player if he scores well enough on the test to gain entry to the academy?

Solution: 50 children who will not become professional players enter the academy (50 "false positives"). Only 1 in 1,000 children who take the test develops into a professional player, and that child will gain entry to the academy.

So the probability that a child, like Bobby, who enters the academy will become a professional tennis player is just 1 in 51, i.e. 1.96%.

It's the same idea as a medical "false positives" problem.

In the equivalent virus version of the problem, a thousand people go to the doctor, where they are all tested for a virus. Only one has the virus and will test positive.

We know that 5% of the 1,000 people will test positive and not have the virus. In other words, there will be 50 "false positives".

So what is the chance that we have the virus if we test positive? In this case, 50 people who do not have the virus test positive. One person who has the virus tests positive. Therefore, the probability we have the virus if we test positive is 1 in 51, i.e. 1.96%.

We can also solve the Bobby Smith problem using Bayes' Theorem, although in this case the solution offered above is simpler. Still, Bayes' Theorem acts as a useful cross-check.

The updated probability that a hypothesis is true after obtaining new evidence, according to the a, b, c formula of Bayes' Theorem, is equal to:

$$ab / \left[ab + c\left(1-a\right) \right]$$

Where,

a is the prior probability, i.e. the probability that a hypothesis is true before the new evidence.

b is the probability of the new evidence if the hypothesis is true.

c is the probability of the new evidence if the hypothesis is false.

In the case of the Bobby Smith problem, the hypothesis is that Bobby will develop into a professional player.

Before the new evidence (the test), this chance is 1 in 1,000 (0.001).

So, a = 0.001.

The probability of the new evidence (entrance to the academy) if the hypothesis is true (Bobby will become a professional player) is 100% since all professional players are graduates of the academy.

So, b = 1.

The probability we would see the new evidence (entrance to the academy) if the hypothesis is false (Bobby will not become a professional player) is 50/999, since 999 of those who took the test will not become professional players, and of these 50 will be admitted to the academy.

So, c = 50 / 999.

Substituting into Bayes' equation gives:

Posterior probability = ab / [ab + c (1 − a)] = 0.001 × 1 / [0.001 × 1 + 50/999 (1 − 0.001)] = 0.001/0.051 = 0.0196

So, using Bayes' Theorem, the chance that Bobby Smith, who gains entry to the academy, will become a top player is not 95% as intuition might suggest, but just 1.96%, as we have shown previously by a different route.

There is, therefore, just a 1.96% chance that Bobby Smith will go on to become a professional tennis player, despite being admitted to the academy.

To look at it another way again, 1,000 children apply for entry to the tennis academy. Of these, 50 will gain entry and never make it to the professional ranks. One will actually become a professional tennis player. So 51 children are admitted to the academy, of which just 1 will make it as a professional. Therefore, the chance of becoming a professional tennis player if you enter the tennis academy is 1 in 51, i.e. 1.96%.

Now for the good news. Bobby Smith was the lucky one. He won the Australian Open under a different name.

1.5.1 Appendix

We can also solve the Bobby Smith problem using the traditional notation version of Bayes' Theorem.

$$P(H|E) = P(E|H).P(H) / \left[P(E|H).P(H) + P(E|H').P(H') \right]$$

Before the new evidence (the test), this chance is 1 in 1,000 (0.001).

So, P(H) = 0.001.

The probability of the new evidence (entrance to the academy) if the hypothesis is true (Bobby will become a professional player) is 100% since all professional players are graduates of the academy.

$$So, \ P(EIH) = 1.$$

The probability we would see the new evidence (entrance to the academy) if the hypothesis is false (Bobby will not become a professional player) is 50/999 since 999 of those who took the test will not become professional players, and of these, 50 will gain entry to the academy.

$$So, \ P(EIH') = 50 / 999.$$

Substituting into Bayes' equation gives:

$$P(HIE) = 0.001 \times 1 / \left[0.001 \times 1 + 50/999 \left(1 - 0.001 \right) \right]$$

$$= 0.001 / 0.051 = 1/51 = 0.0196$$

1.5.2 Exercise

Question 1: Lucy Jones is an excellent school chess player, but only 1 in 1,000 such children goes on to become a professional player.

There is a test, however, that can in this world measure potential. This is provided to a thousand of these aspiring chess professionals, all of whom want to enter the chess academy. Only 2% of those tested will gain entry to the academy but then fail to become professional players.

We also learn that graduation from the academy is a condition of entry to the professional circuit. So we can rule out anyone who does not go to the academy as a future professional chess player.

Lucy scores well on the test and is admitted to the academy.

Without any other information, what is the actual chance now that Lucy will become a professional chess player?

Question 2: The Google Applicants Problem:

Katherine Jane is a graduate who wants to work for Google.

Google has received 1000 graduate applications, including Katherine's, for its five identical entry-level positions.

Five of these applicants will be interviewed and will be successful.

5% of the applicants are interviewed but are unsuccessful.

There are no other positions available. All interviews are conducted before graduates are offered a position.

The only way to get the position at Google is through an interview.

Katherine is offered an interview.

Without any other information, what is the actual chance that Katherine will now secure the position at Google?

- My acknowledgment and thanks to Katherine Fagg for creating the original version of this problem.

1.5.3 Reading and Links

Bayes. 2012. Is your child a football star? 17 July. https://www.massimofuggetta.com/2012/07/17/is-your-child-a-football-star/#.XLXd-GzsY2w

1.6 Bayes and the Broken Window

Let us invent a little crime story in which you are a follower of Bayes and have a friend in a spot of trouble. In this story, you receive a telephone call from the local police station. A police officer says that this friend of many years is helping the police investigate into a case of vandalism of a shop window. It took place in a street close to where she lives at noon that day, which is her day off work. You had heard about the incident earlier but had no good reason at the time to believe that she was in any way linked to it.

She next comes to the telephone and says she has been charged with smashing the shop window, based on the evidence of a police officer who positively identified her as the culprit. There is no other evidence, such as CCTV footage.

She claims mistaken identity.

You must evaluate the probability that she did commit the offence before deciding how to advise her. So the condition is that she has been charged with criminal damage; the hypothesis we are interested in evaluating is the probability that she did it. Bayes' Theorem helps to answer this type of question.

There are three things to estimate.

The first is the Bayesian prior probability (which we represent as "a"). This is the probability you assign to the hypothesis being true before you become aware of the new information. In this case, it means the probability you would assign to your friend breaking the shop window before you received the information that she had been charged.

The second is the probability that the new evidence would have arisen if the hypothesis was correct (which we represent as "b"). In this case, you need to estimate the probability of the police officer identifying your friend if she did break the window.

The third is to estimate the probability that the new evidence would have arisen if the hypothesis was false (which we represent as "c"). In this case, you need to estimate the probability of the police officer identifying your friend if she did *not* break the window.

According to Bayes' Theorem, posterior probability = ab / [ab + c (1 − a)].

So let's apply Bayes' Theorem to the case of the shattered shop window.

Let's start with the prior probability (which we represent as "a").

Well, you have known her for years, and it is totally out of character, but she could have done it. Let us say 5% (0.05). Assigning the prior probability is fraught with problems, as awareness of the new information might easily affect the way you assess the prior information. You need to make every effort to estimate this probability as it would have been before you received the new information.

What about b? This is an estimate of the probability that the police officer would have identified her if she was indeed guilty. If she is guilty, it is easy to imagine how she came to be identified by the police officer. Still, he was not close enough to catch the culprit at the time, which we might bear in mind. Let us estimate the probability to be 80% (0.8).

Let's move on to c. This is an estimate of the probability that the police officer would have identified her if she was not the guilty party, i.e. a false identification. If she did not shatter the window, how likely is the police officer to have wrongly identified her? He might have seen someone of similar age and appearance and jumped to the wrong conclusion, or he may want to identify someone to advance his career. Let us estimate the probability to be 15% (0.15).

Note that b + c does not have to add up to 1.

Once we have assigned these values, Bayes' Theorem can now be applied to establish a posterior probability. The posterior probability is the number in which we are interested. It is the measure of how likely is it that she broke the window, given that she has been identified as the culprit by the police officer.

Given these estimates, we can use Bayes' Theorem to update our probability that the friend is guilty to 21.9% (see Section 1.6.1 "Appendix"), despite assigning reliability of 80% to the police officer's identification.

The clue to the intuitive discrepancy is in the prior probability (or "prior") that you would have attached to the guilt of your friend. If a new piece of evidence now emerges (say a second witness), you should again apply Bayes' Theorem to update to a new posterior probability. In this way, we gradually draw, based on more and more pieces of evidence, ever closer to the truth.

What if the police officer has a reason to identify her either way, in which case b = 1 and c = 1? This implies that the officer would identify the friend as culpable whether she was guilty or not. In this case, it adds no new evidence to the prior probability.

In terms of the equation, ab / [(ab + c (1 − a)] = a / [a + (1 − a)] = a/1 = a.

It is easy to dismiss the implications of this hypothetical case because it was difficult to assign reasonable probabilities to the variables. However, that is what we are doing implicitly when we do not assign numbers. Bayes' Theorem is not at fault for this in any case. It will always correctly update the probability of a hypothesis being true whenever new evidence is identified, based on the estimated probabilities. In some cases, that is not easy, though the approach we adopt to revising our estimates should be better than using intuition to steer a path to the truth.

In many other cases, we do know with precision what the critical probabilities are, and in those cases we can use Bayes' Theorem to identify with precision the revised probability. This often produces startlingly counter-intuitive results.

In seeking to steer the path from ignorance to knowledge, the application of Bayes is the correct method.

1.6.1 Appendix

The calculation and the simple algebraic expression that we have identified in this setting is:

$$ab / \left[ab + c \left(1 - a \right) \right]$$

Where,
   a is the prior probability of the hypothesis (she is guilty) being true. This can also be represented by the notation P (H). In the example, a = 0.05.
   b is the probability the police officer identifies her conditional on the hypothesis being true, i.e. she is guilty. This can also be represented by the notation P (EIH), i.e. probability of E (the evidence) given the hypothesis is true, P (H). In the example, b = 0.8.
   c is the probability the police officer identifies her conditional on the hypothesis not being true, i.e. she is not guilty. This can also be represented by the notation P (EIH'), i.e. probability of E (the evidence) given the hypothesis is false, P (H'). In the example, c = 0.15.

In our example, a = 0.05, b = 0.8, and c = 0.15.

Using Bayes' Theorem, the updated (posterior) probability that the friend is guilty is:

$$ab / \left[ab + c \left(1 - a \right) \right] = 0.04 / \left(0.04 + 0.1425 \right) = 0.04/0.1825$$

$$\text{Posterior probability} = 0.219 = 21.9\%$$

1.6.2 Exercise

You receive a text message from a good friend. The message says that he has been charged with breaking into a local warehouse and stealing some gardening tools. The only evidence is from a neighbour who was passing by and says he recognised him.

Your friend claims he is not guilty and was alone at home at the time.

Which three probabilities do you need to estimate in order to use Bayes' Theorem to evaluate this probability? Choose values for a, b, and c that you think are reasonable in some imaginary scenario, consider why you chose those values, and calculate the result. There is no definitive answer, so only an illustrative solution is provided to the calculation part of this question. For the illustrative solution, we assume a = 0.1, b = 0.8, and c = 0.2.

1.7 The Bayesian Detective Problem

A murder has been committed. There are five suspects, all of whom we consider equally likely to be guilty at the start of the investigation. We know that one of these suspects is the guilty party and must have acted alone.

So, the prior probability of guilt for each of the five possible killers is 20% before finding any new evidence. The names of the suspects are Father Grace, Captain Pepper, Dr. Jones, Miss Golightly, and Professor Wisdom. The victim was Sir Caliban Mackenzie, a famed anthropologist, who was shot in the library while examining a rare first edition of Newton's Principia.

Four hours into the investigation, evidence turns up which eliminates Father Grace. He was celebrating Mass in the chapel at the time of the murder. There are now four remaining suspects, and so the probability that each of the remaining four suspects is guilty rises to 25% (one chance in four).

Two hours later, a new clue arises, which casts some doubt on the alibi of Captain Pepper, whose probability of guilt we now judge to rise from 25% to 40%.

As a result, the probability that one of the other three suspects is guilty falls by 15%, down from a total of 75% to 60%. Since each of the three is equally likely to be guilty, we can now assign each a probability of guilt of 20%, down from 25%.

After 45 minutes, a third clue emerges, which eliminates Dr. Jones. Some very reliable witnesses had spotted her in the congregation at the Mass.

Now that Dr. Jones is out of the frame, what is the best estimate of the revised probability that each of Pepper, Golightly, and Wisdom committed the murder?

One possibility would be to take the 20% probability of guilt we had previously attached to Dr. Jones and divide this equally between the three remaining suspects.

But to do so would be wrong and notably at variance with the toolkit of a Bayesian detective, i.e. a detective who conducts investigations using the Bayesian approach to evidence and probability.

The Bayesian approach considers the prior probability that each suspect is guilty before updating after some new evidence is brought to bear on it. The correct way to adjust the probabilities is in a way that is proportional to the prior probability of guilt of each suspect before Dr. Jones was eliminated. The reason is contained in Bayes' Theorem: P (HIE) = P (EIH) . P (H) / P (E). From the top line of this equation, we see that to update the probability of a hypothesis being true given some new evidence, we must multiply (the probability of the evidence given the hypothesis) by the prior probability of the hypothesis.

Captain Pepper was the prime suspect, with a probability of guilt of 40% before Dr. Jones' elimination. This 40% should be compared to 20% each for Miss Golightly and Professor Wisdom. A Bayesian approach, therefore, is to increase the probability we assign to Pepper's guilt by twice as much as we increase theirs.

So we should now raise the estimate of the probability that Captain Pepper shot Sir Caliban from 40% to 50%. We should increase the probability we assign to Miss Golightly and Professor Wisdom from 20% to 25%.

This is all derived from Bayesian reasoning. We must filter any new evidence through the prior probability of the hypothesis being true before we became aware of the new evidence. This new evidence is Dr. Jones' elimination from the enquiry. This prior probability was twice as great for Captain Pepper as for either of the other remaining suspects.

1.7.1 Epilogue

The estimated 50% probability of guilt was sufficient to persuade the Crown Prosecution Service to haul the Captain before a jury of his peers. In the event the jury convicted him, falling victim to the classic Prosecutor's Fallacy. Like so many juries before them, they confused the probability that someone is guilty in light of the evidence with the likelihood of the evidence arising if they were guilty. The likelihood that Sir Caliban was shot in the library if the Captain was guilty of his murder was naturally rather high, and this led to his conviction. Unfortunately for the Captain, the relevant probability (that he was guilty of murder given that Sir Caliban was shot in the library) was somewhat smaller but bypassed in the jury's deliberations.

Meanwhile, the actual killer, Miss Golightly, got away scot-free. She had concealed a letter with evidence of her involvement in a blackmail plot in a leather-bound volume of Newton's Principia. Unhappily, Sir Caliban had chanced upon it. It left her no option, in her mind, but to use the pistol hidden in the Georgian chest of drawers gracing the back wall of the library. The Captain's appeal was rejected unanimously. He is serving a life sentence. Miss Golightly is living as a tax exile in Belize.

1.7.2 Exercise

A murder has been committed, and there are eight people who could have done it. They are Audrey, Daisy, David, Delilah, Gloria, Joe, Julie, and Tony.

There are no clues, no prior history of which we know. We know that who-ever did it must have acted alone.

So, we consider each suspect equally likely to be guilty at the start of the investigation.

1. What is the prior probability of guilt for each suspect?

2. Now the first clue is found, which eliminates four of the suspects, Audrey, David, Gloria and Julie. What is the new probability that each of the remaining individual suspects, Daisy, Delilah, Joe, and Tony, is guilty?

3. A new clue now arises which casts doubt upon the alibi of Daisy, the owner of the home where the murder was committed. Her probabil-ity of guilt, we now judge, rises to 40%. What is the new probability that each of the other suspects, Joe, Delilah, and Tony, is guilty?

4. The third clue now eliminates Tony. What are the new probabilities of guilt that you, as a Bayesian, will attribute to the remaining sus-pects, Daisy, Delilah, and Joe?

5. Which of these suspects is more likely than not to be guilty of the crime?

1.8 Bayesian Bus Problems

The Bus Problem

Every day, Ted gets the solitary 8 am bus to work. There is no other bus that will get him to his destination.

The bus is early 10% of the time and leaves before he arrives at 8 am.

The bus is late 10% of the time and leaves after 8.10 am.

The rest of the time, the bus departs between 8 am and 8.10 am.

One morning Ted arrives at the bus stop at 8 am, sees no bus, and waits for 10 minutes without the bus coming.

Now, at 8.10 am, what is the probability that Ted's bus will still arrive?

Solution

When Ted arrives at 8 am, there is a 10% chance that his bus will have already left. After he has waited for 10 minutes, he can eliminate the 80% chance of the bus arriving in the period between 8 am and 8.10 am. So only two pos-sibilities remain.

Either the bus has arrived ahead of schedule, or it will leave after 8.10 am. The prior probability that the bus will arrive between 8 am and 8.10 am is 80%.

When the bus does not arrive in that time slot, Ted should distribute the 80% to the other two possibilities in proportion to their individual prior probabilities.

Both outcomes are unusual, but since they are mutually exclusive and equally likely (10% chance of each), the 80% should be shared equally between them, by adding 40% to each.

So, the updated probability that the bus has already left is updated from 10% to 50%, as is the probability that the bus will still arrive. So, there is a 50% chance that he will still catch his bus if he has the patience to wait further, and the same chance that he will wait in vain.

This is all derived from Bayesian reasoning. We must filter any new evidence through the prior probability of the hypothesis being true before we become aware of the new evidence.

The correct way to adjust the probabilities is in a way that is proportional to the prior probability of each option before that option was eliminated. In this example, both prior probabilities were equal. In the exercise to this chapter, that is not the case and the solution must be adapted accordingly.

1.8.1 Exercise

The Bus Plus Problem

Every day, Jack gets the solitary 8 am bus to work. There is no other bus that will get him to his destination.

The bus is early 10% of the time and leaves before he arrives at 8 am.

The bus is late 30% of the time and leaves after 8.10 am.

The rest of the time the bus departs between 8 am and 8.10 am.

One morning Jack arrives at the bus stop at 8 am, sees no bus, and waits for 10 minutes without the bus arriving.

Using Bayesian reasoning, what is the probability that Jack's bus will still arrive?

1.9 Bayes at the Theatre

William Shakespeare wrote the majestic play, Othello, in about 1603. The plot revolves around four central characters: Othello, a General in the Venetian army; his wife, Desdemona; his loyal lieutenant, Cassio; and his trusted ensign, Iago.

A vital element of the play is Iago's plot to convince Othello that Desdemona is having an affair with Cassio. He does this by planting in Cassio's lodgings a treasured keepsake Othello gave to Desdemona, for Othello "accidentally" to find.

If Othello refuses to contemplate any possibility that Desdemona is betraying him, then no evidence could ever change his mind. Othello might concede, however, that there is at least a possibility that Desdemona is betraying him, however small that chance might be. This means that there would exist some level of evidence, however great it would need to be, that would leave him no alternative.

On the other hand, Othello might adopt a more balanced position, trying to assess the likelihood objectively and without emotion. But how? This is where it is easy for Othello to come unstuck. The temptation might be to assume that since it might or might not be true, the likelihood is 50%. This is known as the "Prior Indifference Fallacy". Once Othello falls victim to this fallacy, any evidence against Desdemona could prove devastating.

An alternative would be to find evidence that is logically incompatible with Desdemona's guilt.

What else could Othello turn to? Maybe Bayes' Theorem is the answer.

The (posterior) probability that a hypothesis is true after obtaining new evidence, according to the a, b, c formula of Bayes' Theorem, is equal to:

$$ab / \left[ab + c \left(1 - a \right) \right]$$

Where, a is the prior probability, i.e. the probability that a hypothesis is true before seeing the new evidence; b is the probability of seeing the new evidence if the hypothesis is true; and c is the probability of seeing the new evidence if the hypothesis is false.

In the case of the Othello problem, the hypothesis is that Desdemona is guilty of betraying Othello with Cassio. Before the new evidence (the finding of the keepsake), let us say that Othello assigns a chance of 4% to Desdemona being unfaithful.

So, a = 0.04.

The probability of seeing the new evidence (the keepsake in Cassio's lodgings) if the hypothesis is true (Desdemona and Cassio are conducting an affair) is, say, 50%. There is quite a good chance she would secretly hand Cassio the keepsake as proof of her love for him and not of Othello.

So, b = 0.5.

The probability of seeing the new evidence (the keepsake in Cassio's lodgings) if the hypothesis is false is, say, just 5%. Why would it be there if Desdemona had not been to his lodgings secretly, and why would she take the keepsake along in any case? It could have been stolen and ended up there, but how likely is that?

So, c = 0.05.

Substituting into Bayes' equation gives:

$$\text{Posterior probability} = ab / [ab + c(1 - a)] = 0.294.$$

So, using Bayes' Rule, Othello might assign a probability that Desdemona is guilty of betraying him with Cassio at 29.4%.

If Iago concludes that this is too low a probability for his purposes, his task is to convince Othello to lower his own estimate of "c" from 0.05 to, say, 0.01.

The new Bayesian probability of Desdemona's guilt now becomes:

$$ab / [ab + c(1-a)]$$

a = 0.04 (the prior probability of Desdemona's guilt, as before)

b = 0.5 (as before)

c = 0.01 (down from 0.05)

Substituting into Bayes' equation gives:

$$\text{New probability} = 0.676 = 67.6\%.$$

In contrast, the strategy for Desdemona is to convince Othello that the chance of the keepsake being found in Cassio's place if she were guilty is much lower than 0.5.

For example, she could argue that she would hardly have risked discovery of the keepsake if she were indeed having an affair with Cassio. In other words, she could argue that the presence of the keepsake provides evidence in favour of her innocence. In Bayesian terms, she should try to reduce Othello's estimate of b.

William Shakespeare wrote Othello about a hundred years before Reverend Thomas Bayes was born. That is true. However, the Bard surely was always, in every inch of his being, a true Bayesian. Othello was not, and therein lies the tragedy.

1.9.1 Appendix

The hypothesis is that Desdemona is guilty of betraying Othello with Cassio. Before the new evidence (the finding of the keepsake), let us say that Othello assigns a chance of 4% to Desdemona being unfaithful.

Using traditional notation,

$$P(H) = 0.04$$

The probability we would see the new evidence (the keepsake in Cassio's lodgings) if the hypothesis is true (Desdemona and Cassio are conducting an affair) is, say, 50%.

So, $P(EIH) = 0.5$.

The probability we would see the new evidence (the keepsake in Cassio's lodgings) if the hypothesis is false is, say, just 5%.

So, P (EIH') = 0.05.

Substituting into Bayes' Theorem:

$$P(HIE) = P(EIH) \cdot P(H) / \left[P(EIH) \cdot P(H) + P(EIH') \cdot P(H') \right]$$

$$P(HIE) = 0.5 \times 0.04 / [0.5 \times 0.04 + 0.05 \times 0.96]$$

$$P(HIE) = 0.02 / [0.02 + 0.048] = 0.294$$

Posterior probability = 0.294.

So, using Bayes' Rule and these estimates, the chance that Desdemona is guilty of betraying Othello is 29.4%.

$$\text{If } P(EIH') = 0.01.$$

The new Bayesian probability of Desdemona's guilt now becomes:

$$P(HIE) = 0.5 \times 0.04 / [0.5 \times 0.04 + 0.01 \times 0.96]$$

$$P(HIE) = 0.02 / (0.02 + 0.0096) = 0.02 / 0.0296 = 0.676$$

Updated probability = 0.676 = 67.6%.

1.9.2 Exercise

What are the implications of an increase/decrease in a, b, and c for Othello's perception that Desdemona is having an affair with Cassio?

1.9.3 Reading and Links

Bayes. 2012. Iago's trick. 8 September. https://www.massimofuggetta.com/2012/12/08/iagos-trick/#.XLXijWzsY2w

Shakespeare Birthplace Trust. 2021. Othello: The Moor of Venice. Synopsis and plot overview of Shakespeare's Othello: The Moor of Venice. https://www.shakespeare.org.uk/explore-shakespeare/shakespedia/shakespeares-plays/othello-moor-venice/

1.10 Bayes in the Courtroom

On 9 November 1999, Sally Clark, a 35-year-old solicitor and mother of a young child, was convicted of murdering two of her children. The presiding judge, Mr. Justice Harrison, declared that "we do not convict people in these courts on statistics. It would be a terrible day if that were so". As it turned out, it was indeed a terrible day for Sally Clark and the justice system.

The death of the children was attributed at first to natural causes, probably SIDS ("Sudden Infant Death Syndrome"). Later the Home Office pathologist dealing with the case became suspicious, and Sally Clark was charged with murder and tried at Chester Crown Court. It eventually transpired that essential evidence in her favour had not been disclosed to the defence, but not before a failed appeal in 2000. At a second appeal, in 2003, she was set free, and the case is now recognised as a major miscarriage of justice.

So what went wrong?

A turning point in the trial was the evidence given by a key prosecution witness, who argued that the probability of a child in that environment dying of SIDS was 1 in 8,543. If we are to accept that estimate (which is in fact disputed), the probability of two children dying of SIDS, so he argued, was that fraction squared, or 1 in about 73 million. It's the chance, he argued,

> of backing that long odds outsider at the Grand National ... let's say it's an 80 to 1 chance ... to get to these odds of 73 million you've got to back that 1 in 80 chance four years running ... So it's the same with these deaths ... it's very, very, very unlikely.

The jury convicted her, and the judge sentenced Sally Clark to life in prison.

However, the evidence was flawed, as anyone with a basic understanding of probability would have been aware. It is only correct to multiply probabilities if the probabilities are independent of each other. This would be true only if the cause of death of the first child were independent of the cause of the death of the second child. It assumes no genetic, environmental, or other innocent link between these sudden deaths at all. That is a fundamental error of classical probability.

The other error is somewhat more sinister, in that it is harder to detect the flaw in the reasoning. It is the "Prosecutor's Fallacy", as we have already noted in other sections of this chapter. This is the fallacy of taking the probability of guilt given the available evidence to be the same thing as the probability of the evidence arising given guilt. With particular regard to this case, the following propositions are very different:

1. The probability of observing some evidence (the death of the children) given that a hypothesis is true (that Sally Clark is guilty).

2. The probability that a hypothesis is true (that Sally Clark is guilty) given that we observe some evidence (the death of the children).

These are very different propositions, the probabilities of which diverge widely. Notably, the probability of the former proposition is massively higher than that of the latter. Indeed, the probability of the children dying, given that Sally Clark is a child murderer, is effectively 1 (100%). However, the probability that she is a child murderer given that the children have died is a whole different matter.

Critically, we need to consider the probability that she would kill her children before we are made aware of the sudden deaths. This is the concept of "prior probability", which is central to Bayesian reasoning. We should not view this prior probability through the lens of the later emerging evidence. It should be established on its own merits and then merged using Bayesian updating with the new evidence.

In establishing this prior probability, we need to ask whether there was any other past indication or evidence to suggest that she was a child murderer. Otherwise, the number of mothers who murder their children is very small indeed. Without such evidence, the prior probability of guilt should correspond to something like the proportion of mothers in the general population who serially kill their children. This prior probability of guilt is extremely low. In order to update the probability of guilt, given the evidence of the deceased children, the jury needs to weigh up the relative likelihood of the two competing explanations for the deaths. Which is more likely? Double infant murder by a mother or double SIDS? That is not a question that the jury members, unversed in Bayesian reasoning or conditional probability, seem to have asked themselves. If they did, they reached the wrong conclusion.

More generally, it is likely in any large enough population that one or more cases will occur of something highly improbable in any particular case. Out of the entire population, there is a very good chance that some random family will suffer a case of double SIDS.

In a letter from the President of the Royal Statistical Society to the Lord Chancellor, Professor Peter Green explained the issue succinctly:

> The jury needs to weigh up two competing explanations for the babies' death: SIDS or murder. The fact that two deaths by SIDS is quite unlikely is, taken alone, of little value. Two deaths by murder may well be even more unlikely. What matters is the relative likelihood of the deaths under each explanation, not just how unlikely they are under one explanation.

> (Green, 2002)

To look at the Sally Clark case another way, take the wholly fictional case of Lottie Jones, who is charged with winning the National Lottery by cheating. The prosecution expert gives the following evidence. The probability of winning the Lottery (pick six numbers from 59) jackpot without cheating, he tells the jury, is 1 in 45 million. Lottie won the Lottery. What is the chance she could have done so without cheating in some way? The chance is 1 in 45 million. So she must be guilty. Sounds ridiculous put like that, but it is the same sort of reasoning that sent Sally Clark to prison in real life.

As in the Sally Clark case, the prosecution witness in this fictional parody commits the classic "Prosecutor's Fallacy". He assumes that the probability Lottie is innocent of cheating, given that she won the Lottery, is the same thing as the probability of her winning the Lottery if she is innocent of cheating. The former probability is much higher than the latter unless we have some other indication that Lottie has cheated to win the Lottery. It is an example of how it is likely, in any large enough population, that things will happen that are improbable in any particular case. The probability that needed to be established in the Lottie case was the probability that she would win the Lottery before she did. If she is innocent, that probability is one in tens of millions. The fact that she did win the Lottery does not change that.

In other words, the 1 in 45 million represents the probability that a Lottery entry at random will win the jackpot, not the probability that a player who has won did so fairly!

Lottie just got very, very lucky just as Sally Clark got very, very unlucky.

Sally Clark never recovered from the trauma of losing her children and spending years in prison falsely convicted of killing them. She died on 16 March 2007, of acute alcohol intoxication.

A very different, but related, example of the misapplication of conditional probabilities in the courtroom can be found in the trial of former actor and NFL football star, O.J. Simpson. It was established at his trial that he had repeatedly beaten his wife, Nicole Brown, the murder of whom he was charged with. His defence counsel produced statistics that only one in 2,500 cases of spousal abuse leads to the death of the spouse, which as a statistic was broadly accurate. This conditional probability was, however, irrelevant to the case. The relevant statistic is the probability that a woman's murderer is her partner, if she has been previously physically assaulted by him. In a letter to *Nature* magazine in 1965, the statistician I.J. Good produced an estimate, using some simplifying assumptions, that the latter statistic is about 1 in 3, i.e. about a third of murdered women were murdered by someone who had previously physically assaulted them. Using more detailed statistics, Skorupski and Wainer, in an article in *Significance* in 2015, calculated that the probability that a murdered woman was murdered by a partner who had previously beaten her was 29%, similar to that of Good's estimate.

In other words, Simpson's defence team provided the jury with a broadly accurate but irrelevant probability. Using publicly available US data, Skorupski and Wainer showed that the relevant statistic was over 700 times more likely than the irrelevant statistic.

Specifically, using the letter B as notation to represent "woman battered by her husband, boyfriend or lover", M to represent "woman murdered", and M,B to represent "woman murdered by her batterer", the relevant statistic is P (M,B I M), i.e. the probability that a woman was murdered by her batterer given that she has been murdered. The irrelevant statistic is P (M,B I B), i.e. the probability that a woman is murdered by her batterer given that she has been battered

Using Bayes' Rule, we note that:

$$P(M,B I M) = P(M I \acute{M},B).P(M,B) / P(M)$$

P (M I M,B) = 1, since the probability that a woman has been murdered given that she has been murdered by her batterer is clearly 1.

So, P (M,B I M) = P (M,B) / P (M), which works out to 0.29.

It is, therefore, not only prosecutors who are guilty of some variant of the classic Prosecutor's Fallacy.

1.10.1 Exercise

What is the Prosecutor's Fallacy, using an equation or equations to illustrate your answer? How might this fallacy lead to false convictions?

1.10.2 Reading and Links

Bailer-Jones, C.A.L. 2013, *Statistical Methods SS2013*. Sally Clark Case, 6 March. http://www.mpia.de/~calj/statistical_methods_ss2013/homework/h02_sally_clark_cot_death.pdf

Ben-Israel, A. 2007. Using statistical evidence in courts. A case study. 2 October. http://ben-israel.rutgers.edu/711/Sally_Clark.pdf

Brown, R.J. 2014. Sally Clark – What went wrong? http://www.mathestate.com/Sally%20Clark%20-%20What%20went%20wrong.pdf

Centre for Evidence-Based Medicine. 2018. Taylor, K. The prosecutor's fallacy. 16 July. https://www.cebm.net/2018/07/the-prosecutors-fallacy/

Coles, P. 2013. Archive for Sally Clark. https://telescoper.wordpress.com/tag/sally-clark/

Cornell University. 2018. Bayes' theorem in the court – The prosecutor's fallacy. November 28. http://blogs.cornell.edu/info2040/2018/11/28/bayes-theorem-in-the-court-the-prosecutors-fallacy/

Coursera. The sad story of Sally Clark. An intuitive introduction to probability. University of Zurich. https://www.coursera.org/lecture/introductiontoprobability/the-sad-story-of-sally-clark-bII6g

Disney, M. Cot deaths, Bayes' theorem and plain thinking. http://www2.geog.ucl.ac.uk/~mdisney/teaching/GEOGG121/bayes/COT%20DEATHS.doc

Fenton, N., Neil, M. and Berger, D. 2016. Bayes and the law. *Annual Review of Statistics and Its Applications*, 3, 51–77. https://www.ncbi.nlm.nih.gov/pmc/articles/PMC4934658/

Good, I.J. 1995. When batterer turns murderer. *Nature*, 375, 541.

Green, P. 2002. Letter from the President of the RSS to the Lord Chancellor, 23 January.

Joyce, H. 2002. Beyond reasonable doubt. + *plus magazine*. 1 September. https://plus.maths.org/content/beyond-reasonable-doubt

McGrayne, S.B. Simple Bayesian problems. The Sally Clark case. http://www.mcgrayne.com/disc.htm

Robertson, B., Vignaux, G.A. and Berger, C.E.H. 2016. *Interpreting evidence: Evaluating forensic science in the courtroom*, 2nd edition. Chichester: Wiley.

Royal Statistical Society. 2019. Statistics and probability for advocates: Understanding the use of statistical evidence in courts and tribunals. The Inns of Court College for Advocacy. https://www.icca.ac.uk/wp-content/uploads/2019/11/RSS-Guide-to-Statistics-and-Probability-for-Advocates.pdf

Scheurer, V. October 2009. Understanding uncertainty. Convicted on statistics? https ://understandinguncertainty.org/node/545

Skorupski, W.P. and Wainer, H. 2015. The Bayesian flip. Correcting the prosecutor's fallacy. *Significance*, 12, August, 16–20.

Sally Clark. Wikipedia. https://en.m.wikipedia.org/wiki/Sally_Clark

Prosecutors fallacy. Vaughen, C.S. 23 November 2012. YouTube. https://www.you tube.com/watch?v=nekZF8Gmrug&feature=youtu.be

The Prosecutor's Fallacy. Humphries, M. 6 January 2015. YouTube. https://youtu.be/ V4cMWoGxEwo

2

Probability Paradoxes

In this chapter we examine some classic and some novel problems and paradoxes of probability. Classic paradoxes include the Bertrand's Box Paradox, the Boy–Girl Paradox, and the Monty Hall Problem. The solution to each of these paradoxes presents a conflict with common intuition. We also introduce classic paradoxes involving statistics and data presentation, including Simpson's Paradox, Berkson's Paradox (also known as collider bias), and the Will Rogers Phenomenon. Other classic challenges to intuition explored in this chapter include the Three Prisoners Problem, the Two Envelopes Problem, the Girl Named Florida Problem, and the Birthday Problem. Novel and less well-explored paradoxes that we consider include the Inspection Paradox, the Deadly Doors Problem, and Portia's Challenge.

2.1 The Bertrand's Box Paradox

Someone presents us with three identical boxes. Inside one of the boxes there are two gold coins; inside another box there are two silver coins. The third box holds inside one gold coin and one silver coin. We do not know which coins are in which box.

Now, let us choose a box at random. We reach without looking under the cloth covering the coins and take out one of the coins. Now we can look. It is gold.

So, we can be sure that the box we chose cannot be the box containing the two silver coins. It must be either the box containing two gold coins or the box containing one gold coin and one silver coin. So, the other coin must be either a gold coin or a silver coin.

Given what we now know, what is the probability that the other coin in the box is also gold, and what odds should we take to bet on it?

This is essentially the so-called "Bertrand's Box" Paradox, first proposed by Joseph Bertrand in 1889.

After withdrawing the gold coin, there are only two options left. One is the box containing the two gold coins, and the other is the box containing one gold and one silver coin. It seems intuitively clear that each of these boxes is equally likely to be the one we chose at random, and that therefore the chance it is the box with two gold coins is 1/2, and the chance that it is the box containing one gold and one silver coin is also 1/2. The probability that the other coin is a gold coin must, therefore, be 1/2.

Is this right? In fact, the answer is not 1/2. The probability that the other coin is also gold is 2/3. But this seems counter-intuitive. How can looking at one of the coins make the other coin likely to be of the same type of metal?

To understand why, let us look a little closer. Three equally likely scenarios might have led to us choosing that shiny gold coin. Let us separately label all the coins in the boxes to make this clear.

In the box containing two gold coins, there will be Gold Coin 1 and Gold Coin 2. These are both gold coins, but they are distinct, different coins.

In the box containing the gold and silver coins, we have Gold Coin 3, which is a different coin to Gold Coin 1 and Gold Coin 2. There is also what we might label Silver Coin 3 in the box with Gold Coin 3. This silver coin is distinct and different to what we might label Silver Coin 1 and Silver Coin 2, which are in the box containing two silver coins, which was not selected.

So, here are the equally likely scenarios when we withdrew a gold coin from the box.

a. We chose Gold Coin 1.

b. We chose Gold Coin 2.

c. We chose Gold Coin 3.

We do not know which of these gold coins we withdrew from the box.

If it was Gold Coin 1, the other coin in the box is also gold.

If it was Gold Coin 2, the other coin in the box is also gold.

If it was Gold Coin 3, the other coin in the box is silver.

Each of these possible scenarios is equally likely, so the probability that the other coin is gold is 2/3 and the probability that the other coin is silver is 1/3.

Before withdrawing the gold coin, the chance that the box we had selected was that containing two gold coins was 1/3. By revealing the gold coin, however, we not only excluded the box containing two silver coins but also introduced the new information that we could potentially have chosen a silver coin (if the selected box was that containing one gold and one silver coin) but did not.

That made it more likely (twice as likely) that the box you withdrew the gold coin from was that containing the two gold coins than the box containing one gold and one silver coin.

This is the solution to the Bertrand's Box Paradox.

2.1.1 Exercise

There are three cards. One of the cards has two yellow faces, one has two blue faces and one is yellow on one side and blue on the other.

You are presented with one of these cards, face up on the table. It is blue.

What is the probability that the card is blue on both sides? Explain your reasoning.

2.1.2 Reading and Links

Steemit. 2017. Bertrand's Box problem. https://steemit.com/science/@galotta/bertrand-s-box-problem

Zymergi. 2013. Bertrand's Box paradox. 6 June. http://blog.zymergi.com/2013/06/bertrands-box-paradox.html

Untrammeled Mind. 2018. Bertrand's Box Paradox (with and without Bayes' Theorem). 9 November. https://www.untrammeledmind.com/2018/11/bertrands-box-paradox/

Why Evolution is True. 2018. Bertrand's Box paradox: The answer is 2/3!!! 20 February. https://whyevolutionistrue.wordpress.com/2018/02/20/the-answer-is-2-3/

Bertrand's box paradox. Guestling Bradshaw C of E Primary School. 18 May 2018. YouTube. https://youtu.be/CGMc8B60ZpU

2.2 The Monty Hall Problem

The Monty Hall Problem is a famous, perhaps the most famous, probability puzzle ever to have been posed. It is based on an American game show, Let's Make a Deal, hosted by Monty Hall, and came to public prominence as a question quoted in an "Ask Marilyn" column in *Parade* magazine in 1990.

Suppose you're on a game show, and you're given the choice of three doors. Behind one door is a car: behind the others, goats. The car is randomly placed behind one of the doors, so it's not possible for you to guess based on prior observation or information. You choose a door, say No. 1, and the host, who knows what's behind all the doors, opens another door, say No. 3, which reveals a goat. He then says to you, "Do you want to switch to Door No. 2?" This is not a strategic decision on his part based on knowing that you chose the car, in that he always opens one of the doors concealing a goat and offers the contestant the chance to switch. It is part of the rules of the game.

So should you switch doors?

Consider the probability that you chose the correct door the first time, i.e. No. 1 is the door to the car. What is that probability? Well, it is 1/3 in that you have three doors to choose from, all equally likely.

But what happens to the probability that Door No. 1 is the key to the car once Monty has opened one of the other doors?

This again seems quite straightforward. There are now two doors left unopened, and there is no way to tell behind which of these two doors lies the car. So the probability that Door 1 offers the star prize now that Door 2 (or else Door 3) has been opened would seem to be 1/2. So should you switch? Since the two remaining doors would seem to be equally likely paths to the car, it would seem to make no difference whether you stick with your original choice of Door 1 or switch to the only other door that is unopened.

But is this so? Marilyn Vos Savant, in her "Ask Marilyn" column, declared that you should switch doors to boost your chance of winning the car. This answer was decried by the great majority of the readers who wrote in. They reasoned that once the host opened the door, only two doors remained closed, so the chance that the car was behind either of the doors was identical, i.e. a half. For that reason, whether the contestant switches or sticks to the original choice should make no difference to the chance of winning the car. Vos Savant argued, in contrast, that switching doubled the chance of winning the car.

Let's think it through.

When you choose Door 1, there is a 1 in 3 chance that you have won your way to the car if you stick with it. There is a 2 in 3 chance that Door 1 leads to a goat. On the other hand, if you have chosen Door 1, and it is the lucky door, the host must open one of the two doors concealing a goat. He knows that. You know that. So he is introducing useful new information into the game.

Before he opened a door, there was a 2 in 3 chance that the lucky door was *either* Door 2 or Door 3 (as there was a 1 in 3 chance it was Door 1). Now he is telling you that there is a 2 in 3 chance that the lucky door is *either* Door 2 or Door 3 *but* it is not the door he just opened. So there is a 2 in 3 chance that it is the door he didn't open. So, if he opened Door 2, there is a 2 in 3 chance that Door 3 leads to the car. Likewise, if he opened Door 3, it is a 2 in 3 chance that Door 2 leads to the car. Either way, you are doubling your chance of winning the car by switching from Door 1 (probability of car = 1/3) to whichever of the other doors he did not open (probability of car = 2/3).

It is because the host knows what is behind the doors that his actions, which are constrained by the fact that he can't open the door to the car, introduce valuable new information. Because he can't open the door to the car, he is obliged to point to a door that isn't concealing the car, increasing the probability that the door he doesn't open is the lucky one (from 1/3 to 2/3).

Put simply, if you do not switch, you only win if the door your originally chose leads to the car. That chance is 1/3. If you switch, you win the car if it were behind either of the other doors. That's a 2/3 chance.

Let's itemise the choices.

1. If the car is behind Door 1, and you choose it and stick with your choice, you win the car. Chance = 1/3.

2. If the car is behind Door 2, and you choose Door 1, the host will open Door 3. If you switch, you win. Chance = 1/3.

3. If the car is behind Door 3, and you choose Door 1, the host will open Door 2. If you switch, you win. Chance = 1/3.

   So, you win with a 2/3 chance if you switch, and 1/3 if you stick to your original choice.

Essentially, what is happening is that the odds are against you winning the car when you select your door. With three doors, there's a 2/3 chance that you have chosen a loser. When Monty opens the other door, however, he's eliminating the other loser. So, if you're probably holding a loser, and the other loser has been eliminated, the odds are (2 in 3 chance) that the car is behind the door you didn't select. So, switch!

If this is still not intuitively clear, there is a way of making it more so. Let's say there were 20 doors, with a car behind one of them and goats behind 19 of them.

Now say we choose Door 1. This means that the probability that this is the winning door is 1 in 20. There is a 19 in 20 probability that one of the other doors conceals the car.

Now Monty starts opening one door at a time, taking care not to reveal the car each time. After opening a carefully chosen 18 doors (chosen because they didn't conceal the car), just one door remains. This could be the door to the car, or your original choice of Door 1 could be the path to the car. But your original choice had an original probability of 1/20 of being the winning door. Nothing has changed that, because every time he opens a door he is sure to avoid opening a door leading to the car.

So the chance that the door he leaves unopened points to the car is 19/20. So, by switching, you multiply the probability that you have won the car from 1/20 to 19/20.

In other words, once you make your original selection, the host opens 18 of the remaining doors, each of which he knows doesn't lead to the car. If you stick with your original door, you have a chance of 1/20 of winning the car. If you switch to the last unopened door, you would win if the car is behind any of the other doors (a chance of 19/20).

If he didn't know what lay behind the doors, he could inadvertently have opened the door to the car, so when he does so this adds no new information save that he has randomly eliminated one of the doors. If he randomly opens 18 doors, not knowing what is behind them, and two doors now remain, they each offer a 1 in 2 chance of the car. So you might as well just flip a coin – and hope!

Even when it is explained this way, many people find it impossible to grasp the intuition. So here's the clincher.

Say I have a pack of 52 playing cards, which I lay face down. If you choose the Ace of Spades, you win the car. Every other playing card, you win nothing. Go ahead and choose one. This is now laid aside from the rest of the deck, still face down. The probability that the card you have chosen is the Ace of Spades is 1/52.

Now I, as the host, know exactly where the Ace of Spades is. There is a 51/52 chance that it must be somewhere in the rest of the deck. Now, I carefully turn over the cards in the deck one at a time, taking care never to turn over the Ace, until there is just one card left.

What is the chance that the one remaining card from the deck is this card? It is 51/52 because I have carefully sifted out all the losing cards to leave just one card, the Ace of Spades.

In other words, I have presented you with the 1 card out of the remaining deck of 51 that is the Ace of Spades, assuming that it was not the card you chose in the first place. The chance that the card you chose in the first place was that particular card is 1/52. So the card I have selected for you out of the remaining deck has a probability of 51/52 of being the Ace of Spades.

So should you switch when I offer you the chance to give up your original card for the one that I have filtered out of the remaining 51 cards (taking care each time never to reveal the Ace of Spades)? Of course you should, assuming I know where it's located.

As a final perspective, let's report the results of a challenge posed by *Significance* magazine, published in its October 2013 issue. The challenge was to readers to explain the Monty Hall challenge briefly and clearly. Of those printed, two caught my eye.

The first, offered by reader Peter Button, has echoes of the Bertrand's Box Paradox in its explanation. Instead of labelling the coins, though, he suggests labelling the goats. The idea is that the probability of initially selecting the door concealing the car is 1/3, and similarly for goat A and goat B. Let's say the contestant chooses the door to the car. In that case, sticking with the original choice is the way to win the car. If the contestant chose the door concealing goat A, on the other hand, the host would have no choice but to open the door to goat B. This is because the host *never* opens a door with the car. In this case, the way to win the car is to switch. If goat B is chosen, again the way to win the car is to switch, for the same reason. So, if the contestant chooses to "stick" they would win the car one time out of three. By opting to switch, they would win two times out of three. So the optimal strategy is to switch.

For a neat mix of brevity and clarity, we also have the entry offered by reader Gib Bassett, which goes something like this.

> You choose door A. (i) You stay with door A. *Result*: you win *only* if the car is behind A – 1/3 of the time. (ii) You switch to the "other" door. *Result*: you lose *only* if the car is behind A – 1/3 of the time. Hence you win 2/3 of the time. This is because Monty *never* opens a door with the car. So, if the car is not behind A it will be behind the "other" door.

2.2.1 Appendix

In the standard description of the Monty Hall Problem, Monty can open Door 1 or Door 2 or Door 3. The car can be behind Door 1, Door 2, or Door 3. The contestant can choose any door.

We can apply Bayes' Theorem to solve this.

If P (C) = probability that your chosen door, say Door 1, leads to the car.

P (G) = probability that Monty will choose a door that leads to a goat.

So, P (C) = 1/3.

P (G) = 1, as Monty will never reveal the car.

So, what is the probability that your chosen door, say Door 1, leads to the car, given that Monty opens a door that leads to a goat, i.e.

What is P (C I G)?

P (C I G) = P (G I C) . P (C) / P (G)

P (G) = 1, as Monty will always choose the door leading to a goat.

P (C) = 1/3, as there are three doors to choose from, each equally likely to lead to the car.

P (G I C) = 1, as Monty will always choose a door to a goat if the selected door leads to the car.

So, P (C I G) = 1 . 1/3 / 1 = 1/3. So the probability the original selection leads to the car = 1/3.

So the probability that the other unopened door leads to the car = 2/3.

This relies on Monty always opening a door that leads to a goat, i.e. P (G) = 1 and P (G I C) = 1.

The same logic applies whichever door is selected by the contestant.

So, what happens if Monty does not know where the car is and opens a door randomly?

Now, P (G) = 2/3 as Monty has no idea where the car is, and two of the three doors contains a goat.

What is P (G I C)?

Monty will select a door leading to a goat every time when the contestant has selected the door leading to the car.

So, P (G I C) = 1.

This happens 1/3 of the time; 2/3 of the time he will choose a door to a goat with a probability of 1/2.

In this case, P (G) = 1/3 × 1 + 2/3 × 1/2 = 2/3.

P (C) = 1/3.

So, P (C I G) = 1 . 1/3 / 2/3 = 1/2.

So, switching does not change the probability of winning the car.

2.2.1.1 *Alternative Derivation*

In the standard description of the Monty Hall Problem, Monty can open Door 1 or Door 2 or Door 3. The car can be behind Door 1, Door 2, or Door 3. The contestant can choose any door.

We can apply Bayes' Theorem to solve this.

D1: Monty Hall opens Door 1.

D2: Monty Hall opens Door 2.

D3: Monty Hall opens Door 3.

C1: The car is behind Door 1.

C2: The car is behind Door 2.

C3: The car is behind Door 3.

The prior probability of Monty Hall finding the car behind any particular door is P (C#) = 1/3.

Where, P (C1) = P (C2) = P (C3).

Assume the contestant chooses Door 1 and Monty Hall randomly opens one of the two doors he knows the car is not behind.

The conditional probabilities given the car being behind either Door 1 or Door 2 or Door 3 are as follows.

P (D3 I C1) = 1/2 ... as he is free to open Door 2 or Door 3, as he knows the car is behind the contestant's chosen door, Door 1. We assume he does so randomly.

P (D3 I C3) = 0 ... as he cannot open a door that the car is behind (Door 3) or the contestant's chosen door, so he must choose Door 2.

P (D3 I C2) = 1 ... as he cannot open a door that the car is behind (Door 2) or the contestant's chosen door (Door 1).

These are equally probable, so the probability he will open D3, i.e. P (D3) = (1/2 + 0 + 1) / 3 = 1/2.

So, P (C1 I D3) = P (D3 I C1) . P (C1) / P (D3) = (1/2 × 1/3) / 1/2 = 1/3.

Similarly, P (C1 I D2) = P (D2 I C1) . P (C1) / P (D2) = (1/2 × 1/3) / 1/2 = 1/3.

Therefore, there is a 1/3 chance that the car is behind the door originally chosen by the contestant (Door 1) when Monty opens Door 3 and similarly when Monty opens Door 2.

But P (C2 I D3) = P (D3 I C2) . P (C2) / P (D3) = (1 × 1/3) / 1/2 = 2/3.

Similarly, P (C3 I D2) = P (D2 I C3) . P (D2) = (1 × 1/3) / 1/2 = 2/3.

So, the probability that the car lies behind Door 2 when Monty opens Door 3 is 2/3, and similarly that the car lies behind Door 3 when Monty opens Door 2.

Therefore, there is twice the chance of the contestant winning the car by switching doors after Monty Hall has opened a door (2/3 instead of 1/3).

2.2.2 Exercise

Question 1. As a prize for winning a competition, you're offered a chance by the host to open either a cube or a sphere or a pyramid into one of which he has secretly placed a cheque for £10,000. The others are empty. The host will never open the winning container at this stage. You choose the cube and the sphere is opened. It is empty. You are offered a chance to swap to a different shape before the host reveals where the cheque is located. Should you take the offer or stick to your original choice? Why?

Question 2. As a prize for winning a competition, you're offered a chance by the host to open either a cube or a cylinder or a cone. An independent arbiter has deposited a cheque for £5,000 into one of these, leaving the other containers empty. Only the arbiter knows where the cheque is. You choose the cube and the cylinder, for example, is opened by the host. It is empty. You are now offered a chance to swap to a different shape before the reveal. Should you take the offer or stick to your original choice? Why?

Question 3. As a prize for winning a competition, you're offered a chance by the host to open one of ten boxes, inside one of which there is a prize voucher for a sports car. The others are empty. You choose Box no. 5. You are now offered a chance to either open five further boxes or else let the host open eight of the remaining nine boxes until just two remain, including your original choice. You can then decide to stick or switch. Should you open the extra five boxes or should you ask the host to open the boxes until just two remain? If you choose the latter, should you stick with your original choice or switch?

2.2.3 Reading and Links

Burns, B.D. 2017. Equiprobability principle or "no change" principle? Examining reasoning in the Monty Hall Dilemma using unequal probabilities, COGSCI, 1697-1702.

Caamano-Carrillo, C. and Gonzalez-Navarrete, M. 2020. The Monty Hall problem revisited. 12 April, 1-13. Cornell University. arXiv.2004.05539. https://arxiv.org/pdf/2004.05539.pdf

Crockett, R. 2016. The time everyone "Corrected" the world's smartest woman. Priceonomics. 2 August. https://priceonomics.com/the-time-everyone-corrected-the-worlds-smartest/

Ellis, K.M. 1995. The Monty Hall Problem. https://www.montyhallproblem.com

Frost, J. 2021. The Monty Hall Problem: A Statistical Illusion. https://statisticsbyjim.com/fun/monty-hall-problem/

Frost, J. 2021. Revisiting the Monty Hall Problem with Hypothesis Testing. https://statisticsbyjim.com/hypothesis-testing/monty-hall-problem-hypothesis-testing/

Glen, S. "Monty Hall Problem: Solution Explained Simply" From StatisticsHowTo.com: Elementary Statistics for the rest of us! https://www.statisticshowto.com/probability-and-statistics/monty-hall-problem/

Gruis, L. N. 2020. Card games and the Monty Hall problem. 17 August. https://www.crows.org/blogpost/1685693/354032/Card-Games-and-the-Monty-Hall-Problem

Lawson-Perfect, C. 2019. The big internet math-off 2019, Group 3 – Vicky Neale vs Sophie Carr. Bayes' Theorem. https://aperiodical.com/2019/07/the-big-internet-math-off-2019-group-3-vicky-neale-vs-sophie-carr/

vos Savant, M. 1990. Ask Marilyn. Parade. 16. 9 September. https://web.archive.org/web/20130121183432/http://marilynvossavant.com/game-show-problem/

Open Culture. 2017. The Famously Controversial "Monty Hall Problem" Explained: A Classic Brain Teaser. 3 October. http://www.openculture.com/2017/10/the-famously-controversial-monty-hall-problem-explained-a-classic-brain-teaser.html

Significance. 2013. *Readers' Challenge Report: The Monty Hall Problem*. American Statistical Association/Royal Statistical Society. October, pp. 32–33.

Tubau, E., Aguilar-Lleyda, D. and Johnson, E.D. 2015. Reasoning and Choice in the Monty Hall Dilemma (MHD): implications for improving Bayesian reasoning. Frontiers in Psychology, 31 March, 6:353, 1-11. http://diposit.ub.edu/dspace/bitstream/2445/100384/1/650841.pdf

Uphadyay, R. 2014. Bayes' Theorem – Monty Hall problem. http://ucanalytics.com/blogs/bayes-theorem-monty-hall-problem/

Understanding the Monty Hall problem. Better Explained. https://betterexplained.com/articles/understanding-the-monty-hall-problem/

Monty Hall problem. Wikipedia. https://en.m.wikipedia.org/wiki/Monty_Hall_problem

Monty Hall problem. Numberphile. May 22 2014. YouTube. https://youtu.be/4Lb-6rxZxx0

Monty Hall problem (extended math version). Numberphile2. 23 May 2014. YouTube. https://youtu.be/ugbWqWCcxrg

Monty Hall, Simpson's Paradox. Harvard University. 29 April 2013. YouTube. https://www.youtube.com/watch?v=fDcjhAKuhqQ&feature=youtu.be

The math behind the Monty Hall problem. Bon Crowder presents. November 27 2011. YouTube. https://youtu.be/7WvlPgIjx_M

The Monty Hall problem. MIT OpenCourseWare. February 26 2014. YouTube. https://www.youtube.com/watch?v=UgKrQ2ywVfs&feature=youtu.be

The Monty Hall problem. singingbanana. 19 February 2019. YouTube. https://youtu.be/njqrSvGz8Ps

The Monty Hall problem – Sophie Carr (Big Internet Math-Off 2019). 22 July 2019. YouTube. https://www.youtube.com/watch?v=phb8TOkG_RQ&feature=youtu.be

2.3 The Three Prisoners Problem

First posed in Scientific American in 1959, the Three Prisoners Problem is a variation of the Monty Hall Problem but is a classic of conditional probability in its own right. The problem, or a version of it, is simple to state. There are three prisoners on death row – Arthur, Bob, and Charlie. The warden tells each prisoner that each of their names will be put into a hat and the lucky prisoner, to be chosen randomly, will be pardoned as an act of clemency. The warden knows who has been pardoned, but the prisoners do not.

Arthur asks the warden to name one of the prisoners who will *not* be pardoned. Either way, he agrees that his own fate should not be revealed. If Bob is to be spared, the warden should name Charlie as one of the men to be executed. If Charlie is to be spared, he should name Bob as one of the men to be executed. If it is he, Arthur, who is to be pardoned, the warden should just flip a coin and name either Bob or Charlie as one of the men to be executed.

The warden agrees and names Charlie as one of the men going to the gallows.

Based on this information, what is the probability that Arthur is going to be pardoned, and what is the chance that Bob will instead be pardoned?

Arthur reasons that his chance of being spared before the conversation with the warden was 1/3, as there are three prisoners, and only one of these will be pardoned, chosen by lot. Now, though, he reasons that one, either he or Bob, is to walk free, as he knows that Charlie is not the lucky one. So now Arthur reasons that his chance of being pardoned has risen to 1/2. But is he right?

We can look at it this way. Before talking to the warden, Arthur correctly concludes that his chance of evading the gallows is 1/3. It is either he or Bob or Charlie who will be released, and each has an equal chance, so each has a 1/3 chance of being pardoned.

When Arthur expressly asks the warden to name one of the *other* men who will be executed, Arthur is asking the warden not to name himself, whether he is to be pardoned or not. The warden (as we are told in the question) selects which of the other men to name by flipping a coin. Now, Arthur gains no new information about his fate. The information he does gain is about the fate of Bob and Charlie. By naming Charlie as the condemned man, the warden is ruling out the chance that Charlie is to be pardoned.

So Arthur now knows the chance that Charlie will be spared has decreased from a 1/3 chance before the warden revealed this information to a zero chance after he reveals it.

But his chance of being spared remains unchanged because the warden has not revealed any new information relevant to his fate. New information is a requirement for changing the probability that something will happen or not. So his probability of being pardoned remains at 1/3. The new information he does have is that Charlie is not the lucky man, so the chance that Bob gets pardoned is now 2/3.

Put another way, how can Arthur and Bob receive the same information, but their odds of surviving are so different? It is because, when the warden made his selection, he would never have declared that Arthur was going to die. On the other hand, he might well have declared Bob to be the condemned man. Therefore, the fact that he didn't name Bob provides valuable information as to the likelihood that Bob would be pardoned while telling us nothing as to whether Arthur was. This is an example of the reality that belief updates must depend not merely on the facts observed but also on the method of establishing those facts.

In case there is still any doubt, imagine that there were 26 prisoners instead of 3. Arthur asks the warden not to reveal his own fate but to name in random order 24 of the other prisoners who are to be executed. So what is the chance that Bob will be the lucky one of 26 before the warden reveals any names? It is 1/26, the same chance as each of the other prisoners. Every time, however, that the warden names a dead man walking, say Charlie or Daniel, that reduces their chances to zero and increases the chance of all those left except for Arthur, who has expressly asked not be named, regardless of whether he is to be executed. So it means a lot about Bob's likely fate to learn that the warden has eliminated everyone but Bob given that he had every opportunity to name Bob as one of those going to the gallows. It means nothing that he has not named Arthur because he was expressly told not to, whatever his fate.

In a 26-man line-up, the warden names the condemned men in random order. Once everyone but Bob has been named for execution by the warden, Arthur's chance of surviving stays at 1/26. Bob's chance of being pardoned rises, however, to 25/26. This is even though there are only two remaining prisoners who have not been named for execution by the warden. Would you take 20/1 now that Arthur will be spared? You might, if you were Bob, but you are not getting a good price (the correct odds would be 25/1), and if you win you will not have long to spend it!

2.3.1 Exercise

There are five prisoners on death row. One of them, they are informed, will be pardoned, to be chosen by lot. The warden knows who it is, but the prisoners do not. One of the prisoners asks the warden to name one of the prisoners who will *not* be pardoned but not to name the prisoner making the request, regardless of his fate. If he is not to be pardoned, the warden should randomly reveal one of the other four who will not be pardoned. The warden agrees to the request and names one of the other four prisoners as one of the men going to the gallows. Based on this information, what is the probability that the prisoner who spoke to the warden will now be pardoned?

2.3.2 Reading and Links

Gardner, M. 1959. Mathematical games: Problems involving questions of probability and ambiguity. *Scientific American*, 201(4), October, 174–182.

Gleeson, P. 2017. The paradox of the three prisoners. *Medium.com*. 3 January. https://medium.com/@petergleeson1/the-paradox-of-the-three-prisoners-c8a88fdb67d3

Three prisoners problem. Math Easy Solutions. 11 August 2014. YouTube. https://youtu.be/8vY66MD7nsM

2.4 The Deadly Doors Problem

There are four doors – red, yellow, blue, and green.

Three lead the way to dusty death. One leads the way to fame and fortune. They are assigned in order by your host who draws four balls out of a bag, coloured red, yellow, blue, and green. The first three out of the bag are the colours of the doors that lead to dusty death. The fourth leads to fame and fortune. You must choose one of these doors, without knowing which of them is the lucky door.

Let's say you choose the red door. Since the destinies are randomly assigned to the doors, this means there is a 1 in 4 chance that you are destined for fame and fortune, and a 3 in 4 chance that you are destined for doom.

Your host, Mr. Johnnie, who knows what lies behind each of the doors, opens the yellow door, revealing a door to death. That is part of his job. He always opens a door to death, but not the door to fame and fortune unless it is the last remaining door.

Should you now walk through the red door, the blue door, or the green door?

This is a bit like the Monty Hall Problem, and I have labelled it as Monty Hall Plus.

Common intuition dictates that the chance that the red door leads to fame and fortune is 1 in 4 because there are four doors to choose from, each equally likely to offer the way to fame and fortune. And that's correct. After the yellow door is opened, that probability must increase. Right? After all, once the yellow door is opened, only three doors remain – the red door, the blue door, and the green door. Surely there is an equal chance that fortune beckons behind each of these. If so, the probability in each case is 1 in 3.

Now, the host opens a second door, by the same process. Let's say this time he opens the blue door, which again he reveals to be a death trap. That leaves just two doors. So surely they both now have a 1 in 2 chance. Take your pick, stick or switch. Does it matter?

Yes, it does in fact.

The reason it matters, just as in the standard Monty Hall Problem, is that the host knows where the doors lead. When you choose the red door, there is a 1 in 4 chance that you have won your way to fame and fortune if you stick with it. There is a 3 in 4 chance that the red door leads to death. If you have chosen the red door, and it is the lucky door, the host must open one of the doors leading to a dusty doom. This is valuable information.

Before he opened the yellow door, there was a 3 in 4 chance that the lucky door was *either* the yellow or the blue or the green door. Now he is telling you that there is a 3 in 4 chance that the lucky door is *either* the yellow or the blue or the green door *but* it is not the yellow door. So there is a 3 in 4 chance that it is *either* the blue or the green door. It is equally likely to be either, so there is a 3 in 8 chance that the blue door is the lucky door and a 3 in 8 chance that the green door is the lucky door.

Now he opens the blue door and introduces even more useful information, that it is not the lucky door. So there must be a 3 in 4 chance that it is the green door. So now you can stick with the red door and have a 1 in 4 chance of avoiding a dusty death or switch to the green door and have a 3 in 4 chance of avoiding that fate.

It is because the host knows what is behind the doors that his action introduces new information. Because he can't open the door to fame and fortune, he is increasing the probability that the other unobserved destinies include the lucky one.

If he didn't know what lay behind the doors, he might inadvertently have opened the door to fortune. When he opens a door, therefore, it adds no new information save that he has randomly eliminated one of the doors. If three doors now remain, they each offer a 1 in 3 chance of avoiding a dusty death. If only two doors remain unopened, they each offer in this case a 1 in 2 chance of death or glory. So when the host is as clueless about the lucky door as you are, you might as well roll the dice or toss a coin – and hope!

2.4.1 Exercise

Question 1. You are given the choice of four boxes, labelled 1 to 4. Three are empty, but a prize of £1,000 is inside the fourth. You are asked to select a box. You select box 1.

Your host, Vanessa, knows which box contains the prize and must open one of the boxes that she knows to be empty. She opens box 2 and it is empty.

To maximise your chance of winning the prize, should you switch to box 3, or switch to box 4, or stick with box 1, or does it not matter? Why?

Question 2. Suppose in a separate experiment, you are given the choice of four boxes, coloured black, white, grey, and brown. Three are empty; a prize of £1,000 is inside the fourth. You are asked to select a box. You select the black box.

Your host, Emma, does not know which box contains the prize and must open one of the boxes. She opens the grey box and it is empty.

To maximise your chance of winning the prize, should you switch to the white box, switch to the brown box, stick with the black box, or does it not matter? Why?

2.4.2 Reading and Links

Statistical Ideas. Blogspot. 2015. Games and Monty Hall. 20 June. http://statisticalid eas.blogspot.com/2015/06/games-and-monty-hall.html#!/2015/06/games-and-monty-hall.html

Weinstein, E.W. 2021. "Monty Hall problem". From *MathWorld – A Wolfram Web Resource*. Last updated 18 May. https://mathworld.wolfram.com/MontyHallP roblem.html

2.5 Portia's Challenge

In William Shakespeare's *Merchant of Venice*, potential suitors of young Portia are offered a choice of three caskets: one gold, one silver, and one lead. Inside one of them is a miniature portrait of her. Portia knows it is in the lead casket.

Now, according to her father's will, a suitor must choose the casket containing the portrait to win Portia's hand in marriage. The first suitor, the Prince of Morocco, must choose from one of the three caskets. Each is engraved with a cryptic inscription. The gold casket reads, "Who chooseth me shall gain what many men desire". The silver casket reads, "Who chooseth me shall get as much as he deserves". The lead casket reads, "Who chooseth me must give and hazard all he hath". He chooses the gold casket, hoping to find "an angel in a golden bed". Instead, he finds a skull and a scroll inserted into the skull's "empty eye". The message he reads on the scroll says, "All that glisters is not gold". The Prince beats a hasty exit. "A gentle riddance", says Portia. The next suitor is the Prince of Arragon. "Who chooseth me shall get as much as he deserves", he reads on the silver casket. "I'll assume I deserve the very best", he declares and opens the casket. Inside he finds a picture of a fool with a sharp dismissive note which says, "With one fool's head I came to woo, But I go away with two".

Now let us think about a plot twist where Portia must open one of the other caskets and give Arragon the option of switching his choice of caskets before he opens the silver casket. She is not allowed to indicate where the portrait is and therefore must open the gold casket (she knows it is in the lead casket so can't open that) and show it is not in there. She now asks the Prince whether he wants to stick with his original choice of the silver casket or switch to the lead casket.

Let's imagine that he believes that Portia has no better idea than he has of which casket contains the prize. In that case, should he switch from his original choice of the silver casket to the lead casket? Well, since Portia had no knowledge of the location of the portrait, she might have inadvertently opened the casket containing the portrait, so she adds new information by opening the casket. But if he knows that she is aware of the location of the portrait, her decision to open the gold casket and not the lead casket has doubled the chance that the lead casket contains the portrait compared to his original choice, other things equal. This is because there was just a one-third chance that his original choice (silver) was correct and a two-thirds chance that one of the other choices (gold or lead) was correct. She is forced to eliminate the losing casket of the two (in this case, gold), so the two-thirds chance converges on the lead casket. In this case, should he switch to the lead casket or stay with the silver? It depends on whether other things are in fact equal.

In particular, it depends on how valuable any information contained in the inscriptions is. If he has no faith in the inscriptions to arbitrate, he should switch and improve his chance of winning fair Portia's hand from 1/3 to 2/3. If he thinks, however, that he can unlock some useful information from the inscriptions, the decision is less straightforward.

In summary, the key to the problem is the new information Portia introduced by opening a casket, which she knew did not contain the portrait. By acting on this new information, the Prince can potentially improve his chance of correctly predicting which casket will reveal the portrait from 1 in 3 to 2 in 3. He can do this by switching boxes when given a chance unless he has other information, which makes the opening probabilities different to 1/3 for each casket, such as those cryptic inscriptions. If this information is potentially valuable, or at least if the Prince thinks so, that might meaningfully influence his decision.

2.5.1 Exercise

Potential suitors of Portia are offered a choice of three caskets, one gold, one silver, and one lead. Inside one of them is a miniature portrait of her, which the suitor must correctly identify if he is to win her hand in marriage. Portia knows it is in the lead casket and gives the suitor a chance to select it. The suitor, Bassanio, guesses it is in the gold casket. It is agreed in advance that once a casket is chosen, Portia will reveal one of the other caskets to be empty. She cannot open the lead casket as she knows that contains the portrait, so she opens the silver casket. Portia now asks Bassanio whether he wishes to stick with his original choice or switch to the lead casket. Whatever he now decides, that is his final choice. Should he switch to the lead casket, stick with the gold casket, or does it not matter? Would your answer change if Portia did not know which casket contained the miniature portrait?

2.5.2 Reading and Links

Bellos, A. 2017. Alex Bellos' Monday puzzle. Can you solve it? The mystery of Portia's caskets. *The Guardian*. 13 February. https://www.theguardian.com/science/2017/feb/13/can-you-solve-it-the-mystery-of-portias-caskets

Mackinnon, D. Portia's caskets. https://dmackinnon1.github.io/portia/

Shakespeare Birthplace Trust. 2021. The Merchant of Venice. Synopsis and plot overview of Shakespeare's The Merchant of Venice. https://www.shakespeare.org.uk/explore-shakespeare/shakespedia/shakespeares-plays/merchant-venice/

Understanding the Monty Hall problem. Better Explained. https://betterexplained.com/articles/understanding-the-monty-hall-problem/

2.6 The Boy–Girl Paradox

You meet a man at a sales convention who mentions that he has two children, and you learn that one of them, at least, is a boy. What is the chance that his other child is a girl? It's not a half, as most people think, but 2/3. What now if you see someone in the park accompanied by a child he introduces as his son and mentions that he has one other child? What is the chance that the other child is a girl? 2/3? No, this time it is 1/2. Puzzled? So was a Mr. John Francis, who challenged Marilyn Vos Savant, of Monty Hall fame, who had proposed the equivalent of these answers in response to a question to her "Ask Marilyn" magazine column in 1996. He wrote:

> I have a BA from Harvard, an MBA from the University of Pennsylvania Wharton School, a math SAT score of 800 and a perfect score in the Glazer-Watson critical thinking test, but I'm willing to admit I make mistakes. I hope you will have the strength of character to review your answer to this problem and admit that even a math teacher and the person with the highest IQ in the world can make a mistake from time to time.

Despite his SAT score, she was right, and he was wrong.

Here's a way to look at it. Let's imagine there is a red and a yellow door in a house which you have just entered. You are told that there are two children in the house, and they are each standing, assigned randomly, behind one or other of the doors. So, there are four possibilities:

1. A boy behind each door.
2. A boy behind the red door and girl behind the yellow door.
3. A girl behind the red door and boy behind the yellow door.
4. A girl behind each door.

So the probability that there is a girl behind one of the doors = 3/4. The new information I now impart is that at least one of the doors has a boy behind it. So we can delete option 4. This leaves three equally likely options, two of which include a girl. So the probability that there is a girl behind one of the doors given this new evidence is 2/3.

This is directly comparable to the pairs of options available in a two-child family, assuming for simplicity and illustration that there is an equal chance of any given child being a boy or a girl. The chance of two boys is, therefore, 1/4 (indexed according to some discriminating factor, e.g. by age – an older boy and a younger boy). The chance of two girls is 1/4 (e.g. older girl and younger girl). The chance of a boy and a girl is 1/2 (e.g. an older boy and a younger girl plus an older girl and a younger boy). The discriminating factor does not need to be age, so long as it is discriminating between each element of a pair (e.g. height, alphabetical order of the first name, behind a particular door).

Now consider the case of the man you met at the sales convention, the one with two children, and one at least was a boy. From this information, what is the chance that his other child is a girl? Well, it could be a boy, and it could be a girl. It's 2/3, and the two-door problem gives the reason. In telling you that one of his children was a boy, this was the equivalent of saying that there was a boy behind at least one of the doors but nothing else, so you could exclude option 4 (girl behind both doors), leaving options 1, 2, and 3 alive, with equal probability. Among these options, two also contain a girl and one contains another boy. So the chance his other child is a girl equals 2/3.

Let's state these age-identifying options again. With two children, there are four possible arrangements of younger and older siblings. These are:

Younger boy – older boy
Younger boy – older girl
Younger girl – older boy
Younger girl – older girl

If told that one of the children is a boy, there remain three options.

Younger boy – older boy
Younger boy – older girl
Younger girl – older boy

In two of these cases, the other child is a girl. So the probability that the other child is a girl is 2/3.

What has changed is the context. In terms of the available sample spaces, the additional information that it is not two girls reduces the sample space from four possibilities to three.

A common objection is to argue that Boy–Boy should count as two combinations, just as Boy–Girl does. Similarly, Girl–Girl, it has been argued, should count as two combinations.

In fact, Boy–Boy represents only one possibility, as does Girl–Girl.

Boy–Boy could mean there is a younger boy who has an older brother. It could also mean that there is an older boy who has a younger brother. But these are the same thing. Younger brother–older brother is equivalent to older brother–younger brother. They are two ways of describing the same thing, that there are two brothers, and one is older than the other. So there are not two combinations, but one. Similarly for Girl–Girl.

By way of contrast, Girl–Boy and Boy–Girl represent two distinct possibilities.

Girl–Boy: older girl with a younger brother.
Boy–Girl: older boy with a younger sister.

These are distinct combinations.

Once you eliminate Girl–Girl, therefore, only three possibilities remain, two of which contains a girl. It follows that there is a 2/3 chance that there is a daughter in the family and only a 1/3 chance that there are two sons.

Put another way, if I say that I have an older brother and a younger brother, it is the same event as saying that I have a younger brother and an older brother. If I say that I have an older brother and a younger sister, though, that is a different event to having an older sister and a younger brother. So there are three events here, and two of them include a sister yet only one contains a brother.

The situation is equivalent to learning that there are two coins and being told that at least one of the two coins is heads up. Now what is the probability that the other coin is also Heads? With two coins, there are four possibilities: Heads–Heads, Heads–Tails, Tails–Heads, and Tails–Tails. On being told that at least one of the coins is Heads, we can eliminate Tails–Tails. That leaves Heads–Heads (say left hand and right hand), Heads–Tails (say left hand and right hand), and Tails–Heads (say left hand and right hand). Of these three equally likely possibilities, two contain a Tails as the other element of the binary pair, and one contains a Heads as the other element. So the probability that the other coin is a Tails is 2/3. In each case, there is no distinguishing characteristic offered about the two Heads and about the two Tails.

Look at it this way. Toss two coins one after the other 1,000 times and remove from the data set the combinations that yield two Tails. This leaves all the combinations with at least one Head between the two coins.

The remaining options, each equally likely are:

Heads on first toss – Tails on second toss

Tails on first toss – Heads on second toss

Heads on first toss – Heads on second toss

You have eliminated: Tails on first toss – Tails on second toss.

So, the chance the other coin is Heads is 1/3 (Heads on both tosses). Both other combinations include one Tails.

If you do this experimentally, you will find that the chance that both coins are Heads in this sample is indeed about 1/3. Similarly, take 1,000 two-child families at random and eliminate from the sample all of these families that consist of both girls. Of the remaining families, about 1/3 of them will consist of two boys.

Now, what if you had met the guy in the park accompanied by his son and learned that he had two children? Armed now with this information, you instantly work out the chance that the other child is a girl is not 2/3, but 1/2. Why is the probability different this time? Does it matter how you found out about the boy, whether you were told about him or saw him for yourself? It does. But it wouldn't make any difference whether you actually saw the boy in the park or were simply told that one of the two children (a boy) is in the park and the other child is at home. Now, there are only two possibilities.

1. Boy in the park – Girl at home
2. Boy in the park – Boy at home

So the chance that the other child (the child at home) is a girl is 1/2.

The same reasoning can be found in the example of the doors. If you happen to find out he has a son by seeing the boy, it is like opening a specific door (say the red door) and finding a boy behind it. Being told that a boy is behind a specific door is the same. Either way it leaves two options:

Boy behind red door – Boy behind yellow door

Boy behind red door – Girl behind yellow door

We can rule out the third option: Girl behind red door – Boy behind yellow door.

These two events are (given our starting assumption) equally likely, so the probability that the other child is a girl is 1/2.

The door represents any piece of discriminating information. Location, age, and so on would do as well. The discriminating factor here is in seeing a particular boy, in a particular place, not a generic boy.

Another way to look at it is to state that the older child is a boy. Now, what is the chance that the other child (the younger child) is a girl?

Well, there are now just two possibilities:

Younger girl – older boy

Younger boy – older boy

We can rule out the third option: younger boy – older girl.

The two remaining possibilities are (given our starting assumption) equal, so the probability that the other child is a girl is 1/2.

In terms of coins, it is equivalent to learning that there are two coins and being told that the coin in the right hand is heads up. Now what is the probability that the coin in the left hand is also Heads? With two coins, there are two possibilities – Heads (right hand) and Heads (left hand), or Heads (right hand) and Tails (left hand). Of these two equally likely possibilities, one contains a Tails in the left hand. So the probability that the other coin is Heads is 1/2.

Any distinguishing characteristic about the coins will have the same effect.

Say, for example, there is a shiny coin and a dull coin. Now you learn that the shiny coin lands Heads up. What is the probability that the dull coin also lands Heads up? The coin tosses are independent, and for a fair coin the chance is 1/2. If you are told, however, that one of the coins has landed Heads up, but not which, there are three remaining possibilities. These are:

Heads (shiny) and Heads (dull)

Heads (shiny) and Tails (dull)

Tails (shiny) and Heads (dull)

Of these three possibilities, two involve the other coin being Tails and only one involves the other coin being Heads. Now, the chance that the other coin is Heads is not 1/2, as before, but 1/3.

The same logic applies to any other distinguishing characteristic, such as old coin and new coin.

The act of labelling the elements, using a discriminating characteristic, alters the probabilities.

2.6.1 Appendix

Solutions to the Boy–Girl Paradox using Bayes' Theorem.

What is the probability that the other child is a girl, if he has two children, and you see him with a son?
Assuming that it was equally probable that you would have seen him with the boy or the girl, then the probability of seeing him with a boy is 1/2, and with a girl is 1/2, if he has one of each. The correct calculation now is:

Let G2 be the probability the other child is a girl and P (B1) that he is seen with the boy.

We want to find P (G2|B1).

$$P (G2|B1) = P (B1 \text{ I } G2) . P (G2) / P (B1)$$
$$P (G2|B1) = (1/2 \times 1/2) / 1/2 = 1/4 / 1/2 = 1/2$$

Note that strictly speaking, in our example of the man in the street, this does assume that there is an equal chance that when you met him he would be accompanied by either of his two children. The equivalent assumption in the doors example is that the children are assigned randomly to each coloured door and similarly for the coins. For example, if a boy is rarely assigned to a yellow door, it increases the chance that a boy is behind the red door if you find a girl behind the yellow door.

What is the probability that both children are boys if he has two children and at least one of them is a boy?
Let P (A) be the probability that both children are boys.
Let P (B) be the probability that at least one of the children is a boy.
We want to find P (A I B) = P (B I A) . P (A) / P (B).
Now, P (B I A) = 1, as the probability of there being at least one boy if both children are boys is 100%.
P (A) = 1/4. There are four possible combinations (Boy–Boy, Girl–Girl, Boy–Girl, Girl–Boy), and only one of these involves two boys.

P (B) = 3/4. There are four possible combinations and three of the four involve at least one boy.

Therefore, P (A I B) = 1 × 1/4 / 3/4 = 1/3.

In contrast, the probability that a specific boy (say the first child) is a boy is 1/2.

2.6.2 Exercise

Note that the questions are about probability, not psychology or genetics. Assume for all questions that there is a binary choice of males and females, with an equal number of both in demographics relevant to any question.

1. You overhear a conversation in a bar in which a woman mentions that she has two children and also mentions her daughter. What is the probability that her other child is a boy? Explain your answer.

2. The following week you meet a woman in a park, who is accompanied by her daughter. She mentions that she has two children and that her other child is at home. What is the probability that her other child is a girl? Explain your answer.

3. If you toss a coin twice, which is more likely? Two Heads *or* Heads then Tails?

4. If you toss two coins into the air, which is more likely, that both coins will lands Heads or that one lands Heads and the other Tails?

5. If you have two siblings, which is more likely, that you have two brothers or that you have a brother and a sister?

2.6.3 Reading and Links

Stylianides, N. and Kontou, E. (2020). Bayes Theorem and Its Recent Applications. MA3517 Mathematics Research Journal, March, 1-7. file:///C:/Users/epa3will ilv/Downloads/3488-9410-1-PB.pdf

Boy or girl paradox. Wikivisually. https://wikivisually.com/wiki/Boy_or_Girl_par adox

The Boy or girl probability paradox resolved. Zach Star. April 11 2019. YouTube. https://youtu.be/ElB350w8iJo

2.7 The Girl Named Florida Problem

Suppose that there are two families, each with a child living at home.

Let us assume that one of the children lives in a brick house and the other in a wooden house. Now, what is the probability that both are girls?

There are four possibilities:

1. A boy in both houses.
2. A boy in the brick house and a girl in the wooden house.
3. A girl in the brick house and a boy in the wooden house.
4. A girl in both houses.

So the probability that there is a girl in both houses = 1/4.

This answers the first question. Given the information that there are two children, the chance that both are girls is 1 in 4.

Now, what if we are told that at least one of the children is a girl?

This eliminates option 1, i.e. a boy in both houses, leaving three equally likely possibilities, only one of which is a girl in both houses. So the chance that there is a girl in both houses given that you know that there is a girl in at least one of the houses is 1 in 3.

This is equivalent to asking the probability that both children are girls if you know that at least one of the children is a girl. The answer is 1 in 3.

What if I now tell you that one of the children is a girl called Florida? This is very much a discriminating characteristic which identifies a particular (rather than generic) girl. It is pretty much equivalent to telling you that there is a girl in a particular house, say the brick house. When now asked the probability that there is also a girl in the wooden house, options 1 and 2 (above) disappear, leaving just option 3 (a girl in the brick house and a boy in the wooden house) and option 4 (a girl in both houses). So the probability, given the additional information which identifies or locates one particular girl, is 1 in 2.

In other words, knowing that there is a girl in the brick house, or else knowing that her name is Florida, is like meeting her in the street. If you know there is another child in one of the houses, the chance it is a girl equals 1 in 2. By meeting her, you have identified a feature particular to that individual girl, i.e. that she is standing before you and not at home (or is in the brick house, or is named Florida).

The different information sets can be compared to tossing a coin twice. The possible outcomes are Head–Head (HH), Head–Tail (HT), Tail–Head (TH), and Tail–Tail (TT). If you already know there is "at least" one Head, that leaves HH, HT, and TH. The probability that the remaining coin is a Tail is 2 in 3. If, on the other hand, you identify that the coin in your left hand is a Head, the probability that the coin in your right hand is a Head is now 1 in 2. It is because you have pre-identified a unique characteristic of the coin, in this case its location. Identifying the girl as Florida does the same thing. In terms of two coins, it is like specially marking one of the coins with a blue felt tip pen. You now declare that there are two coins in your hands, and one of them contains a Head with a blue mark on it. Such coins are distinct, perhaps as distinct as girls called Florida. So you are now asked what is the chance that the other coin is Heads (without a blue felt pen mark). Well, there are two possibilities. The other coin is either Heads (almost surely with no blue felt pen mark on it) or Tails. So the chance the other coin is Heads is 1 in 2. Without marking one of the coins, to make it distinct, the chance of the other coin being Heads is 1 in 3. Marking it is the equivalent of saying that the coin in a particular hand is Heads.

Marking the coin with the blue felt tip pen is like pre-identifying a girl (her name is Florida) as opposed to simply declaring that at least one of the children is some generic girl.

In other words, there are four possibilities without identifying either child by a discriminating characteristic.

1. Boy, Boy
2. Girl, Girl
3. Boy, Girl
4. Girl, Boy

If at least one child is a girl, option 1 disappears, and the probability that the other child is a girl is 1 in 3.

If you identify one of the children, say a girl whom you name as Florida, only two of the following four options exist:

1. Boy, Boy
2. Girl named Florida, Girl not named Florida.
3. Boy, Girl named Florida
4. Girl not named Florida, Boy

Options 1 and 4 can be discarded, leaving options 2 and 3. In this scenario, the chance that the other child is a girl (not named Florida) is 1 in 2.

And what if it's a boy named Sue? Same reasoning. The chance that the other child is a boy is different than if it's a boy called Bob.

So what if it's the same family and the girl is called Jane? Well, this is a much more common name than Florida, but it is so unusual for two children in the same family to be called the same name, that it might reasonably be interpreted as a distinguishing characteristic. Put another way, G_{Jane} G_{Jane} is very unusual in the same family. Therefore, the chance that the other child is a girl will be close to a half. This is because the remaining options are: G_{Jane} $G_{Not\ Jane}$, G_{Jane} Boy. One of these options includes two girls.

Finally, what if you find out that one member of the family is born on a Sunday? How much of a distinguishing characteristic is this? Let's assume for simplicity that it's equally likely that a child will be born on any day of the week. In a two-child family, there are four possible children: Girl 1, Girl

2, Boy 1, Boy 2. With the additional day-of-the-week characteristic, there are seven possible combinations for Girl 1, and seven for Girl 2. Similarly for boys.

With the information that at least one of the two children is a girl born on a Sunday, the sample space is composed of Girl 1 born on a Sunday and a boy born on any day of the week (seven possibilities), Girl 1 born on a Sunday and Girl 2 born on any day of the week (seven possibilities), Girl 2 born on a Sunday and a boy born on any day of the week (seven possibilities), and Girl 2 born on a Sunday and Girl 1 born on any day of the week (seven possibilities). This makes a sample space of 28 possibilities. This is nearly correct but note that Girl 1 born on a Sunday and Girl 2 born on a Sunday is the same event as Girl 2 born on a Sunday and Girl 1 born on a Sunday. We shouldn't count the same event twice, so we remove this from the 28 possibilities to give a sample space of 27. Of these 27 possibilities, 13 (14 minus the case where both girls are born on a Sunday) involve a girl born on a Sunday. So, the conditional probability that both children are girls, given that at least one is a girl born on a Sunday, is equal to 13/27.

The more specific the additional information is, the more the sample space is reduced.

The more distinguishing is the characteristic, therefore, the more we diverge from 1/3 to a half, and vice versa.

2.7.1 Appendix

Calculating the probability of two girls given that there are two children, at least one of whom is a girl named Florida.

P (Boy) = 1/2
P (Girl named Florida) = x
P (Girl not named Florida) = 1/2 − x

where x is the percentage of people who are girls named Florida.

If at least one girl is named Florida (GF), there are the following possible combinations, with associated probabilities.

B GF = 1/2 x
GF B = 1/2 x
G GF = x (1/2 − x)
GF G = x (1/2 − x)
GF GF = x²

The cases in bold italics have two girls, and the probability we are seeking to establish, therefore, is the sum of highlighted cases divided by the sum of all cases.

$$= x\left(1/2-x\right)+x\left(1/2-x\right)+x^2 /\left[x\left(1/2-x\right)+x\left(1/2-x\right)+x^2+1/2x+1/2x\right]$$
$$= 1/2\,x-x^2+1/2x-x^2+x^2 /\left[1/2x-x^2+1/2x-x^2+x^2+x\right]$$
$$= x-x^2 /\left(x-x^2+x\right)=x\left(1-x\right)/x\left(2-x\right)=\left(1-x\right)/\left(2-x\right)$$

Assuming that Florida is not a common name, x approaches zero and the answer approaches 1/2. So it turns out that the girl's name is information of relevance.

As x approaches 1/2, the answer converges on 1/3, and the problem reduces to the standard P (GG I G) problem, i.e.

P (GG I G) = P (G I GG) . P (GG) / P (G) ... by Bayes' Theorem
So P (GG I G) = 1 × 1/4 / (3/4) = 1/3.

2.7.2 Exercise

1. You meet your new next-door neighbours, one living on your left and one on your right, who tell you that they each have a child living at home. You learn that one is a son called Vermont, but not which one. What is the chance that the child living in the other house is a girl? Does it matter that the boy has an unusual name?

2. If you learn that at least one of the two children is a girl born in the first week of January, what is the chance that the other child is a girl? Assume it is equally likely that a child will be born in any of the 52 weeks of the year.

3. If you learn that another family has two children, and at least one of these is a boy who was born on a Saturday, what is the chance that the other child is a boy? Assume it is equally likely that a child will be born on any day of the week.

2.7.3 Reading and Links

Downey, A. 2010. Probably overthinking It. 10 November. http://allendowney.blogsp ot.com/2011/11/girl-named-florida-solutions.html
Wallace, D.J. 2017. Untrammeled mind. 31 December. Two-child problem (when one is a child named Florida born on a Tuesday). http://www.untrammeledmind. com/2017/12/two-child-problem-when-one-is-a-girl-named-florida-born-on-a-tuesday/
This may be the most counter-intuitive probability paradox I've ever seen. Can you spot the error? Zach Star. April 7 2019. YouTube. https://youtu.be/bDZieLmya_I

2.8 The Two Envelopes Problem

The Two Envelopes Problem, also known as the Exchange Paradox, is quite simple to state. You are given two identical-looking envelopes, one of which you are informed contains twice as much money as the other. You are asked to select one of the envelopes. You are now given the opportunity, if you wish, to switch it for the other envelope. Once you have decided whether to keep the original envelope or switch to the other envelope, you are allowed to keep the money inside your selection. Should you switch?

Switching does seem like a no-brainer. Note that one of the envelopes (you don't know which) contains twice as much as the other. So, if the first envelope you selected contains £100, for example, the other envelope will contain either £200 or £50. By switching, it seems, you stand to gain £100 or lose £50, with equal likelihood. So the expected gain from the switch is 1/2 (£100) + 1/2 (−£50) = £50 − £25 = £25.

Looked at another way, the expected value of the money in the other envelope = 1/2 (£200) + 1/2 (£50) = £125, compared to £100 from sticking with the original envelope.

You might reason more generally that if X is the amount of money in the selected envelope, the expected value of what is in the other envelope = 1/2 (2X) + 1/2 (X/2) = 5/4 X. Since this is greater than X, it seems like a good idea to switch. So the logic of switching still applies even when you don't know how much money is in the original envelope. As such, it would be correct to switch envelopes even before looking inside the original envelope.

Is this right? Should you always switch?

Look at it this way. If the above logic is correct, then after switching envelopes, the amount of money contained in the other envelope can be denoted as Y.

So by switching back, the expected value of the money in the original envelope = 1/2 (2Y) + 1/2 (Y/2) = 5/4Y, which is greater than Y, following the same reasoning as before. So you should switch back.

But following the same logic, you should switch back again, and so on, indefinitely.

This would be a perpetual money-making machine. Something is clearly wrong here.

One way to consider the question is to note that the total amount in both envelopes is a constant, A = 3X, with X in one envelope and 2X in the other.

If you select the envelope containing X first, you gain 2X − X = X by switching envelopes.

If you select the envelope containing 2X first, you lose 2X − X = X by switching envelopes. So your expected gain from switching = 1/2 (X) + 1/2 (−X) = 1/2 (X − X) = 0.

So which is right? This reasoning or the original reasoning. There does not seem a flaw in either. In fact, there is a flaw in the earlier reasoning, which indicated that switching was the better option. So what is the flaw?

The flaw is in the way that the switching argument is framed, and it is contained in the possible amounts that could be found in the two envelopes. As framed in the original argument for switching, the amount could be £100, £200, or £50. More generally, there could be £X, £2X, or £1/2X in the envelopes. But we know that there are only two envelopes, so there can only be two amounts in these envelopes, not three.

You can frame this as £X and £2X or as £1/2X and £X, but not legitimately as £X, £2X, and £1/2X. By framing it is as two amounts of money, not three, in the two envelopes, you derive the answer that there is no expected gain (or loss) from switching.

If you frame it as £X and £2X, there is a 0.5 chance you will get the envelope with £X, so by switching there is a 0.5 chance you will get the envelope with £2X, a gain of £X. Similarly, there is a 0.5 chance you selected the envelope with £2X, in which case switching will lose you £X. So the expected gain from switching is 0.5 (£X) + 0.5 (−£X) = £0.

If you frame it as £X and £1/2X, there is a 0.5 chance you will get the envelope with £X, so by switching there is a 0.5 chance you will get the envelope with £1/2X, i.e. a loss of £1/2X. Similarly, there is a 0.5 chance you selected the envelope with £1/2X, in which case switching will gain you £1/2X. So the expected gain from switching is 0.5 (−£1/2X) + 0.5 (£1/2X) = £0.

There is demonstrably no expected gain (or loss) from switching envelopes.

In order to resolve the paradox, you must label the envelopes before you make your choice, not after. So envelope one is labelled, say, A, and envelope two is labelled, say, B. A corresponds in advance to, say, £100 and B corresponds in advance to, say, £200, or to £50, but not both. You don't know in advance which of the two defined amounts corresponds to which envelope. So there is no advantage (or disadvantage) in switching in terms of expected value. In summary, the clue to resolving the paradox lies in the fact that there are only two envelopes and these contain two amounts of money, not three.

A related problem is known as the Necktie Paradox. In a version of this paradox, Ed and Harry are at a party and are keen to own the more expensive tie of those they are wearing, but have no idea as to which is actually more expensive. They agree to ask their wives how much each of their ties cost. The proposal is that the man with the more expensive tie gives it to the other as a prize. Ed accepts the bet on the basis that winning and losing are equally likely but if he wins he gains the more expensive tie. If he loses, he only parts with the cheaper tie. The expected value of the wager is positive. Harry uses the same logic to accept the bet. Yet how can the bet be to the advantage of both? To resolve the paradox we need only to note that Ed might have the more expensive tie (say a £20 tie) or the cheaper tie (say a £10 tie) with equal probability. If he has the cheaper tie, he wins the more expensive tie and gains (a £20 tie) from the bet. But if he has the more expensive tie, he must part with that and loses (a £20 tie) from the bet. Since these outcomes are equally likely, what Ed can expect to gain is equal to what he can expect to lose. The expected value of the wager is zero. The same logic applies to Harry. In summary, for both Ed and Harry, what they can expect to win matches what they can expect to lose from the necktie bet.

2.8.1 Exercise

You are presented with two closed boxes, in one of which there will be a prize of a given value. In the other box, there will be a prize of twice that value. Choose one of the boxes. You are now offered the chance to switch your choice to the other box. The problem you face is that you don't know if the box you opened contained the larger or the smaller prize. So what should you do to maximise your expected payout? Stick with the original box, switch, or does it not matter?

What if you could look inside the box first? Would this make any difference to your optimal strategy?

2.8.2 Reading and Links

ThatsMaths. 2018. The two envelopes fallacy. 29 November. https://thatsmaths.com/2018/11/29/the-two-envelopes-fallacy/

Two envelopes problem. Wikipedia. https://en.wikipedia.org/wiki/Two_envelopes_problem

Solve the two envelopes fallacy. Looking glass universe, 23 September 2017. YouTube. https://www.youtube.com/watch?v=OqVFKY504X0&feature=youtu.be

---

2.9 The Birthday Problem

How large should a randomly chosen group of people be, to make it more likely than not that at least two of them share a birthday?

For convenience, assume that all dates in the calendar are equally likely as birthdays and ignore the Leap Year special of 29 February.

Let's examine the intuition. In a group of just two people, the chance that they would share a birthday is just 1/365. There are 365 days in the year and there is only one chance in 365 that the other person would be born on that particular day out of the 365 that your birthday falls on.

If the group is of 366 people, on the other hand, at least one person must share a birthday with someone else.

So, there is a tiny chance that two people will share a birthday in a group of two, but a 100% chance in a group of 366. Intuition might suggest, therefore, that the group size at which there is a 50% chance that two will share a birthday is near the mid-point of these two numbers, say a group of about 180. In fact, the answer is very much smaller.

The first thing to look at is the likelihood that two randomly chosen people would share the same birthday.

Let's call them Felix and Felicity. Say Felicity's birthday is 1 May. What is the chance that Felix shares this birthday with Felicity? Well, there are 365 days in the year, and only one of these is 1 May and we are assuming that all dates in the calendar are equally likely as birthdays. What we call the sample space is, therefore, 365 days and each particular birthday is an "event" in that sample space.

So, the probability that Felix's birthday is 1 May is 1/365, and the chance he shares a birthday with Felicity is 1/365.

So what is the probability that Felix's birthday is not 1 May? It is 364/365. This is the probability that Felix doesn't share a birthday with Felicity.

More generally, for any randomly chosen group of two people, the probability that the second person has a different birthday to the first is 364/365.

With three people, the chance that all three are different is the chance that the first two are different (364/365) multiplied by the probability that the third birthday is different (363/365).

So, the probability that three people have different birthdays = 364/365 × 363/365.

Now, suppose that the room contains four people. What is the chance that at least two of these people share the same birthday?

The probability that all four people have different birthdays = 364 × 363 × 362 / 365 × 365 × 365.

We can then subtract this probability from 1 to establish the probability that at least two of the four share a birthday.

Probability that none of the four people share the same birthday =

$$365 \times 364 \times 363 \times 362 \, / \, 365 \times 365 \times 365 \times 365 = 0.984$$

The probability that at least two share the same birthday = 1 − 0.984 = 0.016.

Similarly, the probability of at least two sharing a birthday increases as n, the number in the room, increases, as below:

n = 16; probability = 0.281

n = 23; probability = 0.505

n = 32; probability = 0.754

n = 40; probability = 0.892

In smaller groups, the chance that two will share a birthday is 11.7% for a group of ten, and of 2.7% for a group of five.

So, the probability that two share a birthday exceeds 0.5 in a room of 23 or more people.

So how large should a randomly chosen group of people be to make it a greater than even chance that at least two of them share a birthday? The answer is 23.

The intuition behind this is quite straightforward if we recognise just how many pairs of people there are in a group of 23 people, any pair of which could share a birthday.

Any group of 23 people contains 253 pairs of people to choose from. Therefore, a group of 23 people generates 253 chances, each of size 1/365, of having at least two people in the group sharing the same birthday.

Another way to look at this is that in a group of two people, there is just one pair of birthdays to consider. In a group of three, there are three pairs to consider, i.e. person 1 and person 2, 2 and 3, and 1 and 3. In a group of four, there are six pairs – 1 and 2, 1 and 3, 1 and 4, 2 and 3, 2 and 4, 3 and 4, and so on. In a group of 23, there are 253 possible pairs who could share a birthday.

The reason for the paradox, therefore, is that the question is not asking about the chance that someone shares your particular birthday or any particular birthday. It is asking whether any two people share any birthday.

The same reasoning applies to balls being randomly dropped into open boxes, such that there is an equal chance that a ball will drop into any of the boxes. If there are 365 such boxes, into which 23 balls are randomly dropped, there is just over a 50% chance that there will be at least two balls in at least one of the boxes. Randomness produces more aggregation than intuition leads us to expect.

A useful rule of thumb for calculating the number of balls or people required for a 50% chance that two or more will share at least one character-istic (such as same box or same birthday) is: "The Square Root of the number of categories times 1.2". So, square root of 365 times 1.2 = 22.93. This means that there is about a 50% chance that in a group of 23 people, two will share a birthday.

The rule of thumb for a 95% chance of the above is to multiply the square root of the number of categories by 1.6. In the case of 365 categories, this equals 30.57. This means that there is about a 95% chance that in a group of 31 people, two will share a birthday. Similarly, if 31 balls are dropped randomly into 365 boxes, all with an equal chance of landing in any box, there is about a 95% chance that two balls will drop into the same box.

The Birthday Problem is, in this way, notable for being a classic example of the Multiple Comparisons Fallacy. This fallacy arises when, in looking at many variables, the number of possible correlations that are being tested is under-estimated. In particular, multiple comparisons arise when a statisti-cal analysis involves multiple simultaneous statistical tests, each of which has the potential to produce a "discovery". For example, with a thousand variables, there are almost half a million (1,000 × 999/2) potential pairs that might appear correlated by chance alone. As such, we may well get a false comparison by chance. This becomes a fallacy when that false comparison is seen as significant rather than a statistical probability.

To summarise the Birthday Problem, in a group of 23 people (assuming all days are equally likely), there is more than a 50% chance that at least two of the group share the same birthday. This seems counter-intuitive since it is rare to meet someone that shares a birthday. Indeed, if you select two random people, the chance that they share a birthday is about 1 in 365. With 23 people, however, there are 253 ($23 \times 22/2$) pairs of people who might have a shared birthday. So by looking across the whole group, we are checking whether any one of these 253 pairings, each of which independently has a tiny chance of coinciding, does indeed match. Because there are so many possibilities of a pair, it makes it more likely than not, statistically, for coincidental matches to arise. For a group of 40 people, it is nearly nine times as likely that at least two share a birthday than that they do not.

The Birthday Problem is, therefore, an example of aggregated coincidences. By stating that in a group of 23 people, there is a 50% chance that at least one pair share a birthday, we are not saying which pair. Similarly, when dropping balls into boxes, we are not stating which balls and which boxes.

The probability that in a group of 23 people, including yourself, at least one of the remaining 22 people will share your birthday is much lower than 50%. The probability $= 1 - P$ (everyone in the group does not share my birthday) $= 1 - (364/365)^{22} = 1 - 0.94 = 0.06$. So, there is only a 6% chance that one of the other members of the group shares your birthday. Even with 366 people in the group, there is only a $1 - (364/365)^{365}$, i.e. 63%, chance that someone will share your birthday, but there is a 100% chance (ignoring Leap Year birthdays) that at least someone will share a birthday with someone else, if we don't specify a particular birthday in advance.

Similarly, when placing one ball in a box out of a choice of 365 boxes and now randomly dropping another 22 balls into these boxes, there is only a 6% chance that one of the other balls will drop into the same box as your ball. Even with 366 balls, there is only a 63% chance of your ball finding a companion ball in its box. There is a 100% chance, however, that some ball will end up in the same box as some other ball if we don't specify the box in advance.

More technically, in a group of 23 people, there are, according to the standard formula, $^{23}C_2$ (called 23 Choose 2) pairs of people.

Generally, the number of ways k things can be chosen from n is:

$$^nC_k = n! / (n-k)! \, k!$$

Here n! (n factorial) is $n \times n - 1 \times n - 2 \ldots$ down to 1. Similarly, for k!

Thus, $^{23}C_2 = 23! \, / \, 21! \, 2! = 23 \times 22 \, / \, 2 = 253$.

These chances have some overlap: if A and B have a common birthday, and A and C have a common birthday, then inevitably so do B and C.

So the probability of at least two people sharing a birthday in a group of 23 is less than 253/365 (69.3%).

The probability that in a group of 23 people, at least two do *not* share a birthday is:

$$(364 / 365)^{253} = 0.4995$$

The odds that at least two of the 23 people share the same birthday = 1 − 0.4995 = 0.505 = 50.5%.

It can be shown that n = 253 for there to be a 50% chance that someone shares your birthday. So, you would need to ask a group of 253 people if anyone shared your birthday for there to be a 50% chance that any of them did.

Similarly, if you place a ball into one of 365 boxes and now randomly drop 253 balls into these 365 boxes, there is about a 50% chance that at least one of the balls will share a box with your ball. This assumes, of course, that any ball is equally likely to end up in any box.

So, the next time you see two football teams line-up with the referee, check out the birthdays. More times than not, you should find a match. It's very unlikely, though, that the match will involve the official (just a 6% chance).

2.9.1 Exercise

1. Using the square root rule-of-thumb method, calculate the number of people required for about a 50% chance that at least two share a birthday. Assume 365 days in a year, all equally likely.

2. Using the square root rule-of-thumb method, calculate the number of balls required for about a 95% chance of at least two dropping randomly into at least 1 of 100 boxes, all equally likely.

3. In a group of 100 people, including yourself, what is the probability that there will be a match to your birthday?

4. In a group of 24 people, how many chances are there of having at least two people in the group sharing a birthday? From this, derive the chance that at least two of these share a birthday. Assume that all dates in the calendar are equally likely as birthdays and ignore the Leap Year, 29 February.

5. What is the probability that in a room of two people, they share the same birthday. Assume that all dates in the calendar are equally likely as birthdays and ignore the Leap Year, 29 February.

2.9.2 Reading and Links

Aczel, A.D. 2016. Chance: A Guide to Gambling, Love, the Stock Market, and Just about Everything Else. 18 May. New York: Thunder's Mouth Press.

Koehrsen, W. 2018. The misleading effect of noise: The misleading comparisons problem. 7 February. https://towardsdatascience.com/the-multiple-compar isons-problem-e5573e8b9578

Science Buddies. 2012. Probability and the birthday paradox. Scientific American. 29 March. https://www.scientificamerican.com/article/bring-science-home-probability-birthday-paradox/

Woodcock, S. 2017. Paradoxes of probability and other statistical strangeness. The Conversation. 4 April. https://theconversation.com/paradoxes-of-probability-and-other-statistical-strangeness-74440

Multiple comparisons fallacy. Logically Fallacious. https://www.logicallyfallacious.com/tools/lp/Bo/LogicalFallacies/130/Multiple-Comparisons-Fallacy

The multiple comparisons fallacy. Fallacy Files. http://www.fallacyfiles.org/multcomp.html

Understanding the birthday paradox. Better Explained. https://betterexplained.com/articles/understanding-the-birthday-paradox/

Birthday problem, Properties of probability. Harvard University. 29 April 2013. https://www.youtube.com/watch?v=LZ5Wergp_PA&feature=youtu.be

Check your intuition: The birthday problem – David Knuffe. Ted Ed. 4 May 2017. YouTube. https://youtu.be/KtT_cgMzHx8

Multiple comparisons. Steve Grambow. 26 March 2013. YouTube. https://youtu.be/EMzcZFtGZZE

The multiple comparisons problem. Sprightly Pedagogue. 5 January 2017. YouTube. https://youtu.be/dzi1CSvzCoU

2.10 The Inspection Paradox

The bus arrives every 20 minutes on average, though sometimes the interval between buses is a bit longer and sometimes a bit shorter. Still, it's 20 minutes taken as an average, or three buses an hour.

You emerge onto the main road from a side lane at some random time and come straight upon the bus stop. How long can you expect to wait for the next bus to arrive?

The intuitive answer is 10 minutes since this is exactly halfway along the average interval between buses. If your usual wait is somewhat longer than this, then you have been unlucky.

But is this right? The *Inspection Paradox* suggests that in most circumstances, you will be quite lucky to wait only 10 minutes for the next bus. Let's examine this more closely. The bus arrives every 20 minutes on average, or three times an hour. But that is only an average. If they do arrive at precise 20 minutes intervals, then your expected wait is indeed 10 minutes (the midpoint of the interval between the bus arrivals). But if there is any variation around that average, things change, for the worse.

Say, for example, that half the time the buses arrive at a 10-minute interval and half the time at a 30-minute interval. The overall average is now 20 minutes. From your point of view, however, it is three times more likely that you'll turn up during the 30 minutes interval than during the 10 minutes interval. Your appearance at the stop is random. As such, it is more likely to take place during a long interval between two buses arriving than during a short interval.

What does this mean for how long you can expect to wait? If you randomly arrive during the long (30 minutes) interval, you can expect to wait 15 minutes. If you randomly arrive during the short (10 minutes) interval, you can expect to wait 5 minutes. But there is three times the chance you will arrive during the long interval, and therefore three times the chance of waiting 15 minutes, as of waiting 5 minutes. So your expected wait is 3 × 15 minutes plus 1 × 5 minutes, divided by 4. This equals 50 divided by 4 or 12.5 minutes.

In summary, the buses arrive on average every 20 minutes but your expected wait time is not half of that (10 minutes) but somewhat more than 10 minutes. This is true in every case except when the buses arrive at exact 20-minute intervals. As the dispersion around the average increases, so does the amount by which the expected wait time exceeds the average wait. This is the *Inspection Paradox* which states that whenever you "inspect" a process, you are likely to find that things take (or last) longer than their "uninspected" average. What seems like the persistence of bad luck is simply the laws of probability and statistics playing out their natural course.

Once made aware of the paradox, it seems to appear everywhere.

For example, take the case where the average class size at an institution is 30 students. If you decide to interview random students from the institution and ask them their class size, you will usually obtain an average rather higher than 30. Let's take a stylised example to explain why. Say that the institution has class sizes of either 10 or 50, and there are equal numbers of each. So the overall average class size is 30. But in selecting a random student, it is five times more likely that he or she will come from a class of 50 students than of 10 students. So for every one student who replies "10" to your enquiry about their class size, there will be five who answer "50". So the average class size thrown up by your survey is 5 × 50 + 1 × 10, divided by 6. This equals 260 /6 = 43.3. So the act of inspecting the class sizes increases the average obtained compared to the uninspected average. The only circumstance in which the inspected and uninspected average coincides is when every class size is equal.

In another example, you visit the library and survey those who are studying there. The survey question is how long they usually stay at the library. The survey answer is likely to be skewed because the students in the sample are likely to include more of those who study for a long time in the library than those who study for a short time.

We can examine the same paradox within the context of what is known as length-based sampling. For example, when digging up potatoes, why does the fork usually go through the very large one? Why does the network connection break down during download of the largest file? It is because these outcomes occur for a greater extension of space or time than the average extension of space or time.

Once you know about the *Inspection Paradox*, the world and our perception of our place in it are never quite the same again.

2.10.1 Exercise

1. You arrive at a friend's home and she takes you into the garden. You know that a train passes the end of the garden approximately every half an hour. Half the trains are scheduled to pass by with an interval of 15 minutes and a half with an interval of 45 minutes. Given that you have no clue when the last train passed by, how long can you expect to wait for the next train?

2. You visit the health club at a random time of the day and ask a random sample how long on average they stay there. Is your survey answer likely to be skewed relative to the actual average length of time people stay at the club?

3. You are the first car to arrive at the traffic light in a foreign land. You have no idea how long the average wait time is between lights, other than that the lights change at regular pre-set intervals. You wait 40 seconds and it turns green. At the next light you arrive second. How long can you expect to wait? More than 40 seconds, less than 40 seconds, or 40 seconds?

2.10.2 Reading and Links

Aczel, A. 2013. On the persistence of bad luck (and good). 4 September. http://blogs. discovermagazine.com/crux/2013/09/04/on-the-persistence-of-bad-luck-and-good/#.XXJL0ihKh3g

Aczel, A.D. 2016. Chance: A Guide to Gambling, Love, the Stock Market, and Just about Everything Else. New York: Thunder's Mouth Press.

Downey, A. 2015. Probably overthinking it. The inspection paradox is everywhere. 18 August. http://allendowney.blogspot.com/2015/08/the-inspection-paradox-is-everywhere.html

The waiting time paradox, or, why is my bus always late? 2018. Pythonic Perambulations. 13 September https://jakevdp.github.io/blog/2018/09/13/waiting-time-paradox/

2.11 Berkson's Paradox

Berkson's Paradox (also known as Berkson's bias or collider bias) is a statistical quirk. It can lead to the appearance of an association between two events or variables which are in fact unrelated. Notably, it can show that two values are negatively correlated in a sample of a population when they are in fact uncorrelated in that population. It arises because of a type of selection bias caused by the observation of some events more than others.

Take the case of a population in which two independent variables, X and Y, are either present or absent. In the population, four possible conditions can be obtained:

X is present, Y is absent; Y is present, X is absent; X is present, Y is present; X is absent, Y is absent.

Berkson's bias occurs when a sample of this population contains three of these categories, but the fourth category (X is absent, Y is absent) is either excluded or under-represented compared to the entire population. In this case, the sample is biased, showing a negative correlation between X and Y when this does not exist in the total population.

Take, for example, the case of a college which admits students based on either musical excellence or sporting excellence. Let us assume for simplicity that there is no link between the two in the total relevant population (say, all students in the country). In other words, let's assume that a musically talented individual is no more or less likely to be gifted at sports. Because the college admits only students who are excellent at music, or excellent at sports, or both, this creates a group or subset of the population which, in this case, displays a negative association between musical and sporting excellence. Those members of the population who are neither excellent at music nor sport are not included in the sample.

This has been shown to have important implications for medical statistics. Say, for example, that a hospital conducts a study that admits patients onto the study who are suffering from either cataracts or diabetes. In this case, there will appear an association between cataracts and diabetes in the set of patients included in the study, which may not appear in the broader population. The paradox arises because the probability of one event happening (cataracts, in this example) is higher in the presence of the other event (diabetes, in this example) because cases in which neither occurs are excluded.

Similarly, take the idea that there is a negative association in our minds between the quality of films based on excellent books and the quality of the books. We can derive one explanation from Berkson's bias. This interpretation is that we remember the instances where the book is excellent or the film is excellent or both. But we forget those cases where both the book and the film were poor. In this case, we find a (spurious) negative correlation between the excellence of the film and the book. The reason is that the bad films/bad books element of the population is not included in the set of films and books under analysis.

Perhaps the most famous example of Berkson's Paradox, proposed by Jordan Ellsberg, was his proposition that "attractive people are jerks". The example works like this. Say that someone only associates with people who are pleasant or attractive or both. That eliminates from the sample pool those who are both unpleasant and unattractive. This leaves a sample which includes attractive people who are unpleasant, and pleasant people who are unattractive, but eliminates those who are neither pleasant nor attractive. So an association is noted between being attractive and being unpleasant, but this is because the unattractive people who are also unpleasant are avoided or ignored. So even if no link exists between attractiveness and unpleasantness in the population, it does in an observed world where the counter-examples who exist in the population are excluded.

More generally, there can be a tendency to construct a sample of the population based on what is present and is of interest. By excluding elements of the population from the sample that are absent and not of interest, this creates a biased sample. This biased sample can indicate the existence of a negative correlation (where X is present, Y is less likely to be present) when this does not exist in the population from which the sample is drawn.

To put it more formally, assume there are two independent events, X and Y. These events are not correlated when observed in nature. If one conditions on the fact that either event X or event Y occurred, however, then if we know that event X did not occur, we know that event Y did occur. This leads to a correlation.

Let's conclude with the appearance of Berkson's Paradox in Covid-19 research. Herbert et al. (2020) highlight a claim that smokers were less likely than non-smokers to be hospitalised with Covid-19. They show how a causal but credible explanation for such a claim is given by Berkson's Paradox. In particular, they note that hospitalised patients are not a random sample of the population but consist disproportionately of those who are older, who are frail, who are smokers, as well as those with Covid-19. For the sake of simplicity and illustration, assume that patients are admitted to hospital for one of two reasons, either for smoking-related illnesses or Covid-19. In this case, tests for Covid on those in hospital are likely to show lower rates of infection among smokers than non-smokers, because the smokers are already hospitalised for smoking-related illnesses unrelated to Covid-19. The non-smokers would be there exclusively because they are suffering from Covid-19. More generally, they argue, if two factors influence being selected into a sample ("collide on selection", hence "collider bias"), they may seem associated when they are in fact independent, or else this might reduce, exaggerate, or reverse an existing association. Ideally, investigators would sample randomly from the target population.

2.11.2 Exercise

Using the idea of collider bias, explain why smokers might appear to be less likely to be hospitalised with pneumonia than non-smokers, even if there is no relationship in the wider population.

2.11.3 Reading and Links

Ellsberg, J. 2014. Why are handsome men such jerks? 3 June. Slate.com https://slate.c om/human-interest/2014/06/berksons-fallacy-why-are-handsome-men-such-j erks.html

Griffith, G., Morris, T.M., Tudball, M.J., Herbert, A., Mancano, G., Pike, L., Sharp, G.C., Palmer, T.M. 2020. Collider bias undermines our understanding of COVID-19 disease risk and severity, Nature Communications, 12 November, 11,5749. Davey Smith, G., Tilling, K., Zuccolo, L., Davies, N.M. and Hemani, G. https://www.nature.com/articles/s41467-020-19478-2

Herbert, A., Griffith, G., Hemani, G. and Zuccolo, L. 2020. The spectre of Berkson's paradox. Collider bias in Covid-19 research. Significance. 29 July. https://rss.onl inelibrary.wiley.com/doi/full/10.1111/1740-9713.01413

Kononovicius, A. 2018. Berkson's paradox. Physics of Risk. 9 October. http://rf.moksl asplius.lt/berkson-paradox/

Shafron, J. 2013. Berkson's paradox explained. Healthcare Economist. 9 July. https://www.healthcare-economist.com/2013/07/09/berksons-paradox-explained/

Simon, C. 2014. Berkson's paradox. Mathemathinking. 5 October. http://corysimon.github.io/articles/berksons-paradox-are-handsome-men-really-jerks/

Woodcock, S. 2017. Paradoxes of probability and other statistical strangeness. The Conversation. 4 April. https://theconversation.com/paradoxes-of-probability-and-other-statistical-strangeness-74440

Are good looking people jerks? AsapSCIENCE. 15 September 2016. YouTube. https://youtu.be/QJ907Aa7TYE

Does hollywood Ruin books? Numberphile. 2018, 28 August 2018. YouTube. https://youtu.be/FUD8h9JpEVQ

2.12 Simpson's Paradox

Simpson's Paradox arises when different groups of frequency data are combined, revealing a different performance rate overall than is the case when examining a breakdown of the performance rate. Ignorance of the implications of Simpson's Paradox can generate false conclusions, which can sometimes be dangerous.

Take the following drugs and their success rate in medical trials over two different days.

	New drug	Old drug
Day 1	63/90 = 70%	8/10 = 80%
Day 2	4/10 = 40%	45/90 = 50%

Overall, new drug = 67% success rate; old drug = 53% success rate.

But the old drug performs better on both days.

So which is the better drug? Clearly, the new drug has a better success rate than the old one, measured over the entire population. The discrepancy with the daily averages can be explained by the relatively good performance of the old drug in Day 1 in the smaller sample and the relatively poor performance of the new drug in Day 2 in the smaller sample.

Take another example. In this trial, there are two groups. The first is a control group of 240 patients supplied with a sugar pill, which is known not to affect the illness under evaluation. The second is a test group of 240 patients supplied with the real drug. The 240 patients are made up of four groups. Group A is elderly adults; Group B is middle-aged adults; Group C is young adults; Group D is children.

Here are the results, with success rate measured by the proportion recovering from the illness within two days of taking the drug:

Those who took the sugar pill

Group A: 20; Group B: 40; Group C: 120; Group D: 60

Success rates are:

Group A: 10%; Group B: 20%; Group C: 40%; Group D: 30%

Overall success rate for those taking the sugar pill = (2 + 8+ 48 + 18) / 240 = 76 / 240 = 31.7%.

Those who took the real drug

Group A: 120; Group B: 60; Group C: 20; Group D: 40

Success rates are:

Group A: 15%; Group B: 30%; Group C: 60%; Group D: 45%

Overall success rate for those taking the real drug = (18+18+12+18) / 240 = 66/240 = 27.5%.

This compares with an overall success rate for those taking the placebo of 31.7%.

So the sugar pill, over the whole sample, produced a higher success rate than the real drug.

Breaking the numbers down, however, reveals a discrepancy.

For the sugar pill

Group A: 10%; Group B: 20%; Group C; 40%; Group D: 30%

For the real drug

Group A: 15%; Group B: 30%; Group C; 60%; Group D: 45%

So, in each group (elderly adults, middle-aged adults, young adults, children), the success rate is greater for those taking the real drug, although in the group as a whole, it is less.

How can we resolve the paradox?
The answer lies in the size and age distribution of each group, which differs between those who received the real drug and those who received the sugar pill. In this study, the group which received the sugar pill consists of a lot more young adults, for example, than the other groups, in contrast with the number taking the real drug. This is important because the natural recovery rates from this illness (as defined in the test) usually are higher in this demographic than the other groups, whether they receive the real drug or the placebo. Again, the elderly (whose recovery rates are normally lower than average) are more heavily represented among those taking the real medication than the placebo.

So which is the better drug? Allowing for the demographic distribution, the real drug would seem to out-perform the sugar pill. The way in which the trials are grouped, however, generates a different result.

2.12.1 Exercise

1. In the 1995/96 baseball seasons, fans could be divided by those who claimed Derek Jeter as the best performing player and those who claimed that title for David Justice. Their respective batting averages in 1995 and 1996 were as follows (a higher batting average is better than a lower batting average).

	1995	1996	Combined
Derek Jeter	12/48 (.250)	183/582 (.314)	195/630 (.310)
David Justice	104/411 (.253)	45/140 (.321)	149/551 (.270)

   In 1997, Jeter averaged 0.291 (190/654), while Justice scored a batting average of 0.329 (163/495).

   Who scores best in each year? Who scores best overall? Who is the better baseball player with the bat?

2. There are two medicines, Super X and Super Y.

   There are two trials of these medicines.

   X: First Trial. Successes 80. Attempts 100.

   Y: First Trial. Successes 78. Attempts 100.

   X: Second Trial. Successes 20. Attempts 40.

   Y: Second Trial. Successes 2. Attempts 5.

Questions:

1. Which is the more successful drug in the first trial?
2. Which is the more successful drug in the second trial?
3. Which is the more successful drug overall?

2.12.2 Reading and Links

Fenton, N., Neil, M. and Constantinou, A. 2015. Simpson's Paradox and the implications for medical trials. 31 August. https://arxiv.org/ftp/arxiv/papers/1912/1912.01422.pdf

Maths: Simpson's Paradox. singingbanana. 21 March 2010, YouTube. https://www.youtube.com/watch?v=wgLUDw8eLB4

Monty Hall, Simpson's Paradox. Harvard University. 29 April 2013. YouTube. https://www.youtube.com/watch?v=0-dFF0w2sPk&app=desktop

Simpson's Paradox. R. Backman. 7 August 2012. YouTube. https://www.youtube.com/watch?v=0-dFF0w2sPk&app=desktop

Simpson's Paradox. Statistics gone bad? Guillaume Riesen. 30 March 2014. YouTube. https://www.youtube.com/watch?v=ZDinnCwP3dg

2.13 The Will Rogers Phenomenon

The Will Rogers Phenomenon occurs when transferring something from one group into another group raises the average of both groups, even though there has been no change in actual values. The name of the phenomenon is derived from a comment made by actor and comedian Will Rogers that "when the Okies left Oklahoma and moved to California, they raised the average intelligence in both states".

In moving a data point from one group into another, the Will Rogers Phenomenon occurs if the point is below the average of the group it is leaving, but above the average of the one it is joining. In this case, the average of both groups will increase.

To take an example, consider six individuals, the remaining life expectancy of whom we assess in turn as 5, 15, 25, 35, 45, and 55.

The individuals with an assessed life expectancy of 5 and 15 years have been diagnosed with a particular medical condition. Those with the assessed life expectancies of 25, 35, 45, and 55 have not. So the mean life expectancy of those with the diagnosed condition is 10 years and those without is 40 years.

If diagnostic medical science now improves such that we can identify the individual with the 25-year life expectancy as suffering from the medical condition (previously this diagnosis was missed), then the mean life expectancy within the group diagnosed with the condition increases from 10 years to 15 years (5 + 15 + 25, divided by 3). Simultaneously, the mean life expectancy of those not diagnosed with the condition rises by 5 years, from 40 years to 45 years (35 + 45 + 55, divided by 3).

So, by moving a data point from one group into the other (undiagnosed into diagnosed), the average of both groups has increased, despite there being no change in actual values. This is because the point is below the average of the group it is leaving (25, compared to a group average of 40), but above the average of the one it is joining (25, compared to a group average of 10). The improvement is, in other words, illusory. The change is simply a consequence of changing the way that the data are grouped.

2.13.1 Exercise

1. Take the following groups of data, A and B.

    A = {10, 20, 30, 40}

    B = {50, 60, 70, 80, 90}

    The arithmetic mean of A is 25, and the arithmetic mean of B is 70.

    Show how transferring one data point from B to A can increase the mean of both.

2. Now take the following example:

    A = {10, 30, 50, 70, 90, 110, 130}

    B = {60, 80, 100, 120, 140, 160, 180}

    By moving the data point 100 from B to A, what happens to the arithmetic mean of A and B?

3. To demonstrate the Will Rogers Phenomenon, does the element which is moved have to be the very lowest of its set, or does it merely have to lie between the arithmetic means of the two sets?

2.13.2 Reading and Links

Feinstein, A. et al. 1985. The Will Rogers phenomenon – Stage migration and new diagnostic techniques as a source of misleading statistics for survival in cancer, *New England Journal of Medicine*, 312(25), 1604–1608.

Inglis-Arkell, E. 2013. The "Will Rogers phenomenon" lets you save lives by doing nothing. 10 October. https://io9.gizmodo.com/the-will-rogers-phenomenon-lets-you-save-lives-by-doi-1443177486

Phillips, B. 2014. Will Rogers phenomenon. Stats Mini Blog. 21 November. https://blogs.bmj.com/adc/2014/11/21/statsminiblog-will-rogers-phenomenon/

Woodcock, S. 2017. Will Rogers paradox. In Paradoxes of Probability and Other Statistical Strangeness. 26 May. https://quillette.com/2017/05/26/paradoxes-probability-statistical-strangeness/

Simple city: Richard Elwes. 2012. The Will Rogers phenomenon., 1 December. https://richardelwes.co.uk/2012/12/01/the-will-rogers-phenomenon/

The Will Rogers effect. UwiSchool. 10 May 2017. YouTube. https://youtu.be/J4NtCsrwq2E

3

Probability and Choice

In this chapter we explore some problems and paradoxes involving choice and reason which have no agreed solution, as well as others with solutions that are somewhat surprising or of particular interest. Examples of problems which defy a unique or consensus solution are Newcomb's Paradox (or Newcomb's Problem), the Sleeping Beauty Problem, the God's Coin Toss Problem, the Keynesian Beauty Contest, and Pascal's Wager (including the Pascal's Mugging Problem). Problems which offer unusual or counter-intuitive insights or perspectives into the world (and the end of the world) include the Doomsday Argument, Benford's Law, and Faking Randomness. A fascinating problem of choice is framed around what is known as the Optimal Stopping Problem, an example of which is the Secretary Problem. We also consider why we seem so often to end up in the slower lane.

3.1 Newcomb's Paradox

You are presented with two boxes. In the transparent box is $1,000. A second box is opaque but contains either $1 million or nothing. Now, you can take away and open just the opaque box and take what is inside, or you can take away both boxes and take the contents inside both. Which should you choose?

Well, if that's all the information you have, it's evident that you should open both boxes. You will not win less than by just opening one of the boxes, but you might win more. Moreover, you will certainly win at least $1,000 instead of a potential zero. So far, so good. But now introduce an additional factor. Before making your decision, you are made aware of a Predictor. The Predictor has a history of being almost always right. When the Predictor has predicted a player will take both boxes, then the opaque box almost always contains nothing. If the Predictor has predicted that the player will take only the opaque box, it almost always contains the million dollars. When you make your final decision the computer's decision has already been made. The contents of the opaque box have already been placed there. You can think of it as a game show where the Predictor informs the organisers of the prediction of what you will do, and the decision about what the opaque box will contain is taken at that point.

This is essentially the basis of what is known as Newcomb's Paradox or Newcomb's Problem, a thought experiment devised by William Newcomb of the University of California and popularised by philosopher Robert Nozick in a paper published in 1969.

So what should you do? Open just the opaque box or open both boxes.

Nozick notes that it is perfectly clear and obvious to almost everyone what should be done. The difficulty is that these people seem to divide almost evenly.

The argument of those who make the case for opening both boxes (the so-called "two-boxers") is that the money has already been deposited at the time you are asked to make your decision. Taking two boxes can't change that, so that's the rational thing to do.

The argument of those who make the case for opening just the opaque box (the so-called "one-boxers") is that the Predictor is very reliable in predicting what people will do. In fact, almost every previous person who has opened two boxes has found the opaque box empty, and almost every previous person who has opened just the opaque box has won the million dollars. On the basis of this evidence, it seems that the sensible thing to do is just to open the opaque box.

In a sense, it can be viewed as a choice between reason, unsupported by evidence, and evidence, unsupported by reason.

Another way of considering the question is to ask whether your choice in some way determines the choice of the Predictor, and thereby the decision as to whether to place the million dollars in the box. Well, there's no time-travelling retro-causality involved. It's a genuine prediction. The Predictor bases its prediction, let's say, on an extremely sophisticated algorithm. It just so happens that the algorithm is uncannily accurate in knowing what people will do.

Look at it this way. The bottom line is that you have a free choice, so why not open both boxes? The problem is that if you are the type of person who is a two-boxer, the Predictor is almost certain to have found this out. If you are the type of person who is a one-boxer, however, the Predictor will almost certainly find that out too.

So, it's not that there is any good reason in itself to open one box rather than two. After all, what you decide now can't change what is already in the box. But there may be a very good reason why you should be the type of person who only opens one box. And the best way to be the type of person who only opens one box is to only open one box. For that reason, the way to win the million dollars is to agree to open just the opaque box and leave the other box untouched.

But why leave behind the $1,000 when the money is either already in the box or not?

That's Newcomb's Paradox. You decide! Are you are a one-boxer or two?

3.1.1 Exercise

Would you open one box or two? Does it matter if it is a different combination of amounts in the two boxes? Does it matter if the Predictor is, or always has been, inerrant in making the predictions?

No solution is provided for this question, as there are no definitive answers. The exercise is for personal or group consideration.

3.1.2 Reading and Links

Bellos, A. 2016. Newcomb's Problem Divides Philosophers. Which Side Are You On? 26 November. https://www.theguardian.com/science/alexs-adventures-in-numberland/2016/nov/28/newcombs-problem-divides-philosophers-which-side-are-you-on?CMP=Share_iOSApp_Other

Bellos, A. 2016. Newcomb's Problem. Which Side Won the Guardian's Philosophy Poll? 30 November. https://www.theguardian.com/science/alexs-adventures-in-numberland/2016/nov/30/newcombs-problem-which-side-won-the-guardians-philosophy-poll

Hickey, G. 2015. The Prisoner's Dilemma and Newcomb's Problem. 5 October. https://www.greghickeywrites.com/prisoners-dilemma-newcombs-problem/

Nozick, R. 1969. Newcomb's Problem and Two Principles of Choice. In: N. Rescher et al. (eds.), *Essays in Honor of Carl G. Hempel*. Springer.

Newcomb's Paradox. Wikipedia. https://en.wikipedia.org/wiki/Newcomb%27s_paradox

Rationally Speaking Podcast. 2015. Kenny Easwaran on "Newcomb's Paradox and the Tragedy of Rationality." 9 August. RS140. http://rationallyspeakingpodcast.org/140-newcombs-paradox-and-the-tragedy-of-rationality-kenny-easwaran/

Transcript of Rationally Speaking Podcast, 2015. Kenny Easwaran on "Newcomb's Paradox and the Tragedy of Rationality." 9 August. RS140. http://rationallyspeakingpodcast.org/wp-content/uploads/2020/11/rs140transcript.pdf

A Problem You'll Never Solve. Vsauce2. 25 February 2019. YouTube. https://www.youtube.com/watch?v=ejUixWn_gE0&feature=youtu.be

Newcomb's Paradox – What Would You Choose. Smart by Design. 20 April 2018. YouTube. https://www.youtube.com/watch?v=B5575Ky0Fz0&feature=youtu.be

Newcomb's Problem and the Tragedy of Rationality. Julia Galef. 12 August 2015. YouTube. https://www.youtube.com/watch?v=2KxJ6eTY9bA&feature=youtu.be

Newcomb's Problem – Evidential & Causal Decision Theory. THUNK. 8 May 2018. YouTube. https://youtu.be/SQw2WmicDYA

3.2 The Sleeping Beauty Problem

Sleeping Beauty volunteers to undergo the following experiment and is told all of the following details. On Sunday she will be sent to sleep. Once or twice during the experiment, Beauty will be woken, interviewed, and sent back to sleep with a drug that makes her forget that awakening.

A fair coin will be tossed on Sunday evening after she goes to sleep, to determine which experimental procedure to undertake: if the coin comes up heads, Beauty will be woken and interviewed on Monday only. If the coin comes up tails, she will be interviewed on Monday and Tuesday. In either case, she will be woken on Wednesday without an interview and the experiment ends.

Any time Sleeping Beauty is interviewed, she is asked, "What is your belief now, as a percentage, that the coin landed heads?"

What should Beauty's answer be?

To one way of thinking about this, the answer is clear. The coin was tossed once before the experiment starts.

Since the coin was tossed just once, and Beauty obtains no further information, her answer should be a 1 in 2 chance that the coin landed heads.

To another way of thinking about it, she is interviewed just once if it landed heads (on Monday) but is interviewed twice if it landed tails (on Monday and Tuesday). She does not know which day it is when she is woken up and interviewed, but from her point of view, there are three possibilities. These are:

1. It landed heads, and it is Monday.
2. It landed tails, and it is Monday.
3. It landed tails, and it is Tuesday.

These three possibilities are of equal likelihood, and two of these involve the coin landing tails and just one for the coin landing heads. So her answer should be a one in three chance that the coin landed heads.

So which answer is correct? The world of probability is by and large divided into those who are adamant that she should go with half (the so-called "halfers") and those who are equally adamant that she should go with a third (the so-called "thirders"). Are they both right, are they both wrong, or somewhere in between?

What if she is told she will be woken 1,000 times if the coin lands tails but only once if the coin lands heads? Now, is her correct estimate of the probability that any particular awakening resulted from the coin landing heads equal to 1 in 1,001? Or should it still be a half?

A way sometimes used to resolve seemingly intractable probability paradoxes is to ask what the fair betting odds should be. At what odds should Beauty be willing to place a bet?

So, if in the original experiment Beauty is offered odds of 3 to 2 (£2 to make a net profit of £3) that the coin landed heads, should she take those odds? If the correct answer is a half, those odds are attractive as the correct odds should be evens (£1 to win a net £1). If the correct answer is a third, those odds are unattractive as the correct odds should be 2 to 1 (£1 to win a net £2).

So what should Beauty do if offered odds of 3 to 2? Bet or decline the bet?

The simplest way to resolve this is to ask what would happen if the experiment is repeated many times and she accepted the odds of 3 to 2. When the coin came up heads, she would be woken just once, placed the £10 bet and made a profit of £15. However, when the coin landed tails she would be woken twice and placed two bets of £10, i.e. a total of £20 and lost both bets.

So her net outcome from this betting strategy would be an average loss of £5. We say that the expected value of her bet is negative.

This suggests that half is a wrong answer as to the probability that the coin landed heads.

At odds of 2 to 1, on the other hand, she would place £10 on the one occasion she would be woken up, i.e. Monday, and would win £20. However, when the coin came up tails, she would lose £10 on Monday and £10 on Tuesday, i.e. £20. If the experiment is repeated many times, therefore, at these odds, she would expect to break even.

This suggests that odds of 2 to 1 are the correct odds, which is consistent with a probability of 1/3.

Applying the "betting test" to this problem, therefore, suggests that Beauty's answer when she is woken up should be that there is a 1 in 3 chance that the coin landed heads.

But how can this be right, when the coin was tossed just once, and we know that the chance of the coin landing heads is a half? If this is the "prior probability" Beauty should assign to the coin landing heads, on what grounds should the probability she assigns change? The only information she acquires is that she has been woken and questioned, but she knew that would happen in advance, so this is not new information. Given that she assigns a prior probability of a half to the coin coming up heads, and she acquires no new information, it is perhaps difficult to see on what grounds she should change her opinion. The posterior probability she assigns (after she receives all new information) should be identical to the prior probability.

This is the kernel of the puzzle, and it is why there is a long-standing and ongoing debate between fervent so-called "halfers" and "thirders".

So the question is whether there is a single correct answer and that one school of thought is wrong. Alternatively, is there no single correct answer and both schools of thought have valid claims?

There is one approach which arguably resolves the problem. To see this, we need to identify the actual "prior probability" that the coin tossed after Beauty goes to sleep is heads.

This depends on the question we are seeking to answer, and what information is available to Beauty before she goes to sleep.

If she is told that a coin will be tossed after she goes to sleep, and nothing else, then her correct estimate that the fair coin will land on heads is a half. This is the answer to a simple question of how likely a fair coin is to land heads with no conditions, i.e. the unconditional probability that the coin will land heads is 1/2.

If she is given the additional information, however, that she will be woken just once if the coin lands heads but twice if it lands tails (albeit she will remember only one of the awakenings), then we are posing a very different question.

The new question she is being asked to answer is to estimate the probability that whenever she awakens, her awakening has resulted from the coin toss landing heads. Since she has just one awakening when the coin lands heads, but two awakenings when it lands tails, the probability that any particular awakening occurred from a heads flip is 1/3, i.e. the conditional probability that the coin landed heads given any individual awakening is 1/3.

By extension, if she is told she will be woken 1,000 times if the coin lands tails but only once if the coin lands heads, then her correct estimate of the probability that any particular awakening resulted from the coin landing heads is 1/1,001.

So the "prior probability" Beauty should assign to the chance of a coin landing heads after any particular awakening is 1/3 within the terms of the experiment, even before she goes to sleep. It is true that she has access to no new information whenever she awakens, but that means her "prior probability" of being awakened by a heads flip remains at 1/3 after she is woken. This is consistent with Bayesian reasoning which states the prior probability of an event will not change unless there is new information.

Given, therefore, that she assigns a prior probability of 1/3 to any particular awakening arising from a heads flip, this should be the answer she gives whenever she wakes.

So the paradox resolves to the question Beauty is being asked to answer. What is the probability that a fair coin will land heads? Answer = 1/2. What is the probability that whenever she is woken the awakening has resulted from a heads flip? Answer = 1/3. She is consistent in these answers both before she goes to sleep and whenever she wakes. In other words, because Beauty knows that she will correctly answer 1/3 whenever she is woken, given the rules of the experiment, of which she is aware, she will answer 1/3 before she goes to sleep.

This, at least, is one way of approaching the classic Sleeping Beauty Problem. Another involves consideration of what is known as the Self-Sampling Assumption (SSA) and the Self-Indication Assumption (SIA). We shall look at that when we consider it in the section titled "The God's Coin Toss Problem".

3.2.1 Exercise

Are you a halfer or a thirder? No solution is provided for this question, as there is no definitive answer. The exercise is for personal or group consideration.

3.2.2 Reading and Links

Downey, A. 2015. Probably Overthinking It. The Sleeping Beauty Problem. 12 June. http://allendowney.blogspot.com/2015/06/the-sleeping-beauty-problem.html

Mutalik, P. 2016. Solution: "Sleeping Beauty's Dilemma." Quanta magazine. 29 January. https://www.quantamagazine.org/solution-sleeping-beautys-dilemma-20160129/

Poundstone, W. 2019. How to Predict Everything. London: Oneworld publications, 102-109.

Sleeping Beauty Problem. Wikipedia. https://en.wikipedia.org/wiki/Sleeping_Beauty_problem

PHILOSOPHY – Epistemology. The Sleeping Beauty Problem. Wireless Philosophy. 27 February 2015. YouTube. https://youtu.be/5Cqbf86jTro

The Sleeping Beauty Problem. Julia Galef. 6 June 2015. YouTube. https://youtu.be/zL52lG6aNIY

3.3 The God's Coin Toss Problem

Imagine a world created by an external Being based on the toss of a fair coin. It's a thought experiment we can call the "God's Coin Toss Problem". In the simplest version, Heads creates a version of the world in which one black-bearded person is created. Let's call that World A. Tails creates a version of the world in which a black-bearded and a brown-bearded person are created. Let's call that World B.

You wake up in the dark in one of these worlds, but you don't know which, and you can't see what colour your beard is, though you do know the rule that created the world. What likelihood do you now assign to the hypothesis that the coin landed Tails and you have been born into World B?

This depends on what fundamental assumption you make about existence itself. One way of approaching this is to adopt what has been called the "Self-Sampling Assumption" (SSA). Essentially, this states that you should reason as if you are randomly selected from all elements that could have been you, what is called your "reference class". So black-bearded and brown-bearded people are in the same reference class, a black-bearded person and a brown billiard ball are not.

Using this assumption, we see ourselves merely as a randomly selected bearded person among the reference class of black- and brown-bearded people. The coin could have landed Heads in which case we are in World A or it could have landed Tails in which case we are in World B. There is an equal chance that the coin landed Heads or Tails, so we should assign a credence of 1/2 to being in World A and similarly for World B. In World B the probability is 1/2 that we have a black beard and 1/2 that we have a brown beard. So far we don't know what colour our beard is.

The light is now turned on and we see that we are sporting a black beard. What is the probability now that the coin landed Tails and we are in World B? Well, the probability we would have a black beard conditional on living in World A is 1, i.e. 100%. This is because we know that the one person who lives in World A has a black beard. The conditional probability of having a black beard in World B, on the other hand, is 1/2. The other inhabitant has a brown beard. So there is twice the chance that we live on World A as World B conditional on finding out that we have a black beard, i.e. a 2/3 chance the coin has landed Heads, and we live in World A.

Let's now say you make a different assumption about existence itself. Your worldview in this alternative scenario is based on what has been termed the "Self-Indication Assumption" (SIA). It can be stated like this. "Given the fact that you exist, you should (other things equal) favour a hypothesis according to which many observers exist over a hypothesis according to which few observers exist".

According to this assumption, you note that there is one hypothesis (the World B hypothesis) according to which there are two observers (one black-bearded and one brown-bearded) and another hypothesis (the World A hypothesis) in which there is only one observer (who is black-bearded). Since there is twice as many observers in World B as World A, then according to the Self-Indication Assumption, it is twice as likely (a 2/3 chance) that you live in World B as World A (a 1/3 chance). This is your best guess while the lights are out. When the lights are turned on, you find out that you have a black beard. The new conditional probability you attach to living in World B is 1/2, as there is an equal chance that as a black-bearded person you live in World B as World A.

So, under the Self-Sampling Assumption, your initial belief that you lived in World B was 1/2, which fell to 1/3 when you found out you had a black beard. Under the Self-Indication Assumption, on the other hand, your initial credence of living on World B was 2/3, which fell to 1/2 when the lights came on.

So which is right and what are the broader implications?

Let us first consider the impact of changing the reference class of the "companion" on World B. Instead of this being another bearded person, you are solely accompanied by a brown billiard ball. In this case, what is the probability you should attribute to living on World B given the Self-Sampling Assumption? While the lights are out, you consider that there is a probability of 1/2 that the coin landed Tails, so the probability that you live on World B is 1/2.

When the lights are turned on, no new relevant information is added as you knew you were black-bearded. There is one black-bearded person on World A, therefore, and one on World B. So the chance that you are in World B is unchanged. It is 1/2.

Given the Self-Indication Assumption, the credence you should assign to being on World B given that your companion is a billiard ball instead of a bearded person is now 1/2 as the number of relevant observers inhabiting World B is now one, the same as on World A. When the lights come on, you learn nothing new, and the chance the coin landed Tails and you are on World B stays unchanged at 1/2.

The choice of underlying assumption has implications elsewhere, most famously regarding the Sleeping Beauty Problem, which we have already considered.

In that problem, Sleeping Beauty is woken either once (on Monday) if a coin lands Heads or twice (on Monday and Tuesday) if it lands Tails. She knows these rules. Either way, she will only be aware of the immediate awakening (whether she is woken once or twice). The question is how Sleeping Beauty should answer if she is asked how likely it is the coin landed Tails when she is woken.

If she adopts the Self-Sampling Assumption, she will answer 1/2. The coin will have landed Tails with a probability of 1/2, and there is no other observer than her. Only if she is told that this is her second awakening will she change her belief that it landed Tails to 1 and that it landed Heads to 0.

If she adopts the Self-Indication Assumption, she has a different world-view in which there are three observation points. According to this assumption, there is one hypothesis (the Heads hypothesis) according to which there is one observer opportunity (Monday awakening) and another hypothesis (the Tails hypothesis) in which there are two observer opportunities (the Monday awakening and the Tuesday awakening). Since there are twice as many observation opportunities in the Tails hypothesis according to the Self-Indication Assumption, it is twice as likely (a 2/3 chance) that the coin landed Tails as that it landed Heads (a 1/3 chance).

Looked at another way, if there is a coin toss that on Heads will create one observer, while on Tails it will create two, then we have three possible observers (first observer on Heads, first observer on Tails, and second observer on Tails, each existing with equal probability), so the Self-Indication Assumption assigns a probability of 1/3 to each. Alternatively, this could be interpreted as saying there are two possible observers (first observer on either Heads or Tails, second observer on Tails), the first existing with probability 1 and the second existing with probability 1/2. So the Self-Indication Assumption assigns a 2/3 probability to being the first observer and 1/3 to being the second observer, which is the same as before. Whichever way we prefer to look at it, the Self-Indication Assumption gives a 1/3 probability of Heads and 2/3 probability of Tails in the Sleeping Beauty Problem.

So what is the problem, if any, with the Self-Indication Assumption? Here the Presumptuous Philosopher Problem has been flagged. It goes like this. Imagine that scientists have narrowed the possibilities for a final theory of physics down to two equally likely possibilities.

The main difference between them is that Theory 1 predicts that the universe contains a million times more observers than Theory 2 does. There is a plan to build a state-of-the-art particle accelerator to arbitrate between the two theories.

Now, philosophers using the Self-Indication Assumption come along and say that Theory 1 is almost certainly correct as we're a million times more likely to exist in the first place with this theory than Theory 2. Indeed, even if the particle accelerator produced evidence that was tens of thousands of times more consistent with Theory 2, we should still hold with the view of the philosophers who are sticking to their assertion that Theory 1 is the correct one.

So we are left with a choice between the Self-Sampling Assumption and the Self-Indication Assumption, and both have their problems. And we need to choose a side. What does our intuition say? And can we rely on that?

3.3.1 Exercise

Do you go along with the Self-Sampling Assumption or the Self-Indication Assumption? No solution is provided, as there is no definitive answer. The exercise is for personal or group consideration.

3.3.2 Reading and Links

Aaronson, S. PHYS771. Lecture 17: Fun with the Anthropic Principle. https://www.scottaaronson.com/democritus/lec17.html

Bostrom, N. 2002. Self-Locating Beliefs in Big Worlds: Cosmology's Missing Link to Observation. Preprint. *Journal of Philosophy* 99 (12), http://philsci-archive.pitt.edu/1625/1/Big_Worlds_preprint.PDF

Problems and Paradoxes in Anthropic Reasoning (Nick Bostrom). PhilosophyCosmology. 7 May 2014. YouTube. https://youtu.be/oinR1jrTfrA

3.4 The Doomsday Argument

Can we demonstrate, purely from the way that probability works, that the human race is likely to go extinct in the relatively foreseeable future, regardless of what humanity might do to try and prevent it? Yes, according to the so-called Doomsday Argument.

Here's how the argument goes. Let's say you want to estimate how many tanks the enemy has to deploy against you, and you know that the tanks have been manufactured with serial numbers starting at 1 and ascending from there. Now let's say you identify the serial numbers on five random tanks, and they all have serial numbers under 10. Even an intuitive understanding of the workings of probability would lead you to conclude that the number of tanks possessed by the enemy is pretty small. On the other hand, if they are identified as serial numbers 2524, 7866, 5285, 3609, and 8009, you are unlikely to be way out if you estimate the enemy has more than 10,000 of them.

Let's say that you only have one serial number to work with, and that it shows the number 18. On the basis of just this information, you would do well to estimate that the total number of enemy tanks is more likely to be 36 than 360, and even more likely than the total tank count being 36,000.

Imagine, in another scenario, that you are made aware that a selected box of numbered balls contains either 10 balls (numbered from 1 to 10) or 10,000 balls (numbered 1 to 10,000), and you are asked to guess which. Before you do so, one is drawn for you. It reveals the number seven. That would be a 1 in 10 chance if the box contains 10 balls, but a 1 in 10,000 chance if it contained 10,000 balls. You would be right based on this information to conclude that the box very probably contains 10 balls, not 10,000.

A similar way of thinking is central to what is known as the mediocrity principle. The idea is to assume mediocrity rather than starting with the assumption that a phenomenon is exceptional, special, privileged, or better. Linked to this is the Copernican principle, the idea in cosmology that we are not privileged or special observers of the universe. It is based on the observation of Nicolaus Copernicus in the sixteenth century that the earth is not located at the centre of the universe.

It was a principle notably used by astrophysicist John Richard Gott when arriving at the Berlin Wall. He asked himself whether, in the absence of other knowledge, there was any reason to believe that the moment in time that he came upon the wall was likely to be any special time in the lifetime of the wall. He decided that there was not and that because any moment was equally likely, therefore, his best estimate was that there was as much time before he visited the wall as there would be for the wall after he visited it. In other words, his best guess as to how long the wall would last going forward was exactly as long as it had already been in existence.

He was also able to go further and estimate its remaining lifetime, to different levels of confidence. If we are visiting the wall a quarter of the way along its existence timeline, he noted, its future would be three times as long as its past. If visiting it three quarters along its timeline, the future would be one third of its past. Because half of its existence is between these two points (75% minus 25% is 50%), there was a 50% chance that it would last a further period between one third and three times its current existence. Based on its age when he observed it in 1969 (8 years), Gott argued that there was a 50% chance that it would fall in between 8/3 years and 8×3 (24) years from then. The wall came down 20 years later. This can be expanded to any specific level of confidence. For example, if visiting the wall 20% along the timeline, its future would be four times as long as its past. If visiting it 80% along its timeline, the future would be a quarter as long as its past. Because 60% of its existence would be between 20% and 80% along the timeline, there was a 60% chance that it would last a further period between four times its past and a quarter of its past (between 32 years and 2 years).

In 1993 Gott applied this method to estimate the time to extinction of the human race, concluding among other things that humans are unlikely to survive long enough to colonise the Galaxy. Gott later gave *The New Yorker* magazine a 95% confidence interval for the closing of 44 Broadway shows based on their opening dates. He was about 95% correct.

It's related to the "Lindy effect", the name of which is derived from a New York delicatessen, famous for its cheesecakes, which was frequented by actors playing in Broadway shows. The Lindy effect was the observation that a Broadway show could expect to last for a further period equal to the length of time it had already been playing. So a show that had been on Broadway for three years could, as a best guess, be expected to last another three years before closing. More generally, the Lindy effect has come to represent the idea that the life expectancy going forward of a non-perishable thing such as a technology or an idea is proportional to its current period of existence, so that every additional period of survival implies a greater future life expectancy.

We can apply this to anything that has a life expectancy determined by chance, where there is a random process determining the timescale for existence. If there is a non-random process at work, with respect (for example) to the biological ageing process, this method is flawed. So this cannot be applied to the life expectancy of a 100-year-old human, since 100 years of age is not some random point in a person's lifetime.

To return to the Copernican principle, in Bayesian terms it can be viewed as Bayes' Rule with an uninformative prior. When we want to estimate how long something will last, in the absence of other knowledge, this principle suggests assuming we are at the mid-point of the timeline.

The final step is to transpose this reasoning to our actual situation here on Earth.

The Doomsday Argument transfers the logic of the laws of probability to the survival of the human race. If we view the entire history of the human race from a timeless perspective, then all else being equal we should be somewhere in the middle of that history. That is, the number of people who live after us should not be too much different from the number of people who lived before us. If the population is increasing exponentially, it seems to imply that humanity has a relatively short time left. Of course, we may have special information that indicates that we aren't likely to be in the middle, but that might simply mitigate the problem, not remove it.

To date there have been, according to indicative estimates, about 110 billion humans on the earth, about 7% of whom are alive today. Projecting demographic trends forward, this makes our best estimate of the termination of the timeline of the human race, as we know it, to be within this millennium.

There is currently no consensus in the debate on the Doomsday Argument, with some proposing that humans will never go extinct. That debate is likely to continue until the end of the world, whenever it comes.

3.4.1 Exercise

1. According to the Doomsday Argument, what is our best estimate (in 2021) of how long the University of Bologna will continue to exist? Is there a caveat raised by the Doomsday argument itself?

2. What spread of years of existence does it have left with a 50% probability? What about with a 60% probability?

Is there a problem with these answers raised by the Doomsday argument itself?

3. If a randomly selected giant tortoise is 20 years old, and we know nothing more about it, what is our best estimate of its future life expectancy?

3.4.2 Reading and Links

Baker, R.D. 2002. Probability Paradoxes: An Improbable Journey to the End of the World. Mathematics Today, December, 185–189.

Barrow, J.D. 2008. How Long Are Things Likely to Survive, Chapter 40. In *100 Essential Things You Didn't Know You Didn't Know*, 109–111. London: Bodley Head.

Bostrom, N. 2002. A Primer on the Doomsday Argument. http://www.anthropic-principle.com/?q=anthropic_principle/doomsday_argument

Farnam Street (fs). The Copernican Principle: How to Predict Everything. https://fs.blog/2012/06/copernican-principle/

Gott, J.R. 1993. Implications of the Copernican principle for Our Future Prospects. *Nature*, 363 (6427): 315–319, https://ui.adsabs.harvard.edu/abs/1993Natur.363..315G/abstract

Kaneda, T. and Haub, C. 2020. How Many People Have Ever Lived on Earth? PRB. 23 January. https://www.prb.org/howmanypeoplehaveeverlivedonearth/

Poundstone, W. 2019. A Math Equation That Predicts the End of Humanity. 5 July. https://www.vox.com/the-highlight/2019/6/28/18760585/doomsday-argument-calculation-prediction-j-richard-gott

Doomsday Argument. Lesswrongwiki. https://wiki.lesswrong.com/wiki/Doomsday_argument

Doomsday Argument. RationalWiki. https://rationalwiki.org/wiki/Doomsday_argument

Doomsday Argument. Wikipedia. https://en.wikipedia.org/wiki/Doomsday_argument

Mediocrity Principle. Wikipedia. https://en.wikipedia.org/wiki/Mediocrity_principle

The Doomsday Argument. When Will Civilization End? Cool Worlds. 16 February 2017. YouTube. https://www.youtube.com/watch?v=84LcUXKNPCY

3.5 When Should You Stop Looking and Start Choosing?

In *The Merchant of Venice*, by William Shakespeare, Portia sets for her suitors a problem to solve to find who is right for her. In the play, there are just three suitors. The suitors are asked to choose between a gold, a silver, and a lead casket, one of which contains a portrait which is the key to her hand in marriage.

Let us change the plot line and base a thought experiment around Portia's quest for love in which she meets the successive suitors in turn, with no test but her own feelings. To make the problem of more general interest, let's say she has a hundred suitors to choose from. Each will be presented to her in random order, and she has three minutes to decide whether he is the one for her. If she turns someone down there is no going back, but the good news is that she is guaranteed not to be turned down by anyone she selects. If she comes to the end of the line and has still not chosen a partner, she will have to take whomever is left, even if he is the worst of the hundred. All she has to go on in guiding her decision is the relative merit of each in the pool of suitors.

Let's say that the first presented to her, whom we shall call No. 1, is a reasonable choice, but she has some doubts. Should she choose him anyway, in case those to follow will be worse? With 99 potential matches left, it seems more than possible that there will be at least one who is a better match than No. 1.

The problem facing Portia is that she knows that if she dismisses No. 1, he will be gone forever, to be betrothed to someone else.

She decides to move on. The second suitor turns out to be far worse than the first, as does the third and fourth. She starts to think that she may have made a mistake in not accepting the first. Still, there are potentially 96 more to see. This goes on until she sees No. 20, whom she prefers to No. 1. Should she now grasp her opportunity before it is too late? Or should she wait for someone even better?

She is looking for the best of the hundred, and this is the best so far. But there are still 80 suitors left, one of whom might be better than No. 20. Should she take a chance?

What is Portia's optimal strategy in finding Mr. Right?

This is an example of an "Optimal Stopping Problem", which is a sequential choice problem. In the classic "Secretary Problem" variation, you are interviewing for a secretary, with your aim being to maximise your chance of hiring the single best applicant out of the pool of applicants. Your only criterion to measure suitability is their relative merit, i.e. who is better than whom. As with Portia's problem, you can offer the post to any of the applicants at any time before seeing any more candidates, but you lose the opportunity to hire that applicant if you decide to move on to the next in line.

This sort of stopping strategy can be extended to anything including the search for a place to live, a place to eat, or the choice of a used car.

In each case, there are two ways you can fail to meet your goal of finding the best option out there. The first is by stopping too early, and the second is by stopping too late.

By stopping too early, you leave the best option out there. By stopping too late, you have waited for a better option that turns out not to exist. So how do you find the right balance?

Let's consider the intuition. The first option is the best yet, and the second option (assuming we are taking the possibilities in random order) has a 50%

chance of being the best yet. Likewise, the tenth option has a 10% chance of being the best to that point.

It follows that the chance of any given option being the best to that point declines as the number of options there have been before increases. So the chance of coming across the "best yet" becomes even smaller as we go through the process.

To see how we might best approach the problem, let's go back to Portia and her suitors and look at her best strategy when faced with different-sized pools of suitors. Can she do better using some strategy other than choosing at some random position in the order of presentation to her?

It can be shown that she can certainly expect to do better than random selection, given that there are more than two to choose from.

We can look at it this way. If she chooses No. 1, she has no information with which to compare the relative merit of her suitors. On the other hand, by the time she reaches No. 3, she must choose him, even if he's the worst of the three. In this way, she has maximum information but no choice.

In the case of No. 2, she has more information than she did when she saw No. 1, as she can compare the two. She also has more control over her choice than she will if she leaves it until she meets No. 3.

So she turns down No. 1 to give herself more information about the relative merit of those available. But what if she finds that No. 2 is worse than No. 1? What should she do? It can be shown that she should wait and take the risk of ending up with No. 3, who could be the worst of the three, as she must do if she leaves it to the last. On the other hand, if she finds that she prefers No. 2 to No. 1, she should choose him on the spot and forego the chance that No. 3 will be a better match.

If there are four suitors, Portia should use No. 1 to gain information on what she should be measuring her standards against and select No. 2 if he is a better choice than No. 1. If he is not, do the same with No. 3. If he is still not better than No. 1, go to No. 4 and hope for the best. The same strategy can be applied to any number of people in the pool.

So, in the case of a hundred suitors, how many should she see to gain information before deciding to choose someone?

This can be generalised to any problem when there is a list of options but it is impractical to recall rejected options for reconsideration. A general problem could exist where the number of options is very large, or where wavering over a choice loses the option to take that choice. Say there are N applicants for a job, where N is very large. The chance of choosing the best applicant at random is $1/N$, but interviewing every single applicant is time-consuming and impractical. Is there a best way of finding the best candidate within practical constraints?

To simplify the problem, let's take the case of three options (candidates) to choose from, which we shall call A, B, and C.

Let's say that option A is better than option B, which is better than option C.

We could see them in six possible orders, with the first to be interviewed on the left and the last to be interviewed on the right:

A,B,C

A,C,B

B,A,C

B,C,A

C,A,B

C,B,A

If our strategy is always to take the first candidate, then this would select the best one (option A) twice out of six times. So we would select the best candidate a third of the time.

If our strategy is instead to use the first candidate as a baseline and then select the first candidate after that who is better than the first, we would select the best candidate in the third (B,A,C), fourth (B,C,A), and fifth (C,A,B) cases. So we would select the best candidate a half of the time.

If our strategy is to eliminate both the first two candidates and select the third, we would choose the best candidate (A) only in the fourth (B,C,A) and sixth (C,B,A) cases. As in the first strategy, the chance of selecting the best again is just a third.

Where there are three candidates, therefore, the strategy of using the first one as a baseline, and selecting the next one after that who is better than the first, is superior to either of the alternative strategies in finding the best candidate.

With four candidates, there are 24 different possible orderings.

In this case, the strategy that most often produces the best candidate is again to reject the first candidate but to choose the next one that is better than the first. This gets the best candidate 11 times out of 24. No other strategy beats this. For comparison, choosing the first candidate (or the last candidate) gives a chance of finding the best candidate of 1/4, which is less than 11/24. Similarly, eliminating the first two candidates and selecting the next who is better than either of these gets the best candidate 5/12 of the time, which is again less than 11/24.

What if there are 100 candidates?

It can be demonstrated that the optimal strategy ("stopping strategy") before converting looking into leaping is 37. See the first 37 of the 100 and reject them, then select the first candidate after that who is better than any of the first 37.

In the case of Portia, she should meet with 37 of the suitors, then choose the first of those to come after who is better than the best of the first 37. By following this rule, she will find the best of the princely bunch of a hundred with a probability, strangely enough, of 37%.

By choosing randomly, on the other hand, she has a chance of 1 in 100 (1%) of settling upon the best.

This stopping rule of 37% applies to any similar decision, such as the Secretary Problem or looking for a house in a fast-moving market. It doesn't matter how many options are on the table. You should always use the first 37% as your baseline and then select the first of those coming after that you find to be better than any of the first 37%.

The mathematical proof is based on the mathematical constant, e (sometimes known as Euler's number) and specifically 1/e, which can be shown to be the stopping point along a range from 0 to 1, after which it is optimal to choose the first option that is better than any of those before. The value of e is approximately equal to 2.71828, so 1/e is about 0.36788 or 36.788%. This has simply been rounded up to 37% in explaining the stopping rule. It can also be shown that the chance that implementing this stopping rule will yield the very best outcome is also equal to 1/e, i.e. about 37%.

If there is a chance that your selection might decide to opt out, the rule can be adapted to give a different stopping rule, but the principle remains. For example, if there is a 50% chance that your selection might opt out, then the 37% rule is converted into a 25% rule. The rest of the strategy remains the same. By doing this, you will have a 25% chance of finding the best of the options, however, compared to a 37% chance if you always get to make the final choice. This is still a lot better than the 1% chance of selecting the best out of a hundred options if you choose randomly. The lower percentage here (25% compared to 37%) reflects the additional variable (your choice might not be available) which adds further uncertainty into the mix. There are other variations on the same theme, where it is possible to go back with a given probability that the option you initially passed over is no longer available.

There is also a rule of thumb which can be derived when the aim is to maximise the chance of selecting a good option, if not the very best. This strategy has the advantage of reducing the chance of ending up with one of the least good options. It is the square root rule, which simply replaces the 37% criterion with the square root of the number of options available. In the case of Portia, she would meet the first 10 of the 100 (instead of 37) and choose the first of the remaining 90 who is better than the best of those 10. Compared to the theoretically best candidate, it has been calculated that in a queue of 100 candidates this method will, on average, get someone about 90% to the top, or 75% in a queue of 10.

Whichever variation you adopt, the numbers will change but the principle stays the same.

The key to the strategy is to balance information with control. The less control you have (because, for example, your selection typically becomes unavailable), the less information you can afford to gather before activating the choice process. If Portia's choice of suitor is quite likely to turn her down, she needs more people to actively choose from, for which she must sacrifice some information. This means seeing fewer suitors before starting the process of deciding. The more control you have (because, for example, your selection is always available), the more information you can afford to gather before actively starting to decide. If Portia is guaranteed her choice, therefore, she can reasonably see more suitors (gather more information) before beginning to actively consider her options.

The optimal strategy might be different if we have some objective standard by which to measure our options. For example, Portia might simply be interested in choosing the richest of the suitors, and she knows the distribution of wealth of all potential suitors. This ranges evenly from the bankrupt suitor to those worth 100,000 ducats.

This means that the upper percentile of potential suitors in the whole population is worth upwards of 99,000 ducats. The lowest percentile is worth up to 1,000 ducats. The 50th percentile is worth between 49,000 and 50,000 ducats.

Now Portia is presented with 100 out of this population of potential suitors, and let's assume that the suitors presented to her are representative of this population.

Say now that the first to be presented to her is worth 99,500 ducats. Since wealth is her only criterion, and he is in the upper percentile in terms of wealth, her optimal decision is to accept his proposal of marriage. One of the next 99 may be worth more than 99,500 ducats, but that isn't the way to bet.

On the other hand, say that the first suitor is worth 60,000 ducats. Since there are 99 more to come, it is a good bet that at least one of them will be worth more than this. If she has turned down all suitors, however, until presented with the 99th, her optimal decision now is to accept him and his 60,000 ducats. In other words, Portia's decision as to whether to accept the proposal comes down to how many potential matches she has left to see. When down to the last two, she should choose him if he is above the 50th percentile, in this case, 50,000 ducats. The more there are to come the higher the percentile of wealth at which she should accept. Most fundamentally, she should never take anyone who is below the average unless she is out of choices. In this version of the stopping problem, the probability that Portia will end up with the wealthiest of the available suitors turns out to be 58%. Indeed, any criterion that provides information on where an option is relative to the relevant population as a whole will increase the probability of finding the best choice of those available.

As such, it seems that if Portia is only interested in the money, she may well be more likely to find it than if she is looking for love.

3.5.1 Exercise

1. There are 64 available luxury apartments in a one-day sell-off, one of which you need to accept. You are offered details of each in turn and must choose to accept or decline within three minutes. If you

decline, you are offered the next on the list. There is no going back. If you accept, it is yours. How many should you assess and reject before starting to consider which of the others to accept? What is the probability that you will have selected the best apartment?

2. What if there was a 50% chance that your selection would turn out to be unavailable? How many should you now assess and reject before starting to consider which of the others to accept? What is the probability now that you have selected the best apartment?

3. Using an established rule of thumb, how many apartments would you assess and reject before starting to consider which of the others to accept in order to maximise the chance of selecting a good option, if not the very best?

4. What is the probability that you would choose the best apartment if you could choose randomly which to choose and obtain that?

3.5.2 Reading and Links

Barrow, J.D. 2008. The Secretary Problem, Chapter 32. In *100 Essential Things You Didn't Know You Didn't Know*, 87–89. London: Bodley Head.

Christian, B. 2016. Optimal Stopping: How to Find the Perfect Apartment, Partner and Parking Spot. Medium.com. July 11. https://medium.com/galleys/opt imal-stopping-45c54da6d8d0

Duso, L. 2020. Math Based Decision Making: The Secretary Problem. How Probability Theory Answers the Question "Accept or Reject?" September 10. https://me dium.com/cantors-paradise/math-based-decision-making-the-secretary-pro blem-a30e301d8489

Hernandez, B. 2010. The Secretary Problem Solution Details. 3 May. https://thebrya nhernandezgame.files.wordpress.com/2010/05/secretary-problem.pdf

Parker, M. 2014. The Secretary Problem. Slate. December 17. https://slate.com/te chnology/2014/12/the-secretary-problem-use-this-algorithm-to-determine-exactly-how-many-people-you-should-assess-before-making-a-new-hire-or-choosing-a-life-partner.html

Smith, D.K. 1997. Mathematics, Marriage and Finding Somewhere to Eat. + plus mag-azine. 1 September. https://plus.maths.org/content/os/issue3/marriage/index

Shakespeare Birthplace Trust. 2021. The Merchant of Venice. Synopsis and plot over-view of Shakespeare's The Merchant of Venice. https://www.shakespeare.org .uk/explore-shakespeare/shakespedia/shakespeares-plays/merchant-venice/

Stats Made Easy. 2017. Reject Love at First Sight Until You Achieve Sufficient Sample Size. 14 February. https://www.statsmadeeasy.net/2017/02/reject-love-at-first-s ight-until-you-achieve-sufficient-sample-size/

Rationally Speaking Podcast. 2016. Tom Griffiths and Brian Christian on "Algorithms to Live By." RS161. 12 June. http://rationallyspeakingpodcast.org/161-algorithm s-to-live-by-tom-griffiths-and-brian-christian/

Transcript of Rationally Speaking Podcast. 2016. Tom Griffiths and Brian Christian on "Algorithms to Live By." RS161. 12 June. http://static1.1.sqspcdn.com/static/ f/468275/27088362/1465770252087/rs161transcript.pdf?token=AZVmgweIqbm ak0lfPO5YbTZiC2o%3D

Algorithms to Live By. Brian Christian & Tom Griffiths. Talks at Google. 12 May 2016.
    YouTube. https://youtu.be/OwKj-wgXteo
The Computer Science of Human Decision Making. Tom Griffiths. TedxTalks. 1
    August 2017. YouTube. https://youtu.be/lOhL-XUQPFE

3.6 Why Do We Always Seem to End Up in the Slower Lane?

Is the line next to you at the check-in at the airport or the check-out at the
supermarket always quicker than the one you are in? Is the traffic in the
neighbouring lane always moving a bit more quickly than your lane? Or
does it just seem that way?

One explanation is to appeal to basic human psychology. For example, is it
an illusion caused by us being more likely to glance over at the neighbouring lane when we are progressing forward slowly rather than quickly? Is
it a consequence of the fact that we tend to look forward rather than back-
ward, so vehicles that are overtaken become forgotten very quickly, whereas
those that remain in front continue to torment us? Do we take more notice or
remember for longer the times we are passed than when we pass others? So
is it an illusion? Or is it more real and fundamental?

Is it true that we really are more often than not in the slower lane? If so,
there may be a reason. Let me explain using an example.

How big is the smallest fish in the pond? You catch 60 fish, all of which
are more than 6 inches long. Does this evidence add support to a hypothesis
that all the fish in the pond are longer than 6 inches? Only if your net is able
to catch fish smaller than 6 inches. What if the holes in the net allow smaller
fish to pass through? We may describe this as a selection effect or an observation bias.

Apply the same principle to your place in the line, or in the lane.

We need to ask, "For a randomly selected person, are the people or vehicles
in the next line or lane actually moving faster?"

Well, one apparent reason why we might be in a slower lane is that there
are more vehicles in it than in the neighbouring lane. This means that more
of our time is spent in the slower lane. In particular, cars travelling at greater
speeds are normally more spread out than slower cars, so that over a given
stretch of road there are likely to be more cars in the slower lane, which
means that more of the average driver's time is spent in the slower lane or
lanes. This is known as an observer selection effect, the key idea being that
observers should reason as if they are randomly selected from the set of all
observers. When making observations of the speed of cars in the next lane,
or the progress of the neighbouring line to the cashier, it is important, therefore, to consider yourself as a random observer and think about the implications of this for your observation.

To put it another way, if you are in a line and think of your present observation as a random sample from all the observations made by all relevant observers, then the probability is that your observation will be made from the perspective that most drivers have, which is the viewpoint of the slower moving queue, as that is where more observers are likely to be. It is because most observers are in the slower lane, therefore, that a typical or randomly selected driver will not only seem to be in the slower lane but actually will be in the slower lane.

If there are, for example, 20 observers in the slower lane and 10 in the equivalent section of the fast lane, there is a 2/3 chance that you are in the slower lane.

So the next time you think that the other lane is faster, be aware that it very probably is.

3.6.1 Exercise

Do you personally spend more time in the slower lane? Whatever your answer, why do you think this is? This exercise is for personal or group consideration. No solution is provided.

3.6.2 Reading and Links

Bostrom, N. 2020. Cars in the Next Lane Really Do Go Faster. + plus magazine. 1 December . https://plus.maths.org/content/os/issue17/features/traffic/index

3.7 Pascal's Wager

Blaise Pascal was a seventeenth-century French mathematician and philosopher, who laid some of the main foundations of modern probability theory. He is also celebrated for his correspondence with mathematician Pierre Fermat, forever associated with Fermat's Last Theorem. School children learning mathematics may have heard of him because of Pascal's Triangle. It is, however, Pascal's Wager, and latterly the Pascal's Mugging Problem, that continues to interest and entertain philosophers. Simply stated, Pascal's Wager can be stated thus: "If God exists and you wager that he does not, your penalty relative to betting that he does exist is enormous. If God does not exist and you wager that he does, your penalty is relatively inconsequential". In other words, there's a lot to gain if it turns out that God does exist and not much lost if not. So, unless it can be proved that God does not exist, you should always side with God's existence and act accordingly.

Put another way, Pascal points out that if a wager were between the equal chance of gaining two lifetimes of happiness and gaining nothing, then a person would be foolish to bet on the latter. The same would go if it were three lifetimes of happiness versus nothing. He then argues that it is simply unconscionable by comparison to bet against an eternal life of happiness for the possibility of gaining nothing. The wise decision is to wager that God exists, since "If you gain, you gain all; if you lose, you lose nothing", meaning one can gain eternal life if God exists, but if not, one will be no worse off in death than by not believing. On the other hand, if you bet against God, you either gain nothing or lose everything.

Intuitively, there seems to be something wrong with this argument. The problem lies in trying to identify what it is. One attempt is known as the "many gods" objection. The argument here is that one can, in principle, come up with multiple different characterisations of a god, including a god that punishes people for siding with his existence. But this assumes that all representations of what God is are equally plausible or probable. Some representations may seem much more plausible than others when the alternatives are investigated. A characterisation that has hundreds of millions of followers, for example, and a strongly developed set of apologetics, would seem more likely to be true than an alternative based on an evil teapot.

If it is indeed more likely that the God of a major established religion exists, relative to the evil teapot religion, the "many gods" objection is weakened. At that point, one needs to take seriously the stratospherically high rewards of siding with belief (at whatever odds one might set for that) compared to the stakes. More generally, dropping or weakening the equal-probability assumption serves to weaken the "many gods" objection.

It's also the case that future rewards tend to be seriously under-weighted by most human decision-makers. While pain suffered in the future will feel just as bad as pain suffered today, most people don't think or behave as if that's so.

Pascal's Wager has, in recent years, been applied to the problem of existential threats like climate change. Let's say, for example, there was only a 1% chance that the planet was on course for catastrophic and irreversible climate disaster if a delay to global imminent action meant passing a point of no return. In that case, not acting now breaches the terms of Pascal's Wager. This has been termed Noah's Law: if an ark may be essential for survival, get building, however sunny a day it is overhead.

Pascal's Mugging is a new twist on the problem. A passing stranger turns up on your doorstep and asks for £10, which they promise to return tomorrow, topped up to a £100. You turn down the deal because you don't believe he will follow through on his promise. This is because you think there is only a very small chance that he will honour any deal you are offered. Let's say, though, that you considered the chance to be 1 in 100 and he offered you £2,000 tomorrow in return for the £10. You work out the expected value of this proposal to be 1/100 times £2,000 or £20 and hand over the tenner. He never comes back and you have, in a way, been intellectually mugged.

Was giving him the £10 irrational? The argument is that for any low probability (short of zero probability) that the promise will be honoured, there is a finite amount that makes it rational to take the bet. However low the probability you assign to the promise being honoured, short of zero, you can be assigned a potential reward which would outweigh it.

It is the Pascal's Wager formulation, however, that is best used to consider the appropriate course of action when confronted more systemically by what may be low-probability, high-downside events. Spending large sums of money and effort on extremely unlikely scenarios might seem irrational, but from Pascal's perspective, it's not at all clear that this is so.

Today, in an age when global existential risk is much higher up the agenda than it was in Pascal's day, it may be time to revisit the lessons to be learned from "The Wager" with renewed urgency.

3.7.1 Exercise

Under what circumstances would you give money to Pascal's Stranger? Is it ever a rational thing to do? There is no solution provided. The exercise is for personal or group consideration.

3.7.2 Reading and Links

Bostrom, N. Pascal's Mugging. pp. 443–444. https://nickbostrom.com/papers/pascal.pdf

BBC Radio 4. In Our Time. Podcast. Pascal. https://www.bbc.co.uk/programmes/b03b2v6m

Rationally Speaking. Podcast. 2017. Amanda Askell on Pascal's Wager and other low risks with high stakes. 6 August. RS190 http://rationallyspeakingpodcast.org/190-pascals-wager-and-other-low-risks-with-high-stakes-amanda-askell/

Transcript of Rationally Speaking Podcast. 2017. Amanda Askell. 6 August. RS190 http://static1.1.sqspcdn.com/static/f/468275/27648050/1502083126473/rs190transcript.pdf?token=xQdh8%2B1IgicYGsJS5D%2Fa%2BB0sFMo%3D

Pascal's Mugging. Apostrophe Philosophy. 8 June 2016. YouTube. https://youtu.be/WT1Z7Y8TkWU

3.8 The Keynesian Beauty Contest

Choose an integer number between 0 and 100. You win a prize if your number is equal or closest to two thirds of the average number chosen by all other participants. What number should you choose?

If you think that the other participants will choose a random number within the range, the average will be 50. Hence you choose 33 (Level 1 rationality). That seems intuitively correct for many people. But hang on. Just as you chose 33, so presumably will other participants, at least on average, based on your same line of reasoning. So if the average number chosen by all participants is 33, then the smart thing to do is to choose 22. This is known as Level 2 rationality.

But do you really think you are smarter than the others? Just as you figured out that 22 is the smart choice, so will others, at least on average. So the super-intelligent thing to do is to choose 15. This is known as Level 3 rationality. But we are heading towards 0 (you get there after 12 iterations). Zero is the only rational choice to make if you don't think you are smarter than the other participants.

You start to get a strong feeling that if you choose 0 you are not going to win the prize. This is because, although you don't think you are necessarily smarter than most, it is reasonable to assume that at least some of the players are not as smart or rational as you. For example, if 10% of players are naive and choose a random number – 50 on average – then the overall average will be 5 and the right answer will be 3. And so on.

Of course, there are plenty more combinations, with varying proportions of players at different levels of "rationality". The higher the winning number, the larger is the percentage of less rational players in the game.

In an experiment conducted with *Financial Times* readers by economist Richard Thaler, made up of 1,476 participants, the winning number was in fact 13. This is roughly consistent with at least the following:

1. All players exhibit Level 3 rationality.

   Or

2. Eighty per cent are fully rational and 20% are totally naive.

   Or

3. Seventy per cent are fully rational and 30% exhibit Level 1 rationality.

John Maynard Keynes, in Chapter 12 of his *General Theory of Employment, Interest and Money*, frames the paradox in terms of the money markets, in a more prosaic way:

> Professional investment may be likened to those newspaper competitions in which the competitors have to pick out the six prettiest faces from a hundred photographs, the prize being awarded to the competitor whose choice most nearly corresponds to the average preferences of the competitors as a whole; so that each competitor has to pick, not those faces which he himself finds prettiest, but those which he thinks likeliest to catch the fancy of the other competitors, all of whom are looking at the problem from the same point of view. It is not a case of choosing those which, to the best of one's judgement, are really the prettiest, not even those which average opinion genuinely thinks the prettiest. We have reached the third degree where we devote our intelligences

to anticipating what average opinion expects the average opinion to be. And there are some, I believe, who practise the fourth, fifth and higher degrees.

In other words, it is those who are able to best out-guess the best guesses of the rest of the crowd who stand to win the prize. Or put another way, the £10 note you spot lying on the floor might well be real after all. Nobody has picked it up yet because they have all assumed that someone else would have picked it up if it were real. You realise that everyone else is thinking like this, and you win yourself a tenner. Let's call that super-rationality.

3.8.1 Exercise

What number would you choose? There is no solution provided to this exercise, as there is no definitive answer. The exercise is for personal or group consideration.

3.8.2 Reading and Links

Bosch-Domenech, A., Montalvo, J.G., Nagel, R. and Satorra, A. 2002. One, Two, Three, ... : Newspaper and Lab Beauty-Contest Experiments. American Economic Review, 92 (5), 1687-1701.

Keynes, J.M. 1936. *The General Theory of Employment, Interest and Money*. London: Macmillan.

Keynes' Beauty Contest. By Richard Thaler in the Financial Times, 10 July 2015. https://www.ft.com/content/6149527a-25b8-11e5-bd83-71cb60e8f08c

Nash Equilibrium Introduction, and the Keynes Beauty Contest. Game Theory Online. 16 October 2013. YouTube. https://youtu.be/-j44yHK0nn4

Strategic Reasoning and the Keynes Beauty Contest Game. Game Theory Online. 16 October 2013. YouTube. https://youtu.be/sVWLrs5wbi4

The Keynesian Beauty Contest. Intermittent Diversion. 11 May 2020. YouTube. https://www.youtube.com/watch?v=j8ZVkVjDPxo

3.9 Benford's Law

Benford's Law is one of those laws of statistics that defies common intuition. Essentially, it states that if we randomly select a number from a table of real-life data, the probability that the first digit will be one particular number is significantly different to it being a different number. For example, the chance that the first digit will be a "1" is about 30%, rather than the intuitive 11% or so, which assumes that all digits from 1 to 9 are equally likely. In particular, Benford's Law applies to the distribution of leading digits in naturally occurring phenomena, such as the population of different countries or the heights of mountains. Now choose a newspaper or magazine with a lot of numbers and circle the numbers that occur naturally. So lengths of rivers and lakes could be included, stock prices, mortality rates, and so on, but not artificial numbers, like telephone numbers. About 30% of these numbers will start with a 1, and it doesn't matter what units they are in. So the lengths of rivers could be denominated in kilometres, miles, feet, centimetres, without it making a difference to the distribution frequency of the digits.

More generally, what are the conditions for Benford's Law (also known as the Newcomb–Benford Law) to hold? One condition is that the quantities should be of the same sort – adding areas of lakes to employee numbers won't work. There shouldn't be any artificial cut-off or limits to the permitted numbers, which should be allowed to take whatever value they wish. This excludes home numbers, for example, or prices of a range of bottles of supermarket wine. The digits should not be allocated by some numbering system such as occurs with postcodes or telephone numbers, and there should be no big spikes around certain numbers. The numbers should not be random either, as the number of each leading digit in random numbers will converge by construct. So, Benford's Law applies to those numbers which are not rigidly constrained or totally random.

The phenomenon can be traced to a note in the *American Journal of Mathematics* in 1881 authored by Simon Newcomb, in which he reported that the pages of logarithms then generally used to perform calculations were much more thumbed at certain leading digits than others; for example, numbers starting with 1 were used much often for calculations than those starting with 9.

Rigorous empirical support for this distribution, however, can be traced to the man after whom the law is named, physicist Frank Benford, in a paper he published in 1938, called "The Law of Anomalous Numbers". In that paper he examined 20,229 sets of numbers, as diverse as baseball statistics, the areas of rivers, numbers in magazine articles, and so forth, confirming the 30% rule for number 1. For information, the chance of throwing up a "2" as the first digit is 17.6% and of a "9" just 4.6%.

This has clear implications for fraud detection. In particular, if declared returns deviate significantly from the Benford distribution, we have a flag for those tackling accounting fraud. Benford's Law has also been used to flag fraud more generally.

If there is a universal law of this kind that governs the digits of numbers that describe natural phenomena, clearly such a law must work regardless of the units that are used. So it should not matter whether the units are inches or yards or centimetres or metres, or any other unit that might be used by some distant reclusive tribe of people. Converting from one unit to another will change the individual numbers but the overall distribution of numbers would still possess the same pattern as previously. This property is known as scale invariance. Benford's Law turns out to indeed be scale invariant and is the only way to distribute digits in a scale invariant way.

To explain the basis of Benford's Law, take £1 as a base. Assume this now grows at 10% per day.

£1.10, £1.21, £1.33, £1.46, £1.61, £1.77, £1.94, £2.14, £2.35, £2.59, £2.85, £3.13, £3.45, £3.80, £4.18, £4.59, £5.05, £5.56, £6.11, £6.72, £7.40, £8.14, £8.95, £9.84, £10.83, £11.92, £13.11, £14.42, £15.86, £17.45, £19.19, £21.11, £23.22, £25.50, £28.10, £30.91, £34.00, £37.40, £41.14, £45.26, £49.79, £54.74, £60.24, £72.89, £80.18, £88.20, £97.02 …

So the leading digits stay a long time in the teens, less in the 20s, and so on through the 90s, and this pattern continues through three digits and so on. Benford noticed that the probability that a number starts with n = $\log_{10}$ (n + 1) − $\log_{10}$ (n), so that:

$$\log_{10} 1 = 0; \log_{10} 2 = 0.301; \log_{10} 3 = 0.4771 \ldots \log_{10} 10 = 1$$

Another way of looking at it is that the percentage of numbers starting with a certain digit, N, is given by $100 \times \log_{10} (1 + 1/N)$.

So, the percentage of numbers with a leading digit of 1 = $100 \times \log_{10} (1 + 1/1) = 30.1\%$.

The percentage of numbers with a leading digit of 2 = $100 \times \log_{10} (1 + 1/2) = 17.6\%$.

Etc.

Leading digit	Probability (%)
1	30.1
2	17.6
3	12.5
4	9.7
5	7.9
6	6.7
7	5.8
8	5.1
9	4.6

Benford's Law can also be used to predict the proportion of digits in the second number, third number, and so on. For the second digit, zero is the most likely digit (12%) with nine the least likely (8.5%). Later digits become progressively more evenly distributed.

As an interesting aside, Benford's Law is related to the Fibonacci sequence which commonly occurs in nature, such as in the spiral pattern of sunflower seeds. The sequence consists of each number being the sum of the two numbers that come before in the sequence. So, it is:

1, 1, 2, 3, 5, 8, 13, 21 …

The ratio of successive numbers in this series tends towards another commonly occurring ratio in nature, the so-called Golden Ratio (1.62) often used in design. The digits of all the numbers in the sequence also tend to conform to Benford's Law.

3.9.1 Exercise

Choose a source with a lot of numbers, and now identify the numbers that occur naturally, such as stock prices. So lengths of rivers and lakes could be included, but not artificial numbers like telephone numbers. Calculate how many numbers start with a 1, 2, 3, 4, 5 6, 7, 8, and 9. How closely do they follow Benford's Law? No solution is provided. This is an exercise designed for personal or group activity.

3.9.2 Reading and Links

ACFE. Association of Certified Fraud Examiners. 2018. Using Benford's Law to Detect Fraud, pp. 46–57. https://www.acfe.com/uploadedFiles/Shared_Content/Products/Self-Study_CPE/UsingBenfordsLaw_2018_final_extract.pdf

Benford, F. 1938. The Law of Anomalous Numbers. *Proceedings of the American Philosophical Society*, 78 (4): 551–572.

Gill, J. 2019. What Is Benford's Law and Why Do Fraud Examiners Use It? 16 May. ACFE Insights. https://www.acfeinsights.com/acfe-insights/what-is-benfords-law

Misal, D. 2019. The Power of Benford's Law in Detecting Financial Fraud. 2 November. Praxis. https://analyticsindiamag.com/the-power-of-benfords-law-in-detecting-financial-fraud/#:~:text=Accounting%3A%20The%20idea%20behind%20detecting,rescue%20to%20identify%20the%20fraud.&text=Transaction%2Dlevel%20accounting%20data%20for,2.

Revolutions 2009. Statistical Analysis Suggests Possible Fraud in Polling Data. 6 October. https://fivethirtyeight.com/features/comparison-study-unusual-patterns-in/.

Silver, N. 2009. Strategic Vision Polls Exhibit Unusual Patterns, Possibly Indicating Fraud. FiveThirtyEight. 25 September. https://fivethirtyeight.com/features/strategic-vision-polls-exhibit-unusual/

Silver, N. 2009. Comparison Study: Unusual Patterns in Strategic Vision Polling Data Remain Unexplained. FiveThirty Eight. 26 September. https://fivethirtyeight.com/features/comparison-study-unusual-patterns-in/

Silverstein, S. 2014. How Forensic Accountants Use Benford's Law to Detect Fraud. Business Insider. 10 December. https://www.businessinsider.com/benfords-law-to-detect-financial-fraud-2014-12?r=US&IR=T#:~:text=Benford's%20Law%20can%20be%20used,the%20numbers%20have%20been%20manipulated.

Stalcup, C. 2010. Benford's Law. How a Simple Misconception can Trip up a Fraudster and How a Savvy CFE Can Spot it. Fraud magazine. https://www.acfe.com/uploadedFiles/Shared_Content/Products/Self-Study_CPE/UsingBenfordsLaw_2018_final_extract.pdf

Benford's Law. DataGenetics. http://datagenetics.com/blog/march52012/index.html

Benford's Law. Nagwa English, 30 November 2018. YouTube. https://youtu.be/EcN9GkdNvyQ

Benford's Law Explained: a response to Numberphile. TheHue's SciTech. 19 October 2013. YouTube. https://youtu.be/Az3kXCPZpYs

Benford's Law explanation. Sequel to mysteries of Benford's Law. Algebra II. Khan Academy. 23 August 2011. YouTube. https://www.youtube.com/watch?v=SZUDoEdjTzg&feature=youtu.be

Brady's Videos and Benford's Law – Numberphile. 21 January 2013. YouTube. https://youtu.be/VbtNy54ya9A

Number 1 and Benford's Law – Numberphile. 20 January 2013. YouTube. https://youtu.be/XXjlR2OK1kM

Vi and Sal talk about the mysteries of Benford's Law. Logarithms. Algebra II. Khan Academy. 23 August 2011. YouTube. https://www.youtube.com/watch?v=6KmeGpjeLZ0&feature=youtu.be

What is Benford's Law? StatisticsHowTo. 15 July 2016. https://www.statisticshowto.datasciencecentral.com/benfords-law/

3.10 Faking Randomness

Dr. Theodore P. Hill used to ask his mathematics students at the Georgia Institute of Technology to go home and either toss a coin 200 times and record the results or fake 200 results. When the results were handed in, he would almost always be able to distinguish the flippers from the fakers. The reason is that most people under-estimate how common a long sequence of heads or of tails really is.

To illustrate, ask someone to toss a fair coin 32 times. Which of the following rows of coin toss patterns is more likely to result if they actually do toss the coins and record them accurately, and which is likely to be the fake?

HTTHTHTTHHTHTHHTTTHTHTTHTHHTTHHT

OR

HTTHTHTTTTTHTHTTHHHHHTTHTHTHHTHHT

In both cases, there are 15 heads and 17 tails.

But would we expect a run of five heads or a run of five tails in the series? The answer is yes.

This can be demonstrated as follows:

The probability of 5 heads in a row = 1/32.

The probability of *not* getting 5 heads in a row from a particular run of 5 coin tosses = 31/32.

But there are 28 opportunities for a run of 5 in 32 tosses.

The chance of *not* getting 5 heads in a row from 28 runs of 5 coin tosses = $(31/32)^{28} = 41.1\%$.

Therefore, the probability of getting 5 heads in a row from 28 runs of 5 coin tosses = 58.9%.

Similarly for tails.

The probability of 5 heads *or* 5 tails in a row = 1/32 + 1/32 = 1/16.

The probability of *not* getting 5 heads *or* 5 tails in a row from a particular run of 5 coin tosses = 15/16.

The chance of *not* getting 5 heads *or* 5 tails in a row from 28 runs of 5 coin tosses = $(15/16)^{28}$ = 16.4%.

Therefore, the probability of getting 5 heads *or* 5 tails in a row from 28 runs of 5 coin tosses = 83.6%.

The probability of 4 heads in a row = 1/16.

The probability of *not* getting 4 heads in a row from a particular run of 4 coin tosses = 15/16.

Probability of *not* getting 4 heads in a row from 29 runs of 4 coin tosses = $(15/16)^{29}$ = 15.4%.

Therefore, the probability of getting 4 heads in a row from 29 runs of 4 coin tosses = 84.6%.

Similarly for tails.

The probability of 4 heads *or* 4 tails in a row = 1/16 + 1/16 = 1/8.

The probability of *not* getting 4 heads *or* 4 tails in a row from a particular run of 4 coin tosses = 7/8.

The chance of *not* getting 4 heads *or* 4 tails in a row from 29 runs of 4 coin tosses = $(7/8)^{29}$ = 2.1%.

Therefore, the probability of getting 4 heads *or* 4 tails in a row from 29 runs of 4 coin tosses = 97.9%.

Now, consider the series of coin tosses above. The first series has no run of heads (or tails) longer than three. The second series has a run of five tails and of four heads.

The second series is likely, therefore, to be the genuine one and the first one to be the fake.

3.10.1 Exercise

An experiment produces the following sequence of heads and tails from 32 flips of a coin. Which is more likely to be the genuine sequence?

THHTHTHHTTTHTHTTHHHTHTHHTHTTHHTT

Or

HTTHTTHHHHHHTHTTTTTTHTHTHTHTTHHT

3.10.2 Reading and Links

Barrow, J.D. 2012. Chapter 40: Faking It. In *100 Essential Things You Didn't Know You Didn't Know about Sport*, pp. 106–108. London: Bodley Head.

Benjamin, D. 2019. Chapter 2: Errors in Probabilistic Reasoning and Judgment Biases. In *Handbook of Behavioral Economics: Applications and Foundations 1*, Vol. 2, pp. 69–186, Elsevier. https://www.sciencedirect.com/science/article/pii/S2352 239918300228?casa_token=hP1q5z3zbKAAAAAA:RSt6-KtBAb0Kt0sGpS7 IZwlHjmg0Cnvpa1msfhNTICj0eEucr9_9q5UV6_fhPXRU9ndYx2U9fzs

Browne, M.D. 1998. Following Benford's Law, or Looking Out for No. 1. New York Times. 1 August. http://www.rexswain.com/benford.html

4

Probability, Games, and Gambling

In this chapter we consider the application of probability to games and gambling. Some of these problems date back to the seventeenth and eighteenth centuries, including the Chevalier's Dice Problem, the Problem of Points, and the Newton–Pepys Problem. The first two led to Blaise Pascal and Pierre Fermat laying the foundations of modern probability theory, while the third is a fascinating example of the use of classic probability theory to solve problems. We explore a range of other systems, biases, and anomalies, ranging from the potential to exploit the well-established favourite-longshot bias in betting markets, to a consideration of the so-called martingale betting system. We examine the use of the Poisson distribution in modelling football scores and the use of card counting as a winning blackjack strategy. We learn the optimal way to reach a target sum at the casino and how much to bet whenever (if ever) the edge is in our favour. We explore the Expected Value Paradox and conclude with an introduction to spread betting and the use of options.

4.1 The Chevalier's Dice Problem

What is probability? It is the logic of uncertainty. In classic terms, it can represent relative frequency (the proportion of times something occurs relative to the number of times it could occur). It can also represent a degree of belief. The number we assign to the probability of something occurring can range from 0 (for example, the chance of rolling a 14 with a pair of normal dice) to 0.5 (the chance of getting a Head from tossing a fair coin) to almost 1 (that the sun will rise tomorrow).

An example of probability as a degree of belief is to assign a probability to the existence of life of some form on a moon of Saturn. There either is or isn't, but currently we don't know either way for sure. So there are two possibilities (is or isn't). Because there are only two possibilities, however, this doesn't make the probability one half. For example, what is the probability of intelligent life on a moon of Saturn? Again, there are only two possibilities, but it can't be the same as before, given that it is less likely that there's intelligent life than any kind of life. So an equal likelihood assumption needs explicit justification in any individual case.

DOI: 10.1201/9781003083610-4

To consider relative frequency, we can introduce the terms "sample space" and "events". Say you roll a single die. How many possible outcomes are there? The answer is six. The set of all these possible outcomes is known as the sample space. So the sample space is made up of each individual possible outcome. An event is a subset of the sample space. In the case of a single die, therefore, the sample space is 1, 2, 3, 4, 5, 6. Each individual number constitutes an "event".

If we roll two dice, there are 36 possible outcomes. The probability of an event, P (A), can be defined as: "In how many of all possible outcomes does A occur divided by the number of all possible outcomes (size of the sample space)".

If we toss a coin twice, the sample space is the four possible outcomes – HH, HT, TT, TH.

Say that P (A) is the probability of two heads from two tosses of a fair coin or a successive toss of two fair coins. A coin is fair if, on a single toss, the probability of tails equals the probability of heads.

The probability of the first head is 1/2. The probability of the second head is also 1/2.

According to the laws of probability, we can multiply the probability of two independent events.

This can be written as: P (head $\cap$ tail) = P (head) $\times$ P (tail)

So, P(A) = 1/2 $\times$ 1/2 = 1/4.

For P (A) = 1/4, we are assuming that outcomes are equally likely. This is the assumption of symmetry. Coins have no memory, so a coin landing tails does not make tails (or heads) any more likely on the next toss – similarly for heads.

We also assume a finite sample. If the denominator, the sample space, was infinite, the definition of probability would be meaningless.

When two or more possible outcomes are mutually exclusive, we can add the probabilities.

This can be written as: P (head $\cup$ tail) = P (head) + P (tail).

For example, probability of a head *or* a tail (mutually exclusive) on a single fair coin = P (head) + P (tail) = 1/2 + 1/2 = 1.

When two events are independent but not necessarily mutually exclusive, we can add the probabilities then subtract the overlap of the probabilities. This is known as the inclusion–exclusion principle.

For example, the probability of a head when tossing a coin then rolling a six on a die = P (head) + P (six) – P (head *and* six) = 1/2 + 1/6 – (1/2 $\times$ 1/6) = 8/12 – 1/12 = 7/12.

This can be written as: P (head $\cup$ six) = P (head) + P (six) – P (head $\cap$ six).

The reason for subtracting the overlap is that when we add P (head) to P (six), the probability of "both head *and* six" is already included in the probability of head and in the probability of six. Put another way; we have counted "the probability of both" twice. So we need to subtract "the probability of both" to avoid over-counting (counting it twice).

For an example of relative frequency, symmetry, and finite sample spaces in action, we can turn to a true story about the New York gambling-house operator, the Butch, who made his fortune booking dice games. In 1952 he was challenged by a big-time gambler known as the Brain to a simple wager. The bet was an even money proposition that the Butch could throw a double-six in 21 rolls of two dice. We can assume symmetry – the dice were not loaded or biased in any way. All numbers were equally likely to come up.

On the face of it, the edge seems to be with Butch. After all, there are 36 possible combinations that could come up when throwing two dice, from 1-1, 1-2, 1-3, to 6-4, 6-5, and 6-6. Intuition might suggest, therefore, that 18 throws should give you a 50–50 chance of throwing any one of these combinations, including a double-six. In 21 throws, the chance of a double-six should, therefore, be more than 50–50. On this basis, the Butch accepted the even money bet at $1,000 a roll. After 12 hours of rolling, the Brain was $49,000 up, at which point the Butch called it a day, sensing that something was wrong with his strategy.

The Brain had, in fact, profited from a classic probability puzzle known as the Chevalier's Dice Problem, which can be traced to the seventeenth-century French gambler and bon vivant, Antoine Gombaud, better known as the Chevalier de Méré. The Chevalier would agree even money odds that in four rolls of a single die he would get at least one six. At even money odds, a bet of £x will win £x if the bet comes off and lose £x if the bet loses.

His logic seemed impeccable. The Chevalier reasoned that since the chance that a 6 will come up in any one roll of the die is 1 in 6, then the chance of getting a 6 in 4 rolls is 4/6, or 2/3, which is a good bet at even money.

In fact, it is straightforward to show that this reasoning is faulty. If it were correct, then we would calculate the chance of a 6 in 5 rolls of the die as 5/6, and therefore the chance of a 6 in 6 rolls of the die would be 6/6 = 100%, and in 7 rolls, 7/6!!! Something is, therefore, clearly wrong here.

Still, even though his reasoning was faulty, he continued to make a profit by playing the game at even money. To see why, we need to calculate the true probability of getting a six in four rolls of the die. The key idea here is that the number that comes up on each roll is independent of any other rolls, i.e. dice have no memory. Since each event is independent, we can (according to the laws of probability) multiply the probabilities.

The probability of a 6 followed by a 6, followed by a 6, followed by a 6, is: $1/6 \times 1/6 \times 1/6 \times 1/6 = 1/1296$.

So what is the chance of getting at least one 6 in 4 rolls of the die?

Since the probability of getting a 6 in any one roll of the die = 1/6, the probability of *not* getting a 6 in any one roll of the die = 5/6.

The chance of *not* getting a 6 in 4 rolls of the die is:

$$5/6 \times 5/6 \times 5/6 \times 5/6 = 625/1296$$

The chance of getting at least one 6 is 1 minus this, i.e. $1 - (625/1296) = 671/1296 = 0.5177$, which > 0.5.

So, the odds are still in favour of the Chevalier, since he agrees even money odds on an event with a probability of 51.77%.

This was all very well as long as it lasted, but eventually the Chevalier decided to branch out and invent a new, slightly modified game. In the new game, he asked for even money odds that a pair of dice, when rolled 24 times, will come up with a double-6 at least once. His reasoning was the same as before and quite similar to the reasoning employed by the Butch. If the chance of a 6 on one roll of the die is 1/6, then the chance of a double-6 when two dice are thrown $= 1/6 \times 1/6$ (as they are independent events) $= 1/36$.

So, reasoned the Chevalier, the chance of at least one double-6 in 24 throws is: $24/36 = 2/3$.

So this is a very profitable game for the Chevalier. Or is it? No, it isn't, and this time Monsieur Gombaud paid for his faulty reasoning. He started losing. In desperation, he consulted the mathematician and philosopher, Blaise Pascal. Pascal derived the correct probabilities as follows:

The probability of a double-6 in one throw of a pair of dice $= 1/6 \times 1/6 = 1/36$.

So the probability of *no* double-6 in one throw of a pair of dice $= 1 - 1/36 = 35/36$.

The probability of no double-6 in 24 throws of a pair of dice $= 35/36 \times 35/36 \ldots 24$ times $= 35/36$ to the power of 24, i.e. $(35/36)^{24} = 0.5086$.

The probability of at least one double-6 is 1 minus this, i.e. $1 - 0.5086 = 0.4914$, i.e. less than 0.5. Under the terms of the new game, the Chevalier was betting at even money on a game that he lost more often than he won. It was an error that the Butch was to repeat almost 300 years later! We would say that the edge against the Chevalier is equal to 0.5086 minus $0.4914 = 0.0172 = 1.72\%$.

In the "pair of dice" game (24 throws), the Chevalier's edge =

$$49.14\% - 50.86\% = -1.72\%$$

What if the Chevalier had changed the game to give himself 25 throws?

Now, the probability of throwing at least one double-6 in 25 throws of a pair of dice is:

$$1 - (35/36)^{25} = 0.5055$$

In this "pair of dice" game (25 throws), the Chevalier's edge =

$$50.55\% - 49.45\% = 1.1\%$$

These odds, at even money, are in favour of the Chevalier, but this probability is still lower than the probability of obtaining one "6" in four throws of a single die.

In the single-die game, the Chevalier has a house edge of 51.77% − 48.23% = 3.54%.

A better game for the Chevalier would have been to offer even money that he could get at least one run of 10 heads in a row in 1024 tosses of a coin. The derivation of this probability is similar in method to the Dice Problem.

First, we need to determine the probability of 10 heads in 10 tosses of a fair coin.

The odds are: $1/2 \times 1/2 \times 1/2 \times 1/2 \times 1/2 \times 1/2 \times 1/2 \times 1/2 \times 1/2 \times 1/2$.

$$\text{Odds} = (1/2)^{10} = 1/1024, \text{ i.e. } 1023/1.$$

So what is the probability of *no run* of 10 heads in one attempt?

This is: $1 - 1/1024$.

In 1024 attempts, it is:

$$\left(1 - 1/1024\right)^{1024}$$

The probability of *no runs of ten heads* = $(1023/1024)^{1024} = 37\%$.

So the probability of *at least* one run of 10 heads = 63%.

Now assume you have already had 234 attempts out of 1024, without a run of 10 heads, what is your chance now of getting 10 heads?

Probability of *no runs of ten heads* in remaining 790 attempts = $(1023/1024)^{790}$ = 46%.

So the probability of at least one success = 54%.

The Chevalier could have played either of these games and expected to come out ahead. But the game would have taken a long time. He preferred the shorter game, which produced the longer loss.

Until he was put right by Pascal.

More importantly, though, the Chevalier's question led to a correspondence between Blaise Pascal and Pierre de Fermat, most of which has survived, which led to the foundations of modern probability theory.

Out of this correspondence emerged quite a few jewels, one of which has become known as the "Gambler's Ruin" problem.

This is an idea set in the form of a problem by Pascal for Fermat, subsequently published by Christiaan Huygens (*On Reasoning in Games of Chance*, 1657) and formally solved by Jacob Bernoulli (*Ars Conjectandi*, 1713).

One way of stating the problem is as follows. If you play any gambling game long enough, will you eventually go bankrupt, even if the odds are in your favour, if your opponent has unlimited funds?

For example, say that you and your opponent toss a coin, where the loser pays the winner £1. The game continues until either you or your opponent has all the money. Suppose you have £20 to start and your opponent has £40. What are the probabilities that (a) you and (b) your opponent will end up with all the money?

The answer is that the player who starts with more money has more chance of ending up with all of it. The formula is:

$$P_1 = n_1 / (n_1 + n_2)$$

$$P_2 = n_2 / (n_1 + n_2)$$

Where n_1 is the amount of money that Player 1 starts with, and n_2 is the amount of money that Player 2 starts with, and P1 and P2 are the probabilities that Player 1 or Player 2, your opponent, wins.

In this case, you start with £20 of the £60 total, and so have a 20 / (20 + 40) = 20/60 = 1/3 chance of winning the £60; your opponent has a 2/3 chance of winning the £60. But even if you do win this game, and you play the game again and again, against different opponents, or the same one who has borrowed more money, eventually you will lose your entire bankroll. This is the case even if the odds are in your favour. Eventually, you will meet a long-enough bad streak to bankrupt you. In other words, infinite capital will overcome any finite edge. This is one version of the "Gambler's Ruin" problem, and many gamblers over the years have been ruined because of their unawareness of it.

4.1.1 Exercise

1. When the Butch played the Brain at dice, he offered even money odds that in 21 throws of two dice he would get at least one double-6. He reasoned that the true probability was 21/36, which gave him an advantage at even money odds. What is the true probability of throwing a double-6 in 21 throws of the dice, and what was the edge in percentage terms (if any) against the Butch?

2. You and your opponent toss a coin, where the loser pays the winner £10. The game continues until either you or your opponent has all the money. Suppose you have £100 to start and your opponent has £400. What are the probabilities that (a) you and (b) your opponent will end up with all the money?

3. In the novel *Bomber*, Len Deighton claims that a World War II pilot had a 1 in 50 or 2% chance of being shot down on each mission. So he is "mathematically certain" to be shot down if he flies 50 missions. Is this correct? If not, what is the true probability?

4. What is the probability of tossing a tail with a fair coin followed by an even number on a fair die?

5. Suppose there are five red marbles and five yellow marbles in a bag. While your back is turned, your friend selects one marble at random from the box and puts it in her pocket, so you don't know what colour it was. You now select a marble at random from the nine left

in the box. What is the probability that the marble you have selected is yellow?

6. Ted and Alice take turns tossing their own coin. If the faces match (heads–heads or tails–tails) the game goes on. The game ends when the faces don't match (heads–tails or tails–heads). Ted wins if it lands heads–tails and Alice wins if it lands tails–heads. There is one twist in the game. The coins are not fair but weighted to land heads 99% of the time, and tails 1% of the time. Now, who is more likely to win the game, Ted or Alice?

7. There are two containers, a cube and a cone, into one of which I have placed two plain chocolates and in the other one plain chocolate and one nut chocolate. If I randomly put my hand into a container and draw out a plain chocolate, what is the probability that I took it from the cube? Give one simple intuitive solution and one using Bayes' Theorem.

8. During World War II, planes returning to base after a mission were checked for bullet holes. You are asked to look at the statistics and decide which parts of the plane to focus on reinforcing. In order, you find that most hits were in the wings, then the tail, then the centre of the body of the plane, and then the engine. For practical reasons, it's not possible to reinforce everywhere. Where do you recommend reinforcing as a priority?

4.1.2 Reading and Links

Bernoulli, J. 2005. [1713]. *The Art of Conjecturing, Together with Letter to a Friend on Sets in Court Tennis (English translation)*, translated by Edith Sylla, Baltimore: Johns Hopkins University Press. https://books.google.com/?id=-xgwSAjTh34C&dq=edith+dudley+sylla

DeMéré's Paradox. ProofWiki. https://proofwiki.org/wiki/De_M%C3%A9r%C3%A9%27s_Paradox

Huygens, C. 1714. *The value of chances in games of fortune*. English translation. https://math.dartmouth.edu/~doyle/docs/huygens/huygens.pdf

Miller, B. 2020. How 'survivorship bias' can cause you to make mistakes. *Worklife*. 29 August. Survivorship Bias. https://www.bbc.com/worklife/article/20200827-how-survivorship-bias-can-cause-you-to-make-mistakes

One gambling problem that launched modern probability theory. Introductory Statistics. 12 November 2010. https://introductorystats.wordpress.com/2010/11/12/one-gambling-problem-that-launched-modern-probability-theory/

The Decision Lab. Why do we misjudge groups by only looking at specific group members? The Survivorship Bias, explained. https://thedecisionlab.com/biases/survivorship-bias/

Weisstein, E.W. 2021 de Méré's problem. From *MathWorld* – A Wolfram Web Resource. Last updated 18 May. https://mathworld.wolfram.com/deMeresProblem.html

Weisstein, E.W. 2021 "Gambler's ruin." From *MathWorld* – A Wolfram Web Resource. Last updated 18 May. https://mathworld.wolfram.com/GamblersRuin.html

Survivorship Bias. Wikipedia. https://en.wikipedia.org/wiki/Survivorship_bias

BBC Radio 4. 2008 In our time. Podcast. 29 May. Probability. https://www.bbc.co.uk/programmes/b00bqf61

4.2 The Pascal–Fermat "Problem of Points"

What is the fair split of stakes in a game which is interrupted before its conclusion? This was the problem posed in 1654 by French gambler Chevalier
de DeMéré to philosopher and mathematician Blaise Pascal. Pascal shared it
in famed correspondence with mathematician Pierre de Fermat (best known
these days perhaps for Fermat's Last Theorem). It has come to be known as
the Problem of Points.

The question had been addressed by the Franciscan friar and Leonardo
Da Vinci collaborator, Luca Bartolomeo de Pacioli, father of the double-entry
system of bookkeeping. Pacioli's method was to divide the stakes in proportion to the number of rounds won by each player to that point. There is a
problem with this method, however. What happens, for example, if only a
single round of many has been played? Should the entire pot be allocated to
the winner of that single round? In the mid-1500s, Venetian mathematician
and founder of the theory of ballistics, Niccolo Fontana Tartaglia, proposed
basing the division on the ratio between the size of the lead and the length of
the game. But this method is not without its problems. For example, it would
split the stakes in the same proportion in a 1 to 100 contest whether one
player was ahead by 40-30 or by 99-89, although the latter situation is hugely
more advantageous to the leader than the former.

The solution adopted by Pascal and Fermat defied the prevailing intuition
by basing the division of the stakes not on the history of the interrupted
game to that point as on the possible ways the game might have continued
were it not interrupted. In this method, a player leading by 6-4 in a game to
10 would have the same chance of winning as a player leading by 16-14 in
a game to 20, so that an interruption at either point should lead to the same
division of stakes. What is important in the Pascal–Fermat solution is not the
number of rounds each player has yet won, but the number of rounds each
player still needs to win.

Take another example. Suppose that two players agree to play a game of
coin-tossing to win £32, and the winner is the first player to win four times.

If the game is interrupted when one of the players (Player 1) is ahead of the
other (Player 2) by two games to one, how should the £32 be divided fairly
between the players?

In Fermat's method, imagine playing another four games. Outcomes of
each coin-tossing game are equally likely and are represented as P (won by
Player 1) and Q (won by Player 2).

The possible outcomes of the next four games are as follows:

PPPP; PPPQ; PPQP; PPQQ

PQPP; PQPQ; PQQP; PQQQ

QPPP; QPPQ; QPQP; QPQQ

QQPP; QQPQ; QQQP; QQQQ

The probability that Player 1 would have won is 11/16 (in bold) = 68.75%.

The probability that Player 2 would have won is 5/16 = 31.25%.

The method can be generalised to any game of chance which ends before the game is complete.

4.2.1 Appendix

Pascal proposed an alternative method which dispenses with the need to consider possible steps after the game had already been won. In doing so, he was able to devise a relatively simple formula which would solve all possible Problems of Points without needing to go beyond the point at which the game resolves to the advantage of one or other of the players. This is based on Pascal's Triangle demonstrated below.

```
        1 =                         1
     1    1 =                       2
   1   2   1 =                      4
   1  3   3   1 =                    8
   1  4  6   4  1 =                 16
   1  5  10   10   5  1 =           32
   1  6  15  20  15   6  1 =        64
   1  7  21  35  35  21  7  1 =    128
```

Each of the numbers in Pascal's Triangle is the sum of the adjacent numbers immediately above it.

If the game is interrupted 2-1, after three games, in a one to four contest, Player 1 has to win two more games and Player 2 three more games. This adds up to five, so we select the row of five (1, 4, 6, 4, 1). The resolution is (1 + 4 + 6) / 16 to (4 + 1) / 16, i.e. 11/16 to Player 1 and 5/16 to Player 2.

This is the same solution as derived from Fermat's method.

Now, consider the probability that Player 1 would win if leading 3-2 in a game in which the first player to win four games is the outright winner. If Player 1 wins the next coin toss, he goes ahead 4-2 and wins outright (value = 1). There is a 50% chance of this. There is a 50% chance of Player 2 winning the coin toss, however, in which case the game is level (3-3). If the game is level, there is a 50% chance of Player 1 winning and a 50% chance of Player 2 winning.

So the expected chance of Player 1 winning when leading 3-2 = 50% × 1 + 50% × 50% = 50% + 25% = 75%. Expected chance of Player 2 winning = 25%.

Using Pascal's Triangle provides the same solution. Player 1 needs to win one more game; Player 2 needs to win two more games. The resolution is read off from the third row of the Triangle, i.e. (1 + 2) / 4 = 3/4 to Player 1 and 1/4 to Player 2.

What is the probability that Player 1 would win if leading 3-1 in a game in which the first player to win four games is the outright winner? If Player 1 wins the next coin toss, he goes ahead 4-1 and wins outright (value = 1). There is a 50% chance of this. There is a 50% chance of Player 2 winning the coin toss, however, in which case the game goes to 3-2. We know that the expected chance of Player 1 winning if ahead 3-2 is 75% (derived above).

So the expected chance of Player 1 winning when leading 3-1 = 50% × 1 + 50% × 75% = 50% + 37.5% = 87.5%. Expected chance of Player 2 winning = 12.5%.

Using Pascal's Triangle provides the same solution. Player 1 needs to win one more game; Player 2 needs to win three more games. The resolution is read off from the fourth row of the Triangle, i.e. (1 + 3 + 3) / 8 = 7/8 to Player 1 and 1/8 to Player 2.

Fermat's method and Pascal's method yield the same solution, by different routes, and will always do so in determining the correct division of stakes in an interrupted game of this nature.

4.2.2 Exercise

Suppose that two players agree to play a game of coin-tossing repeatedly to win £32, and the winner is the first player to win four times.

1. The game is interrupted when the match stands at two games each, after the first player is at one time leading 2-0. Determine how the £32 should be divided fairly between the players, if the match stopped at that point.

2. What if, instead of stopping, the match continues and the first player now wins the fifth game. How should the prize be divided?

4.2.3 Reading and Links

Blaise Pascal – 2020. The story of mathematics. Luke Mastin. https://www.storyofm athematics.com/17th_pascal.html

BBC Radio 4. 2008. In our time. Podcast. 19 September. Pascal. https://www.bbc.co. uk/programmes/b03b2v6m

Can You Solve the Problem that Inspired Probability Theory? (Problem of the Points). MindYourDecisions. 31 January 2016.. YouTube. https://youtu.be/C_nV3cVNjog

Pascal's Problem of Points Video. Weida L.20 March 2018. YouTube. https://youtu.be/ uPLNdat2Xi4

4.3 The Newton–Pepys Problem

One of the most celebrated pieces of correspondence in the history of probability, and gambling, involves an exchange of letters. The letters were between perhaps the greatest diarist of all time, Samuel Pepys, and perhaps the greatest scientist of all time, Sir Isaac Newton.

The six letters exchanged between Pepys in London and Newton in Cambridge related to a problem posed to Newton by Pepys about gambling odds. The interchange took place between 22 November and 23 December 1693. The ostensible reason for Mr. Pepys' interest was to encourage the thirst for truth of his young friend, Mr. Smith. Whether Sir Isaac believed that tale, we shall never know. The real reason was later revealed in a letter written to a confidante by Pepys indicating that he was about to stake £10, a considerable sum in 1693, on a wager.

The first letter to Newton introduced Mr. Smith as a fellow with a "general reputation ... in this towne (inferiour to none, but superiour to most) for his maistery [of] ... Arithmetick".

What emerged has come down to us as the aptly named Newton–Pepys Problem.

Essentially, the question came down to this:

Which of the following has the greatest chance of success?

A. Six fair dice are tossed independently and at least one "6" appears.

B. Twelve fair dice are tossed independently and at least two "6"s appear.

C. Eighteen fair dice are tossed independently and at least three "6"s appear.

Pepys was convinced that C had the highest probability and asked Newton to confirm this.

Newton chose A as the highest probability, then B, then C, and produced his calculations for Pepys, who wouldn't accept them.

So who was right? Newton or Pepys?

Well, let's see.

The first problem is the easiest to solve.

What is the probability of A?

Probability that one roll of a die produces a "6" = 1/6.

So the probability that one roll of a die does *not* produce a "6" = 5/6.

So the probability that six independent rolls of a die produces *no* "6" = $(5/6)^6$.

So the probability of *at least* one "6" in 6 rolls = $1 - (5/6)^6 = 0.6651$.

This is a formal solution to Part 1 of the Newton–Pepys Problem.

So far, so good.

The probability of problem B and the probability of problem C are more difficult to calculate and involve the use of the binomial distribution, though Newton derived the answers from first principles, by his method of "Progressions".

Both methods give the same answer but using the more modern binomial distribution is easier.

So let's do it along the way by introducing the idea of so-called "Bernoulli trials".

The nice thing about a Bernoulli trial is that it has only two possible outcomes.

Each outcome can be framed as a "yes" or "no" question (success or failure).

Let the probability of success = p.

Let the probability of failure = 1 – p.

Each trial is independent of the others but identical and the probability of the two outcomes remains constant for every trial.

An example is tossing a coin. Will it land heads?

Another example is rolling a die. Will it come up with a "6"?

Yes = success (S); No = failure (F).

Let the probability of success, P (S) = p; the probability of failure, P (F) = 1 – p.

So to the question: How many Bernoulli trials are needed to get to the first success?

This is straightforward, as the only way to need exactly five trials, for example, is to begin with four failures, i.e. FFFFS.

Probability of this = $(1 - p)(1 - p)(1 - p)(1 - p) p = (1 - p)^4 p$.

Similarly, the only way to need exactly six trials is to begin with five failures, i.e. FFFFFS.

Probability of this = $(1 - p)(1 - p)(1 - p)(1 - p)(1 - p) p = (1 - p)^5 p$.

More generally, the probability that success starts on trial number n =
$(1 - p)^{n-1} p$

This is a geometric distribution. This distribution deals with the number of trials required for a single success.

But what is the chance that the first success takes *at least* some number of trials, say 12 trials?

The solution to this is straightforward. The only time you will need *at least* 12 trials is when the first 11 trials are all failures, i.e. $(1 - p)^{11}$.

In a sequence of Bernoulli trials, the probability that the first success takes *at least* n trials is $(1 - p)^{n-1}$.

Let's take a couple of examples.

The probability that the first success (heads on coin toss) takes at least three trials (tosses of the coin) = $(1 - 0.5)^2 = 0.25$.

The probability that the first success (heads on coin toss) takes at least four trials (tosses of the coin) = $(1 - 0.5)^3 = 0.125$.

But so far we have only learned how to calculate the probability of one success in so many trials.

What if we want to know the probability of two, or three, or however many successes?

To take an example, what is the probability of exactly two "6"s in five throws of the die?

To determine this, we need to calculate the number of ways two "6"s can occur in five throws of the die and multiply that by the probability of each of these ways occurring.

So, probability = number of ways something can occur multiplied by the probability of it occurring in each way.

How many ways can we throw two "6"s in five throws of the die?

Where S = Success in throwing a "6", F = Fail in throwing a "6", we have:

SSFFF; SFSFF; SFFSF; SFFFS; FSSFF; FSFSF; FSFFS; FFSSF; FFSFS; FFFSS

So there are ten ways of throwing two "6"s in five throws of the dice.

More formally, we are seeking to calculate how many ways two things can be chosen from 5. This is known as "5 Choose 2", written as:

$$^{5}C_2 = 10$$

More generally, the number of ways k things can be chosen from n is:

$^{n}C_k = n! / (n - k)! \, k!$

n! (n factorial) = $n \, (n - 1) \, (n - 2) \ldots 1$

k! (k factorial) = $k \, (k - 1) \, (k - 2) \ldots 1$

Thus, $^{5}C_2 = 5! / 3! \, 2! = 5 \times 4 \times 3 \times 2 \times 1 / (3 \times 2 \times 1 \times 2 \times 1) = 5 \times 4/ (2 \times 1) = 20/2 = 10$.

So what is the probability of throwing exactly two "6"s in five throws of the die, in each of these ten cases? p is the probability of success and $1 - p$ is the probability of failure.

In each case, the probability = $p \cdot p \cdot (1 - p) \cdot (1 - p) \cdot (1 - p)$.

$$= p^2(1-p)^3$$

Since there are $^{5}C_2$ such sequences, the probability of exactly two "6"s =

$$10 \, p^2(1-p)^3$$

Generally, in a fixed sequence of n Bernoulli trials, the probability of exactly r successes is:

$$^{n}C_r \times p^r(1-p)^{n-r}$$

This is the binomial distribution. Note that it requires that the probability of success on each trial be constant. It also requires only two possible outcomes.

So, for example, what is the chance of exactly three heads when a fair coin is tossed five times?

$$^5C_3 (1/2)^3 \times (1/2)^2 = 10/32 = 5/16$$

And what is the chance of getting exactly two 6s when a fair die is rolled five times?

$$^5C_2 \times (1/6)^2 \times (5/6)^3 = 10 \times 1/36 \times 125/216 = 1250/7776 = 0.1608$$

So let's now use the binomial distribution to solve the next two parts of the Newton–Pepys Problem.

1. What is the probability of *at least* two 6s with 12 dice?
2. What is the probability of *at least* three 6s with 18 dice?

Probability of at least two 6s with 12 dice

Probability of at least two 6s with 12 dice is equal to "1" minus the probability of no 6s minus the probability of exactly one 6.

This can be written as:

P $(x \geq 2) = 1 - P(x = 0) - P(x = 1)$
P $(x = 0)$ in 12 throws of the dice = $(5/6)^{12}$
P $(x = 1)$ in 12 throws of the dice = $^{12}C_1 \cdot (1/6)^1 \cdot (5/6)^{11}$
$^nC_k = n! / (n - k)! \, k!$
So $^{12}C_1 = 12! / (12 - 1)! \, 1! = 12! / 11! \, 1! = 12$
So, P $(x \geq 2) = 1 - (5/6)^{12} - 12 \cdot (1/6) \cdot (5/6)^{11}$
$= 1 - 0.112156654 - 2 \cdot (0.134587985) = 0.887843346 - 0.26917597$
$= 0.618667376 = 0.619$ (to 3 decimal places)

This is a formal solution to Part 2 of the Newton–Pepys Problem.

Probability of at least three 6s with 18 dice

The probability of at least three 6s with 18 dice is equal to "1" minus the probability of no 6s minus the probability of exactly one 6 minus the probability at exactly two 6s.

This can be written as:

$P (x \geq 3) = 1 - P (x = 0) - P (x = 1) - P (x = 2)$

$P (x = 0)$ in 18 throws of the dice $= (5/6)^{18}$

$P (x = 1)$ in 18 throws of the dice $= {}^{18}C_1 . (1/6)^1 . (5/6)^{17}$

$^nC_k = n! / (n - k)! \, k!$

So ${}^{18}C_1 = 18! / (18 - 1)! \, 1! = 18$

So $P (x = 1) = 18 . (1/6)^1 . (5/6)^{17}.$

$P (x = 2) = {}^{18}C_2 . (1/6)^2 . (5/6)^{16}$

$^{18}C_2 = 18! / (18 - 2)! \, 2! = 18!/16! \, 2! = 18 . (17/2)$

So $P (x = 2) = 18 . (17/2) (1/6)^2 (5/6)^{16}.$

So $P (x = 3) = 1 - P (x = 0) - (P (x = 1) - P (x = 2)).$

$P (x = 0) = (5/6)^{18}$

$= 0.0375610365$

$P (x = 1) = 18 . 1/6 . (0.0450732438) = 0.135219731$

$P (x = 2) = 18 . (17/2) (1/36) (0.0540878926) = 0.229873544$

So $P (x \geq 3) = 1 - 0.0375610365 - 0.135219731 - 0.229873544.$

$P (x \geq 3) = 0.597345689 = 0.597$ (to 3 decimal places)

This is a formal solution to Part 3 of the Newton–Pepys Problem.

So, to re-state the Newton–Pepys Problem:

Which of the following has the greatest chance of success?

A. Six fair dice are tossed independently and at least one "6" appears.

B. Twelve fair dice are tossed independently and at least two "6"s appear.

C. Eighteen fair dice are tossed independently and at least three "6"s appear.

Pepys was convinced that C had the highest probability and asked Newton to confirm this.

Newton chose A, then B, then C, and produced his calculations for Pepys, who wouldn't accept them.

So who was right? Newton or Pepys?

According to our calculations, what is the probability of A? 0.665.

What is the probability of B? 0.619.

What is the probability of C? 0.597.

Sir Isaac's solution was right. Samuel Pepys was wrong, a wrong compounded by refusing to accept Newton's solution. How much he lost gambling on his misjudgment is mired in the mists of history. Fortunately for us, the Newton–Pepys Problem is not lost and continues to tease our brains to this very day.

4.3.1 Exercise

Mr. Jacky Thomas and Mr. John Donelon are equally matched opponents at tennis, so each has a 1 in 2 chance of winning any particular match.

If Mr. Thomas plays Mr. Donelon, what is the probability that he will win exactly five games out of eight?

4.3.2 Reading and Links

Newton and Pepys. DataGenetics. http://datagenetics.com/blog/february12014/in dex.html

The Newton-Pepys Problem. Nicholson, E. 26 August 2014.. YouTube. https://youtu. be/7HSmXPxMipU

4.4 Staking to Reach a Target Sum

John needs £216 to pay off an urgent debt but has only £108 available. This is unacceptable to the lender and is as good as nothing. He decides to try to win the money at the roulette wheel.

So what is his best strategy? The answer might seem a little surprising.

Take the case of a single-zero roulette wheel. There are 36 slots plus the zero, and the payout to a winning bet is 35/1, while the chance of winning is 1 in 37 (so the payout should be at odds of 36/1). The way to look at it is that the house edge is equal to the proportion of times the ball lands in the zero slot, which is 1/37 or 2.7%. This edge in favour of the house is the same whatever individual bet we make.

So let's see what happens when John goes for the "bold" play and stakes the entire £108 on red. In this case, 18 times out of 37 (statistically speaking), or 48.6% of the time, John can cash his chips immediately for £216. Of course, he is only doing this once, so this 48.6% should be interpreted as the probability that he will win the £216.

An alternative "timid" strategy is to divide his money into 18 equal piles of £6 and be prepared to make successive bets on a single number. He does this until he either runs out of cash or one bet (at 35 to 1) yields him £210 plus his stake = £216.

To calculate the odds of success using this strategy, calculate the chance that all the bets lose. So any single bet loses with a probability of 36 in 37. So the chance that all 18 bets lose = $(36/37)^{18}$ = 0.61. Therefore, the probability that at least one bet wins = 1 − 0.61 = 0.39. The chance that he will achieve his target has been reduced, therefore, from 48.6% to 39% by substituting the timid strategy for the bold play.

There are many alternative staking strategies that might put John over the top, but none of them can make it more probable that he will achieve his target than the boldest play of them all – the full amount on one spin of the wheel.

Yes, that's right. In unfavourable games (house edge against you), bold play is best and timid play is worst. Always place the fewest bets you need to reach your target.

The intuition behind bold play is simple. Since the odds are in favour of the house, the strategy that minimises your exposure to that advantage, by minimising the number of games you play, is the optimal one. We can also look at it another way. If you split your bank between red and black, betting an equal amount on red and black, you would break even most of the time (36 times in 37). When it lands on the zero, however, with a probability of 1/37, you lose your entire bank. The more often you play the game, the more likely it is that the zero comes up.

4.4.1 Exercise

1. You need £432 to pay off an urgent debt but have only a bank of £216 available. This is unacceptable to the lender and is as good as nothing. You decide to try to win the money at the roulette wheel. It is a single-zero wheel.

   a. What is the probability that you will win the target sum if you place all your bank on one spin of the wheel?

   b. Is there any alternative staking strategy that can increase your chance of winning enough to pay off your debt?

2. You play a game in which ten balls, numbered 1 to 10, are placed into a lottery draw machine and randomised. A ball is now delivered out of the machine. If it is numbered 1 to 4, you win a point. If the ball is numbered 5 to 10, your opponent wins a point. The ball is now replaced, and the game can take place again. At the end of the session of play, you must be ahead in points to win the match. Would you be more likely to win if the match is made up of only one game, or of three games, or does it make no difference? What is the actual probability that you would win at least 2 out of 3 games played?

4.4.2 Reading and Links

Dubins, L.E., and Savage, L.J. 1960. Optimal gambling systems. 19 October. https://www.ncbi.nlm.nih.gov/pmc/articles/PMC223086/pdf/pnas00211-0067.pdf
StackExchange. Mathematics. How to win at roulette? https://math.stackexchange.com/questions/98981/how-to-win-at-roulette

4.5 The Favourite-Longshot Bias

In a betting market characterised by a market-maker, the odds offered to bettors are designed so that the market-maker secures a profit overall, i.e. the average bettor trades at a loss. For this reason, we might consider information efficiency to mean that no bettor is able systematically to derive above-average returns, or that no one trading strategy yields a better return than another, except by incurring correspondingly greater costs (broadly defined).

For example, if it can be shown that bets placed at a given odds level tend to yield a significantly higher return than equivalent bets at a different odds level, this would constitute prima facie evidence of information inefficiency in the sense that a trading strategy exists which can yield greater returns than another. In conventional financial markets, on the other hand, information efficiency is more closely related to the absence of opportunities to secure abnormal returns.

This brings us directly to the so-called "favourite-longshot bias", which is the well-established tendency in most betting markets for bettors to bet too much on "longshots" (events with long odds, i.e. low-probability events) and to relatively under-bet "favourites" (events with short odds, i.e. high-probability events). This is strangely counter-intuitive as it seems to offer a betting strategy that can expect to generate above-average returns. Assume, for example, that Mr. Jon Miller and Mr. Stan Stiller, who know nothing about the horses other than the odds, both start with £1,000. Now Mr. Miller places a level £10 stake on 100 horses quoted at 2 to 1 against (£10 to win a net £20). Meanwhile, Mr. Stiller places a level £10 stake on 100 horses quoted at 20 to 1 against (£10 to win a net £200).

Who is likely to end up better off at the end? Surely the answer should be the same for both. Otherwise, either Mr. Miller or Mr. Stiller would seem to be doing something very wrong. So let's take a look.

The *Ladbrokes Flat Season Pocket Companion* for 1990 provides evidence for British flat horse racing between 1985 and 1989. Still, the same sort of pattern applies for any set of years we care to choose, or (with a few rare exceptions) pretty much any sport, anywhere.

The table conveniently presented in the Companion shows that not 1 out of 35 favourites starting at 1/8 or shorter lost between 1985 and 1989. This is a rapid return of between 4% and 12.5%, a substantial rate of interest. Therefore, the shorter the odds, the better the return. The group of "white-hot" favourites (odds between 1/5 and 1/25) were successful in 88 out of 96 races for a 6.5% profit. The following table examines other odds groupings. Note, however, that in 1990 there was a tax levied on betting stakes or returns, so the actual return was somewhat less than shown in these before-tax figures.

Odds	Wins	Runs	Profit	%
1/5 – 1/2	249	344	+£1.80	+0.52

4/7 – 5/4	881	1,780	–£82.60	–4.64
6/4 – 3/1	2,187	7,774	–£629	–8.09
7/2 – 6/1	3,464	21,681	–£2,237	–10.32
8/1 – 20/1	2,566	53,741	–£19,823	–36.89
25/1 – 100/1	441	43,426	–£29,424	–67.76

An interesting argument advanced by Robert Henery in 1985 is that the favourite-longshot bias is a consequence of bettors discounting a fixed fraction of their losses, i.e. they underweight their losses compared to their gains, and this causes them to bias their perceptions of what they have won or lost in favour of longshots. The rationale behind the Henery hypothesis is that bettors will tend to explain away and therefore discount losses as atypical or unrelated to the judgment of the bettor. They might be viewed, for example, as "near wins" or the outcome of "fluke" events. In the Appendix (Section 4.5.1) we show how discounting losses can produce the observed favourite-longshot bias.

If the Henery Hypothesis is correct as a way of explaining the favourite-longshot bias, the bias can be explained, therefore, as the natural outcome of bettors' pre-existing perceptions and preferences. There is little evidence, however, that the market offers opportunities for players to earn consistent profits. Still, they certainly do much better (lose a lot less) by a blind level-stakes strategy of backing at shorter odds instead of at longer odds. Intuitively, we would think that people would wise up and switch their money away from the less favourable longshots to the more favourable favourites. In that case, the forces of supply and demand should mean that favourites would become less good value, as their odds would shorten, and longshots would become better value as their odds would lengthen. But it doesn't happen, despite a host of published papers pointing this out, as well as the Ladbrokes Pocket Companion.

What explanations exist for the persistence of the favourite-longshot bias? One explanation is based on consumer preference for risk. The idea here is that bettors are risk-loving and so prefer the risky series of long runs of losses followed by the odd big win to the less risky strategy of betting on favourites that will win more often albeit pay out less for each win. Such an assumption of risk-love by bettors runs contrary, however, to conventional explanations of financial behaviour, which tend to assume people like to avoid risk. It's alternatively been argued that bettors are not risk-lovers but skewness-lovers, which would also explain a preference for backing longshots over favourites, seminally by Golec and Tamarkin (1998). This is consistent with what is termed "prospect theory" whereby people derive utility from gains and losses measured relative to some reference point. The possibility of gaining a lot from a single bet (positive skew) could prove attractive to bettors, especially if they over-estimate the chance that the event (the longshot winning) will occur. This could also be the consequence of what is known as "narrow framing", which occurs when an individual evaluates an event separately from other concurrent events. While longshots offer lower expected returns than favourites measured over all bets, the narrow framing of a single bet as an isolated case could encourage bettors to seek a single big win from a small stake, ignoring the bigger picture of how much they stand to win/lose over all bets.

Another explanation that has been proposed for the bias is based on the existence of unskilled bettors in the context of high margins and other costs of betting which deter more skilled agents. These unskilled bettors find it more difficult to arbitrate between the true win probabilities of different horses and so over-bet those offered at longer odds. One test of this hypothesis is to compare the size of the bias in person-to-person betting exchanges (characterised by lower margins) and bookmaker markets (higher margins). The bias was indeed lower in the former (Smith et al., 2006), a finding which is at least consistent with this theory.

So far, it should be noted that these are all demand-side explanations, i.e. based on the behaviour of bettors. Another explanation of at least some of the bias is the idea that odds-setters defend themselves against bettors who potentially have superior information to bookmakers by artificially squeezing odds at the longer end of the market, or they might squeeze longer odds because it is less noticeable to do so than at shorter odds. Even so, the favourite-longshot bias continues to exist in the so-called "pari-mutuel" markets, in which there are no odds-setters, but instead a pool of all bets which is paid out (minus fixed operator deductions) to winning bets. To the extent that the favourite-longshot bias cannot be fully explained by this odds-squeezing explanation, we can classify the remaining explanations as either preference-based or perception-based. Risk-love is an example of a preference-based explanation, as is skewness-love. Discounting of losses or other explanations based on a poor assessment of the true probabilities can be categorised as perception-based explanations.

The favourite-longshot bias has even been found in online poker, notably in lower-stake games (Vaughan Williams et al., 2016). In that context, the evidence suggests that it was misperception of probabilities rather than risk-love that offered the best explanation for the bias.

In conclusion, the favourite-longshot bias is a well-established market anomaly in betting markets, which can be traced in the published academic literature as far back as the late 1940s (Griffith, 1949). Explanations can broadly be divided into demand-based and supply-based, preference-based and perceptions-based. A significant amount of modern research has been focused on seeking to arbitrate between these competing explanations of the bias by formulating predictions as to how data derived from these markets would behave if one or other explanation was correct. See, for example, Vaughan Williams and Paton, 1997; Vaughan Williams et al., 2016. A compromise position, which may or may not be correct, is that all these explanations have some merit, the relative merit of each depending on the market context.

Before concluding this section, we turn to another betting bias, known as the "Gambler's Fallacy". The Gambler's Fallacy is the proposition that people, instead of accepting an actual independence of successive outcomes, are influenced in their perceptions of the next possible outcome by the results of the preceding sequence of outcomes – for example, dice rolls and spins of a wheel. Dek Terrell (1994, p. 309) states it this way: "The 'gambler's fallacy' is the belief that the probability of an event is decreased when the event has occurred recently, even though the probability of the event is objectively known to be independent across trials". The existence of a "Gambler's Fallacy" can be traced to laboratory studies, lottery-type games, and lotteries.

Clotfelter and Cook (1993) found, in a study of a Maryland numbers game, a significant fall in the amount of money wagered on winning numbers in the days following the win, an effect which did not disappear entirely until after about 60 days. This game was, however, characterised by a fixed-odds payout to a unit bet, and so the Gambler's Fallacy had no effect on expected returns.

In pari-mutuel games, on the other hand, the pool of all bets is divided among winning bets (minus commission), so the return to a winning number is linked to the amount of money bet on that number. As such, the operation of a systematic bias against certain numbers will tend to increase the expected return on those numbers. Does this make a difference to the Gambler's Fallacy?

Terrell (1994) investigated one such system, the New Jersey State Lottery. In a sample of 1,785 drawings, he constructed a subsample of 97 winners which repeated as a winner within a 60-day cut-off point.

The expected payout increased by 28% one day after winning and decreased from this level by c. 0.5% each day after the number won, returning to its original level 60 days later.

The size of the Gambler's Fallacy, while significant, was less than that found by Clotfelter and Cook in their fixed-odds numbers game.

In other words, the size of the Gambler's Fallacy was greater in a game where its existence didn't matter than it was in a game where it did matter.

4.5.1 Appendix

Let's look more closely at how the Henery odds transformation works.

If the true probability of a horse losing a race is q, then the true odds against winning are $q / (1 - q)$.

For example, if the true probability of a horse losing a race (q) is 3/4, the chance that it will win the race is 1/4, i.e. $1 - 3/4$. The odds against it winning are: $q / (1 - q) = 3/4 / (1 - 3/4) = 3/4 / (1/4) = 3/1$.

Henery now applies a transformation whereby the bettor will assess the chance of losing not as q, but as Q which is equal to fq, where f is the fixed fraction of losses not discounted by the bettor. In other words, the bettor discounts a constant fraction $1-f$ of losses, thus believing that the chance of losing is fq, where q is the true chance of losing.

If, for example, $f = 3/4$, and the true chance of a horse losing is 1/2 ($q = 1/2$), then the bettor will rate subjectively the chance of the horse losing as $Q = fq$.

So $Q = 3/4 \cdot 1/2 = 3/8$, i.e. a subjective chance of winning of 5/8.

So the perceived (subjective) odds of winning associated with the true (objective) odds of losing 50% (evens, i.e. q = 1/2) is 3/5 (60%), i.e. odds-on.

This is derived as follows:

$$Q/(1-Q) = fq/(1-fq) = 3/8/(1-3/8) = 3/8/(5/8) = 3/5$$

If the true probability of a horse losing a race is 80%, so that the true odds against winning are 4/1 (q = 0.8), then the bettor will assess the chance of losing not as q, but as Q which is equal to fq, where f is the fixed fraction of losses not discounted by the bettor.

If, for example, f = 3/4, and the true chance of a horse losing is 4/5 (q = 0.8), then the bettor will rate subjectively the chance of the horse losing as Q = fq.

So Q = 3/4 × 4/5 = 12/20, i.e. a subjective chance of winning of 8/20 (2/5).

So the perceived (subjective) odds of winning associated with the true (objective) odds of losing 80% (4 to 1, i.e. q = 0.8) is 6/4 (40%).

This is derived as follows:

$$Q/(1-Q) = fq/(1-fq) = 12/20/(1-12/20) = 12/8 = 6/4$$

To take this to the limit, if the true probability of a horse losing a race is 100%, so that the true odds against winning are ∞ to 1 against (q = 1), then the bettor will again assess the chance of losing not as q, but as Q which is equal to fq, where f is the fixed fraction of losses not discounted by the bettor.

If, for example, f = 3/4, and the true chance of a horse losing is 100% (q = 1), then the bettor will rate subjectively the chance of the horse losing as Q = fq.

So Q = 3/4 × 1 = 3/4, i.e. a subjective chance of winning of 1/4.

So the perceived (subjective) odds of winning associated with the true (objective) odds of a 100% chance of losing (∞ to 1, i.e. q = 1) is 3/1 (25%).

This is derived as follows:

$$Q/(1-Q) = fq/(1-fq) = 3/4/(1/4) = 3/1$$

Similarly, if the true probability of a horse losing a race is 0%, so that the true odds against winning are 0 to 1 against (q = 0), then the bettor will assess the chance of losing not as q, but as Q which is equal to fq, where f is the fixed fraction of losses not discounted by the bettor.

If, for example, f = 3/4, and the true chance of a horse losing is 0% (q = 0), then the bettor will rate subjectively the chance of the horse losing as Q = fq.

So Q = 3/4 × 0 = 0, i.e. a subjective chance of winning of 1.

So the perceived (subjective) odds associated of winning with the true (objective) odds of losing of 0% (0 to 1, i.e. q = 0) is also 0/1.

This is derived as follows:

$$Q/(1-Q) = fq/(1-fq) = 0/1 = 0/1$$

This can all be summarised in a table.

Objective odds (against)	Subjective odds (against)
Evens	3/5
4/1	6/4
Infinity to 1	3/1
0/1	0/1

We can now use these stylised examples to establish the bias.

In particular, the implication of the Henery odds transformation is that, for a given f of 3/4, 3/5 is perceived as fair odds for a horse with a 1 in 2 chance of winning.

In fact, a £100 wagered at 3/5 yields £160 (3/5 × £100, plus stake returned) half of the time (true odds = evens), i.e. an expected return of £80.

A £100 wagered at 6/4 yields £250 (6/4 × £100, plus the stake back) one fifth of the time (true odds = 4/1), i.e. an expected return of £50.

A £100 wagered at 3/1 yields £0 (3/1 × £100, plus the stake back) none of the time (true odds = infinity to 1), i.e. an expected return of £0.

It can be shown that the longer the odds the lower is the expected rate of return on the stake, although the relationship between the subjective and objective probabilities remains at a fixed fraction throughout.

Now onto the over-round (OR).

The same simple assumption about bettors' behaviour can explain the observed relationship between the over-round (sum of win probabilities minus 1) and the number of runners in a race, n.

If each horse is priced according to its true win probability, then the over-round = 0. So in a six horse race, where each has a 1 in 6 chance, each would be priced at 5 to 1, so none of the loss probability is shaded by the bookmaker. Here the sum of probabilities = (6 × 1/6) − 1 = 0. Note that an over-round of 0 (as derived here) is often presented in the format of 1 or 100%, so that an over-round of 0.5 is often presented as 1.5 or 150%. This does nothing to change the argument but we shall here use 0 in this demonstration to represent zero over-round.

If only a fixed fraction of losses, f, is counted by bettors, the subjective probability of losing on any horse is f (qi), where qi is the objective probability of losing for horse i, and the odds will reflect this bias, i.e. they will be shorter than the true probabilities would imply. The subjective win probabilities in this case are now 1 − f (qi), and the sum of these minus 1 gives the over-round.

Where there is no discounting of the odds, the over-round (OR) = 0, i.e. n times correct odds minus 1. Assume now that $f = 3/4$, i.e. 3/4 of losses are counted by the bettor.

If there is discounting, then the odds will reflect this, and the more the runners the bigger will be the over-round.

So in a race with 5 runners, q is 4/5, but $fq = 3/4 \times 4/5 = 12/20$, so subjective win probability $= 1 - fq = 8/20$, not 1/5. So OR $= (5 \times 8/20) - 1 = 1$.

With 6 runners, $fq = 3/4 \times 5/6 = 15/24$, so subjective win probability $= 1 - fq = 9/24$. OR $= (6 \times 9/24) - 1 = (54/24) - 1 = 1_{1/4}$.

With 7 runners, $fq = 3/4 \times 6/7 = 18/28$, so subjective win probability $= 1 - fq = 10/28$. OR $= (7 \times 10/28) - 1 = 42/28 = 1_{1/2}$.

If there is no discounting, then the subjective win probability equals the actual win probability, so an example in a 5-horse race is that each has a win probability of 1/5. Here, OR $= (5 \times 1/5) - 1 = 0$. In a 6-horse race, with no discounting, subjective probability $= 1/6$. OR $= (6 \times 1/6) - 1 = 0$.

Hence, the over-round is linearly related to the number of runners, assuming bettors discount a fixed fraction of losses (the "Henery Hypothesis").

4.5.2 Exercise

1. Using the method demonstrated in the Appendix, calculate the subjective odds (against) in this table assuming that f, the fixed fraction of losses undiscounted by the bettor, is a half.

   Objective odds (against)   Subjective odds (against)

   a. Evens
   b. 4/1

2. Does the favourite-longshot bias imply for a particular event that the expected return to a unit stake is greater for a series of bets at shorter odds or longer odds?

3. Assuming no other information, does the favourite-longshot bias imply that the expected return to a unit stake will be greater for a series of bets on 4/1 favourites or on 4/1 longshots?

4. Assuming the existence of a favourite-longshot bias in a racetrack betting market that you are trading, what can you expect by backing the favourite in every race?

   a. At least an above-average return
   b. An abnormal return
   c. Both
   d. Neither

5. Assuming no other information, in which of the following would you expect the return to betting on a series of ten favourites in a row to be greatest?

a. Horse racing

b. Football

c. Tennis

6. Can knowledge of the Gambler's Fallacy be used in principle to improve your chances of winning the UK National Lottery main draw?

7. Can knowledge of the Gambler's Fallacy be used in principle to improve the expected return to a £2 ticket on the UK National Lottery main draw?

4.5.3 Reading and Links

Clotfelter, C., and Cook, P.J. 1993. The 'gambler's fallacy' in lottery play, *Management Science*, 39, 12, 1521–1525. https://www.nber.org/system/files/working_papers/w3769/w3769.pdf

Golec, J. and Tamarkin, M. 1998. Bettors love skewness, not risk, at the horse track. *Journal of Political Economy*, 106(1), 205–225.

Griffith, R.M. 1949. Odds adjustments by American horse-race bettors. *American Journal of Psychology*, 62(2), 290–294. https://www.jstor.org/stable/pdf/1418469.pdf?casa_token=jYYMVAnyE08AAAAA:SDLUVkL9J8EkFqtyJp_5kuucBWsS7tEu_B-uVz2_hsNHIzjXa9Wi9nJ_5hFJulmoVSWLModLHshKRS3WfJja0HjccSP1XCutMzIcTMFHBkoohFJbCp4

Henery, R.J. 1985. On the average probability of losing bets on horses with given starting price odds. *Journal of the Royal Statistical Society. Series A (General)*, 148(4), 342–349. https://www.jstor.org/stable/2981894?seq=1#page_scan_tab_contents

Paton, D. and Vaughan Williams, L. 2008. Do betting costs explain betting biases? *Applied Economics Letters*, 5(5), 333–335. https://www.tandfonline.com/doi/pdf/10.1080/758524413?casa_token=ZJzhh11lXXoAAAAA:n-H9hlSk-w0R9Q4C_QS31EG0-4em3QTf7ptOz47XWXtl8qMNyC3w8I8dLxb2dK0_EyWzmRFNvkuh1w

Smith, M.A., Paton, D., and Vaughan Williams, L. 2006. Market efficiency in person-to-person betting. *Economica*, 73, 292, 673–689. https://onlinelibrary.wiley.com/doi/pdf/10.1111/j.1468-0335.2006.00518.x?casa_token=ZL7pql3NsoQAAAAA:BADN5ZtZdFbY1dej1Lnyw7m7nkIBxMThhVKN6nyq3R7ARTkSRw8WTrjkDGQj4Gzt_50i13OzioPbsXvx

Smith, M.A., Paton, D., and Vaughan Williams, L. 2009. Do bookmakers possess superior skills to bettors in predicting outcomes? *Journal of Economic Behavior and Organization*, 71(2), 539–549. https://www.sciencedirect.com/science/article/pii/S0167268109000833?casa_token=rdqW7jp7sGQAAAAA:QA6ztSJpe8xcyQEIRXdsrdln8psADoFfysXJPlWnLJVUy748g3c8XNkWh1UZqXdZK3hNNTA1vnk

Smith, M.A., and Vaughan Williams, L. 2010. Forecasting horse race outcomes: New evidence on odds bias in UK betting markets. *International Journal of Forecasting*, 26(3), 543–550. https://www.sciencedirect.com/science/article/pii/S0169207009002155?casa_token=HgkJGg5vfCEAAAAA:vZF0xwzI0wieWYDgCU4fCa2tllDF87xjqasaLY5hDuodWTVeTuTVQrBu5nPZppa4FMuv-u-pF5Q

Terrell, D. 1994. A test of the gambler's fallacy: Evidence from pari-mutuel games. *Journal of Risk and Uncertainty*, 8, 309–317. https://www.jstor.org/stable/pdf/41760730.pdf?casa_token=FHGbFRegbXYAAAAA:bbqJPWybiSpGlx3lbUobMya_VyiJvbU1WBBht8RqLMuklgLT-PdF4ytc0wTonMC1NCxlu_NX99--MRG-QULRAqRJKIMJwXo78YdQwmFQ_OKiijX9QpM

Vaughan Williams, L. 1999. Information efficiency in betting markets: A survey. 51(1), 1–39. https://onlinelibrary.wiley.com/doi/pdf/10.1111/1467-8586.00069?casa_token=9Ph0y2DicOgAAAAA:nWvz0aY1OXwqEbOqp8M4xCEQu04DJjDJjrCK2pfv0hMVqAmUqMcISRHQf2qo-TLAERxArVnolL86nZUH

Vaughan Williams, L. 2003. *Betting to Win: A Professional Guide to Profitable Betting.* Oldcastle Books Ltd.

Vaughan Williams, L. 2003. *The Economics of Gambling.* London: Routledge.

Vaughan Williams, L. 2005. *Information Efficiency in Financial and Betting Markets,* ed. L. Vaughan Williams. Cambridge University Press.

Vaughan Williams, L. 2012. *The Economics of Gambling and National Lotteries.* The International Library of Critical Writings in Economics. Cheltenham: Edward Elgar.

Vaughan Williams, L. and Paton, D. 1996. Risk, return and adverse selection: A study of optimal behaviour under asymmetric information. *Rivista di Politica Economica,* 11–12, 63–81.

Vaughan Williams, L. and Paton, D. 1997. Why is there a favourite-longshot bias in British racetrack betting markets? *Economic Journal,* 107, 150–158. http://bit.ly/2P0Z8u5

Vaughan Williams, L. and Paton, D. 1998. Why are some favourite-longshot biases positive and others negative? *Applied Economics,* 30(11), 1505–1510. https://www.tandfonline.com/doi/pdf/10.1080/000368498324841?casa_token=EkyXRb-6vJgAAAAA:Qh6OCaUjXp-dZcn3QAOrqEeTx_W5nS-fCFQdsuwNuNomVK2RAM2fyoK7XXDuOZOv3RIQjCweqvKcPA

Vaughan Williams, L. and Siegel, D. 2013. *The Oxford Handbook of the Economics of Gambling.* New York: Oxford University Press.

Vaughan Williams, L., Sung, M., Fraser-Mackenzie, P., Peirson, J., and Johnson, J.E.V. 2016. Towards an understanding of the origins of the favourite-longshot bias: Evidence from online poker, a real-world natural laboratory. *Economica,* 85, 338, 360–382. https://onlinelibrary.wiley.com/doi/pdf/10.1111/ecca.12200?casa_token=qg2-SdM-C18AAAAA:UuyhES9uPGraMnF1oULlwbnMhlAh5WyUcUx0HTEkV4uILT2EdnFzhIknn5hNO8qivrONYYciVe8N4Wg

4.6 The Poisson Distribution

The Poisson distribution, named after Simeon Denis Poisson, arises in the context of events which occur infrequently but at some average rate. It was famously used in the nineteenth century to model the number of Prussian cavalry officers kicked to death by horses in various Army regiments over 20 years.

More recently, it has been used to model the number of goals scored in football matches.

If we use X to denote the relevant quantity, suppose its average value is μ. Then X will have a Poisson distribution when it can take any of the values 0, 1, 2, 3 ..., and the probability that X takes the value r is:

$$e^{-\mu} \mu^r / r!$$

For example, let $\mu = 2.5$. Then the probabilities for the Poisson are:

Value of r	0	1	2	3	4	5	6	7+
Probability	.082	.205	.257	.214	.134	.067	.028	.014

Here's an example of the type of problem that can be addressed using the Poisson.

Question: The average number of homes sold by a local estate agent is two homes per day. What is the probability that exactly three homes will be sold tomorrow?

This is a Poisson experiment in which we know the following:

$\mu = 2$; since two homes are sold per day, on average.

$r = 3$; since we want to find the chance that three homes will be sold tomorrow.

e = a constant equal to approximately 2.71828.

We plug these values into the Poisson formula as follows:

$P(r;\mu) = (e^{-\mu})(\mu^r) / r!$

$P(3;2) = (2.71828^{-2})(2^3) / 3!$

$P(3;2) = (0.13534)(8) / 6$

$P(3;2) = 0.180$

The probability that three homes will be sold tomorrow is 18%.

A feature of football matches is that goals are infrequent. Even the best teams at top level score an average of under two goals a match.

The simplest plausible model of the scoring assumes that goals occur at essentially random times, at some average rate that depends on the team, the opposition, and who is playing at home. To the extent that this assumption is right, we can use the Poisson distribution. Here is a sample (using rounded percentages) of the distribution to illustrate.

• Goals	0	1	2	3	4 or more
• Average = 0.8	45%	36%	14%	4%	1%
• Average = 1.2	30%	36%	22%	9%	3%
• Average = 1.6	20%	32%	26%	14%	8%
• Average = 2.0	14%	27%	27%	18%	14%

This distribution can be used to estimate the probability of a win, loss, or draw for a team, or for a particular scoreline, to the extent that the Poisson is a good representation of the distribution of goals in a football match.

Take a game in which the home side is expected to score 1.6 goals, on average, while the corresponding estimate for the away side is 1.2 goals.

Assuming that the number of goals scored by the two sides is independent, the probability that the game ends in any particular score is easily derived from the table.

For example, the chance of a 0-0 draw is (20%) × (30%) = 6%.

The chance of a 1-1 draw = (32%) × (36%) = 11.5%.

By adding scores that lead to a draw, we estimate that the chance of the match ending in a draw is about 25%.

Aggregating similarly for all scores that lead to a home win sums to 48%.

The chance of an away win = 27%.

A useful addition to the sports forecasting toolbox!

4.6.1 Exercise

The average number of homes sold by a local estate agent is two homes per day. What is the probability that exactly four homes will be sold tomorrow?

4.6.2 Reading and Links

STAT TREK. Poisson distribution. https://stattrek.com/probability-distributions/poisson.aspx

An Introduction to the Poisson Distribution. jbstatistics. 30 October 2013.. YouTube. https://www.youtube.com/watch?v=jmqZG6roVqU

The Poisson Distribution: An Introduction (Fast Version). jbstatistics. 26 May 2012. YouTube. https://www.youtube.com/watch?v=8x3pnyYCBto&app=desktop

Lecture 11: The Poisson Distribution. Statistics 110. Harvard University. 29 April 2013. YouTube. https://www.youtube.com/watch?v=TD1N4hxqMzY&feature=youtu.be

4.7 Card Counting

It is said that on returning from a day at the races, a certain Lord Falmouth was asked by a friend how he had fared. "I'm quits on the day", came the triumphant reply. "You mean by that", asked the friend, "that you are glad when you are quits?" When Falmouth replied that indeed he was, his companion suggested that there was a far easier way of breaking even, and without the trouble or annoyance. "By not betting at all!" The noble Lord said that he had never looked at it like that and, according to legend, gave up betting from that very moment.

While this may well serve as a very instructive tale for many, Ed Thorp, writing in 1962, took a rather different view. He had devised a strategy, based on probability theory, for consistently beating the house at blackjack (or "21"). In his book *Beat the Dealer: A Winning Strategy for the Game of Twenty-One*, Thorp presents the system. On the inside cover of the dust jacket, he claims that "the player can gain and keep a decided advantage over the house by relying on the strategy".

The rules of blackjack are simple. To win a round, the player has to draw cards to beat the dealer's total and not exceed a total of 21. Because players have choices to make, most obviously as to whether to take another card or not, there is an optimal strategy for playing the game. The precise strategy depends on the house rules, but generally speaking it pays, for example, to hit (take another card) when the total of your cards is 14 and the dealer's face-up card is 7 or higher. If the dealer's face-up card is a 6 or lower, on the other hand, you should stand (decline another card). This is known as "basic strategy".

While the basic strategy will reduce the house edge, it is not enough to turn the edge in the player's favour. That requires exploitation of the additional factor inherent in the tradition that the used cards are not shuffled back into the deck. This means that by counting which cards have been removed from the deck, we can re-evaluate the probabilities of particular cards or card sizes being dealt moving forward. For example, a disproportionate number of high cards in the deck is good for the player, notably because in those situations where the rules dictate that the house (though not the player) is obliged to take a card (for example, on a total of 16), a plethora of remaining high cards increases the dealer's exposure to going bust (exceeding a total of 21).

Thorp's genius was in devising a method of reducing this strategy to a few simple rules which could be understood, memorised, and made operational by the average player in real time. As the book blurb puts it, "The presentation of the system lends itself readily to the rapid play normally encountered in the casinos". Essentially, all that is needed is to attach a tag to specific cards (such as +1 or −1) and then add or subtract the tags as the cards are dealt. Depending on the net score concerning the cards dealt, it is easy to estimate whether the edge is with the house or the player. This system is called keeping a "running count".

There are variations on this theme, but the core strategy and original insights hold. The problem now became one familiar to many successful horseplayers – how to get your money on before being closed down.

4.7.1 Exercise

When playing blackjack at the tables, with cards not shuffled back into the deck, will basic strategy generate a profit? Will card counting generate a profit?

4.7.2 References and Links

Shackleford, M. 2019. Card Counting. The Wizard of Odds. 21 January. https://wizardofodds.com/games/blackjack/card-counting/introduction/

Shackleford, M. 2019. Introduction to the High-Low Card Counting Strategy. The Wizard of Odds. 13 August. https://wizardofodds.com/games/blackjack/card-counting/high-low/

Shackleford, M. 2019. 4-Deck to 8-Deck Blackjack Strategy. The Wizard of Odds. 13 August. https://wizardofodds.com/games/blackjack/strategy/4-decks/

Shackleford, M. 2019. The Ace/Five Count. The Wizard of Odds. 21 January. https://wizardofodds.com/games/blackjack/appendix/17/

Thorp, E.O. (1966). Beat the Dealer: A Winning Strategy for the Game of Twenty-One. Vintage Books.

Card counting. Wikipedia. https://en.wikipedia.org/wiki/Card_counting

Blackjack Expert Explains How Card Counting Works. WIRED. 6 March 2017. YouTube. https://youtu.be/G_So72lFNIU

4.8 Can the Martingale Betting System Guarantee a Profit?

The basis of the martingale betting system is a strategy in which the gambler doubles the bet after every loss, so that the first win would recover all previous losses plus a profit equal to the original stake. The martingale strategy has been applied to roulette in particular, where the probability of hitting either red or black is near to 50%.

Take the case of a gambler who wagers £2 on Heads, at even money, and so profits by £2 if the coin lands Heads and loses £2 if it lands Tails. If he loses, he doubles the stake on the next bet, to £4, and wins £4 if it lands Heads, minus £2 lost on the first bet, securing a net profit over both bets of £2 (£4 − £2). If it lands Tails again, however, he is £6 down, so he doubles the stake in the next bet to £8. If it lands Heads he wins £8, minus £6 lost on the first two nets, securing a net profit over the three bets of £2 (£8 − £6). This can be generalised for any number of bets. Whenever he wins, the gambler secures a net profit over all bets of £2.

The strategy is essentially, therefore, one of chasing losses. In the above example, the loss after n losing rounds is equal to $2 + 2^2 + 2^3 + \ldots + 2^n$.

So the strategy is to bet in the next round $2 + 2^2 + 2^3 + \ldots + 2^n + 2$.

In this way, the profit whenever the coin lands Heads is 2.

To look at it another way, for any positive integer n, it can be shown that the sum of numbers $1 + 2 + 4 + 8 + \ldots + n = 2n − 1$. So, for example, $1 + 2 = 2 \times 2 − 1 = 3$; $1 + 2 + 4 + 8 = 2 \times 8 − 1 = 15$. There will always be a profit of a unit stake to any doubling strategy when it generates a win.

For a gambler with infinite wealth, and hence an infinite number of coin tosses to generate heads eventually, the martingale betting strategy has indeed been interpreted as a sure win.

However, the gamble's expected value remains zero (or less than zero) because the small probability of a very large loss exactly balances out the expected gain. In a casino, the expected value is negative due to the house edge. There is also conventionally a house limit on bet size.

The martingale strategy fails, therefore, whenever there is a limit on earnings or on bets or bet size, as is the case in the real world. It is only with infinite or boundless wealth, bet size, and time that it could be argued that the martingale becomes a winning strategy.

A variant on the martingale system is the "Devil's Shooting Room Paradox", traceable to Eckhardt (1997) and explored in Leslie (1998). In this paradox, a group of people are ushered into a room and told by the Devil that he will shoot everyone in the room if he throws a double-six. He also declares that over 90% of those who enter the room will be shot. How can both these declarations be true? Because while there is only a 1 in 36 chance that any particular room of people will fall victim, each time he ushers a group of people into the room, the size of the group is over ten times bigger than the previous one. It can be shown that this generates a probability of being shot of over 90%, and at the same time the chance that any one particular group falls victim to the Devil is 1 in 36.

The only way that the Devil can guarantee that 90% of those who enter the room will be shot, however, is if there is an infinite supply of people to fill the room. As such, this is a variant of the martingale paradox.

Infinity is also central to other famous gambling paradoxes, notably the St. Petersburg Paradox, proposed by Nicolas Bernoulli and published by his cousin, Daniel Bernoulli. To state the paradox, imagine tossing a coin until it lands heads up and suppose that the payoff grows exponentially according to the number of tosses you make. If the coin lands heads up on the first toss, then the payoff is £2. If it lands tails on the first toss, you receive £1. If it lands heads on the second toss, the payoff is £4; if it takes three tosses, the payoff is £8; and so forth, ad infinitum. You can play as many rounds of the game as you wish.

Now the odds of the game ending on the first toss is 1/2; of it ending on the second toss is $(1/2)^2 = 1/4$; on the third $(1/2)^3 = 1/8$, etc., so your expected win from playing the game $= (1/2 \times £1) + (1/2 \times £2) + (1/4 \times £4) + (1/8 \times £8) + \dots$, i.e. £0.5 + £1 + £1 + £1 … = infinity. It follows that you should be willing to pay any finite amount for the privilege of playing this game. Yet it seems irrational to pay very much at all.

According to this reasoning, any finite stake is justified because the eventual payout increases infinitely through time, so you must end up with a profit whenever the game ends. Yet most people are only willing to pay a few pounds, or at least not much more than this. So is this yet further evidence of our intuition letting us down? That depends on why most people are not willing to pay much. It is true, of course, that you will, if you play an infinite number of rounds of the game, win an infinite amount. In the real world we inhabit finite time and resources, however, so the expected value is not an infinite sum but a finite sum of several terms, with a finite payout. In this world, the expected value of the bet is now a much more constrained sum.

It can also be explained by distinguishing between money and the value put on money, which declines as additional sums are added to one's pile.

This is the idea of diminishing marginal utility of wealth, proposed by Daniel Bernoulli, the man who made the paradox famous. Related to this is the concept of the geometric mean, which is the square root of two possible outcomes multiplied together. So the geometric mean of £5 and £80 is the square root of £400, i.e. £20. The arithmetic mean of £5 and £80 is half the sum of the amounts, i.e. £42.50. Similarly, the geometric mean of £1 and £1 million is about £1,000, while the arithmetic mean is about £500,000. Bernoulli suggests that reasonable people might actually seek to maximise the geometric mean of outcomes.

Ultimately, though, the utility-based explanation is surplus to needs. In a world of finite resources, the expected value of such a bet is a finite sum with a finite payout, and it can be totally rational to pay a relatively modest sum to play the game.

4.8.1 Appendix

The probability of losing three fair coin tosses = 1/8.

The probability of losing n times = $1/2^n$.

Total loss with starting stake of 2, with 3 losses of coin toss = 2 + 4 + 8 = 14.

So martingale strategy suggests a bet of 14 + 2 = 16.

Loss after n losing rounds = $2 + 2^{2+} \ldots + 2^n$.

So martingale bet = $(2 + 2^2 + \ldots + 2^n) + 2 = 2^{n+1}$.

This strategy always wins a net 2.

This strategy, of always betting to win more than lost so far, works in principleregardless of the odds or whether they are fair. This holds so long as there is no finite stopping point at which the next martingale bet is unavailable (such as a maximum bet limit) or can't be afforded.

So, let us assume that everyone has some number of losses such that they don't have enough money to pay a stake large enough for the next round that it would cover the sum of the losses to that point. Call this run of losses n.

n differs across people and could be very high or very low.

The probability of losing n times = $1/2^n$.

Using a martingale +2 strategy, the player wins 2 if able to play on and then wins.

So, the player wins 2 with a probability of $(1 - 1/2^n)$.

Total losses after n losing bets = $(2 + 2^2 + \ldots + 2^n) = (2^{n+1} - 2)$.

Expected gain is equal to the probability of not folding times the gain plus the probability of folding times the loss.

Expectation = $(1 - 1/2^n) \cdot 2 - 1/2^n (2^{n+1} - 2)$.

$= 2 - 2/2^n - 2 + 2/2^n = 0$

So the expected gain in a fair game for any finite number of bets is zero using the martingale system, but it is positive if the system can be played to infinity. The increment per round need not be 2, but could be any number, x. The net gain to a winning bet is this number, x.

The intuitive explanation for the zero expectation is that the player (take the simplest case of an increment per round of 2) wins a modest gain (2) with a large probability $(1 - 1/2^n)$ but with a small probability $(1/2^n)$ makes a disastrous loss $(2^{n+1} - 2)$.

More generally, for an increment of x:

Expectation $= (1 - 1/x^n) \cdot x - 1/x^n (x^{n+1} - x)$

$= x - x/x^n - x + x/x^n = 0$

The mathematical paradox remains. In the case where on the nth round, the bet is 2^n, the martingale expectation $= 1/2 \times 2 + 1/4 \times 2^2 + 1/8 \times 2^3 + \ldots = 1 + 1 + 1 + 1 \ldots \infty$.

Yet the actual expectation, when the odds are fair, in all realistic cases $= 0$.

If the odds are tilted against the bettor, so that, for example, the bettor wins less if a fair coin lands Heads than he loses if it lands Tails, the expected gain in a finite series of coin tosses is less than zero, but the same principle applies.

4.8.2 Exercise

Show that whenever the player using the martingale strategy wins, the net gain overall to the player is positive.

Show how this can be reconciled with the fact that the expected value of a martingale strategy in a fair game of Heads/Tails, with a starting stake of x, is zero.

4.8.3 Reading and Links

Baker, R.D. 2002. Probability paradoxes: An improbable journey to the end of the world. *Mathematics Today*, December, 185–189.

Eckhardt, W. 1997. A shooting-room view of doomsday. The Journal of Philosophy, 94(5), 244–259. https://www.jstor.org/stable/pdf/2564582.pdf?casa_token=Vm MZNgUeRK4AAAAA:ml-2kYrmnaiAokDrLIyZKD034E4Oax0wRwcl9_6 FXz3xLcZ3Ty7DmuEK_qPhrvSE6rUHffAuodvl7BNzn9AFm-hMme-tP-POXK5 rcQunj_FWR-RioY3x

Leslie, J. 1998. The End of the World: The Science and Ethics of Human Extinction. London: Routledge.

Can You Solve this Gambling Paradox? Looking Glass Universe. 8 January 2012.. YouTube. https://youtu.be/t8L9GCophac

Why the Martingale Betting System Doesn't Work. Looking Glass Universe. 15 January 2017. YouTube. https://youtu.be/Ry3B9hJbBfk

4.9 How Much Should We Bet When We Have the Edge?

How much should we bet when we believe the odds are in our favour. The answer to this question was first formalised in 1956 by the daredevil pilot, recreational gunslinger, and physicist John L. Kelly, Jr. at Bell Labs. The so-called Kelly criterion is a formula employed to determine the optimal size of a series of bets when we have the advantage, in other words when the odds favour us. It takes account of the size of our edge as well as the adverse impact of volatility. In other words, even when we have the advantage, we can still go bankrupt along the way. This can happen by staking too much on any individual bet or series of bets.

Essentially, then, the Kelly strategy is to wager a proportion of our capital, equivalent to our advantage at the available odds. So if we are offered even money, and we back heads, and we are certain that the coin will come down heads, we have a 100% advantage. So the recommended wager is the total of our capital. If there is a 60% chance of heads and a 40% chance of tails, our advantage is now 20%, and we are advised to stake accordingly. This is a simplified representation of the literature on the Kelly criterion and its derivatives, but the bottom line is clear. As well as being able to gauge when we have the advantage, it is important to know how much to stake when we do. But it's not easy unless we can accurately identify that advantage.

Put more technically; the Kelly criterion is the fraction of capital to wager to maximise compounded growth of capital. The system is not the only proportional betting system. There are numerous potential proportional systems, such as always betting 5% of your bankroll, or 25%. What is special about the Kelly system is that it grows wealth faster than any other.

What Kelly seeks to address is that even when there is an edge, beyond some threshold larger bets will result in lower compounded return because of the adverse impact of volatility. The Kelly criterion defines the threshold and indicates the fraction that should be wagered to maximise compounded return over the long run (F), which is given by:

$$F = Pw - (Pl / W)$$

where
   F = Kelly criterion fraction of capital to bet
   W = Amount won per amount wagered (i.e. win size divided by loss size)
   Pw = Probability of winning
   Pl = Probability of losing

When win size and loss size are equal, W = 1, and the formula reduces to:

$$F = Pw - Pl$$

For example, if Jane loses £1,000 on losing trades and gains £1,000 on winning trades, and 60% of all trades are winning trades, the Kelly criterion indicates an optimal trade size equal to 20% (0.60 – 0.40 = 0.20). As another example, if Mike wins £2,000 on winning trades and loses £1,000 on losing trades, and the probability of winning and losing are both equal to 50%, the Kelly criterion indicates an optimal trade size equal to 25% of capital: 0.50 – (0.50/2) = 0.25.

Proportional over-betting is more harmful than under-betting. For example, betting half the Kelly will reduce compounded return compared to a full Kelly strategy by 25%, while betting double the Kelly will eliminate 100% of the gain. It can perform well in the case of lucky streaks, but all the gain will eventually be eliminated. Betting more than double the Kelly will result in an expected negative compounded return, regardless of the edge on any individual bet. The Kelly criterion implicitly assumes that there is no minimum bet size. This assumption prevents the possibility of total loss. If there is a minimum trade size, as is the case in most practical investment and trading situations, then ruin is possible if the amount falls below the minimum possible bet size.

So should we bet the full amount recommended by the Kelly criterion? In fact, betting the full amount recommended by the Kelly formula may be unwise. Notably, an accurate estimation of the advantage of the bets is critical; if we over-estimate the advantage by more than a factor of two, the Kelly strategy will cause a negative rate of capital growth, and this is easily done. So, a full Kelly betting strategy may be a "rough ride" (Benter, 2008), and a fractional Kelly betting strategy might be substituted. A fractional strategy is where we bet some fraction of the recommended Kelly bet, such as a half or a third.

Essentially, the Kelly system is driven by the "law of large numbers", the idea that a positive edge on the spin of a wheel, say, will turn ever more surely into an actual gain the more times the wheel is spun. It is a way, therefore, of managing a bankroll such that the bettor stays in the game long enough to take advantage of the law of large numbers.

Ironically, John Kelly himself died in 1965, never having used his own criterion to make money.

So that's the Kelly criterion. In a nutshell, the advice is only to bet when you believe you have the edge, and to do so using a stake size related to the size of the edge. Mathematically, it can mean betting a fraction of your capital equal to the size of your advantage. So, if you have a 20% edge at the odds, bet 20% of your capital. In the real world, however, we need to allow for errors that can creep in, like uncertainty as to the size of any actual edge that we may have at the odds. So, unless we're happy to risk a very bumpy ride, and we have total confidence in our judgment, a preferred strategy may be to stake a defined fraction of that amount, known as a fractional Kelly strategy.

4.9.1 Exercise

1. If Steve is offered even money on a heads/tails bet and knows that the chance of heads is 70%, the Kelly criterion indicates an optimal trade size equal to x% of capital. Calculate x.

2. If Tom wins £1,000 on winning trades and loses £1,000 on losing trades, and 60% of all trades are winning trades, the Kelly criterion indicates an optimal trade size equal to ... % of capital?

3. If Ceinwen wins £2,000 on winning trades and loses £1,000 on losing trades, and the probability of winning and of losing both equal 50%, the optimal trade size, according to the Kelly criterion, equals to ... % of capital?

4. If Nora wins £2,000 on winning trades and loses £1,000 on losing trades, and 60% of all trades are winning trades, the optimal trade size, according to the Kelly criterion, equals to ... % of capital?

5. In what circumstance would the Kelly criterion indicate an optimal trade size of 100% of capital?

4.9.2 Reading and Links

Kelly, J.L. 1956. A new interpretation of information rate. *Bell System Technical Journal*, 35(4), 917–926. https://www.princeton.edu/~wbialek/rome/refs/kelly_56.pdf

LessWrong. The Kelly Criterion. 15 October 2018. https://www.lesswrong.com/posts/BZ6XaCwN4QGgH9CxF/the-kelly-criterion

Poundstone, W. 2005. Fortune's Formula. The Untold Story of the Scientific System that Beat the Casinos and Wall Street. New York: Hill and Wang.

Quant Channel. 2015. Introduction to the Kelly Criterion. 5 October 2015. YouTube. https://youtu.be/d4yzXbdq2DA

Understanding Kelly Criterion Sbr Justin. 2 August 2008. YouTube. https://youtu.be/IyATmCJf4fc

Benter, W. 2008. Computer based horse race handicapping and wagering systems: A report. In *Efficiency of Racetrack Betting Markets*, Eds. Hausch, D.B., Lo, V.S.Y. and Ziemba, W.T. Chapter 19, pp. 183–198, London: World Scientific Publishing Co. Pt. Ltd. https://www.gwern.net/docs/statistics/decision/1994-benter.pdf

4.10 The Expected Value Paradox

To illustrate the Expected Value Paradox, let us propose a coin-tossing game, in which you gain 50% of what you bet if the coin lands Heads and lose 40% if it lands Tails. What is the Expected Value of a single play of this game?

The Expected Value can be calculated as the sum of the probabilities of each possible outcome in the game times the return if that outcome occurs.

Say, for example, the unit stake for each play of the game is £10. In this case, the gain if the coin lands Heads is 50% × £10 = £5, and the loss if the coin lands Tails is 40% × £10 = £4.

In this case, the Expected Value (given a fair coin, with 0.5 chance of Heads and 0.5 chance of Tails) = 0.5 × £5 − 0.5 × £4 = £0.5, or 50 pence.

So the Expected Value of the game is 5%. This is the positive net expectation for each play of the game (toss of the coin).

Let's see how this plays out in an actual experiment in which 100 people play the game. What do we expect would be the average final balance of the players?

The expected gain from the 50 players tossing Heads = $50 \times £5 = £250$.

The expected loss from the 50 players tossing Tails = $50 \times £4 = £200$.

So, the net gain of over 100 players = $£250 - £200 = £50$.

The average net gain of the 100 players = $£50/100 = £0.5$, or 50 pence.

Expected Value = $0.5 \times £1.5 + 0.5 \times 60p. = £1.05$. As above, this is an expected gain of 5%.

Regardless of how many coin tosses the group throws, the Expected Value is positive.

Take now the case of one person playing the game through time. Say there are four coin tosses, each for a stake of £10.

From four coin tosses, our best estimate is two Heads and two Tails.

Expected value for two Heads and two Tails = $£10 \times 1.5 \times 1.5 \times 0.6 \times 0.6$.

Expected value goes from £10 to £15 to £22.50 to £13.50 to £8.10. This is a net loss.

If we throw the same number of Heads and Tails after tossing the coin N times, we would expect more generally to earn the following.

$1.5^{N/2} \times 0.6^{N/2} \times$ unit stake = $(1.5 \times 0.6)^{N/2} \times$ unit stake = $0.9^{N/2} \times$ unit stake

Eventually, all the stack used for betting is lost.

Herein lies the paradox. When many people play the game a fixed number of times, the average return is positive, but when a fixed number of people play the game many times, they should expect to lose money.

This is a demonstration of the difference between what is termed "time averaging" and "ensemble averaging".

Thinking of the game as a random process, time averaging is taking the average value as the process continues. Ensemble averaging is taking the average value of many processes running for some fixed amount of time.

Let's say that we wish to determine the most visited parts of a city. In this case, we could take a snapshot in time of how many people are in neighbourhood A, how many in neighbourhood B, etc. Alternatively, we could follow a particular individual or a few individuals over a period of time and see how often they visit neighbourhood A, neighbourhood B, etc. The first analysis (the ensemble) may not be representative over a period of time, while the second (time) may not be representative of all the people.

An ergodic process is one in which the two types of statistics give the same results. In an ergodic system, time is irrelevant and has no direction. Say, for example, that 100 people rolled a die once, and the total of the scores is divided by 100. This is finite-time, ensemble averaging. Now, take the case of a single person rolling a die 100 times, and the total scored is divided by 100. This is time averaging.

An implication of ergodicity is that the result of ensemble averaging will be the same as time averaging.

And here is the key point: In the case of ensemble averages, it is the size of the sample that eventually removes the randomness from the sample. In the case of time averages, it is the time devoted to the process that removes randomness.

In the dice rolling example, both methods give the same answer, subject to errors. In this sense, rolling dice is an ergodic system.

However, if we now bet on the results of the dice rolling game, wealth does not follow an ergodic system. If a player goes bankrupt, he or she stays bankrupt, so the time average of wealth can approach zero over time as time passes, even though the ensemble value of wealth may increase.

As a new example take the case of 100 people visiting a casino, with a certain amount of money. Some may win, some may lose, but we can infer the house edge by counting the average percentage loss of the 100 people. This is the ensemble average. This is different from one person going to the casino 100 days in a row, starting with a set amount. The probabilities of success derived from a collection of people do not apply to one person. The first is the "ensemble probability", and the second is the "time probability" (the second is concerned with a single person through time).

Here is the key point: no individual person has sure access to the returns of the game or market without infinite pockets and an absence of the so-called "uncle points" (the point at which they need, or feel the need, to exit). To equate the two is to confuse ensemble averaging with time averaging.

If the player/investor has to reduce exposure because of losses or maybe retirement or other change of circumstances, that player/investor's returns will be divorced from those of the market or the game. The essential point is that success first requires survival. This applies to an individual in a different sense to the ensemble.

So where does the money lost by the non-survivors go? It gets transferred to the survivors, some of whom tend to scoop up much or most of the pool, i.e. the money is scooped up by the tail probability of those who keep surviving, which may just be by blind good luck, just as the non-survivors may have been forced out of the game/market by blind bad luck.

The so-called Kelly approach to investment strategy, discussed in a separate chapter, is an investment approach which seeks to respond to the survivor issue.

Say, for example, that the probability of Heads from a coin toss is 0.6, and Heads wins a dollar, but Tails (with a probability of 0.4) loses a dollar. Although the Expected Value of this game is positive, if the response of an investor in the game is to stake all their bankroll on Heads on each toss of the coin, the expected time until bankroll bankruptcy is just $1 / (1 - 0.6) = 2.5$ tosses of the coin.

The Kelly strategy to optimise the growth rate of the bankroll is to invest a fraction of the bankroll equal to the difference in the likelihood of winning and losing.

In the above example, it means we should in each game bet a fraction of x equal to 0.6 – 0.4, i.e. 0.2 of the bankroll.

If we bet all our bankroll on each coin toss, we will most likely soon lose the bankroll. This is balanced out over all players by those who with low probability win a large bankroll.

In trying to maximise Expected Value, the probability of bankroll bankruptcy soon gets close to one. It is better to invest, say, 20% of bankroll in each game and maximise long-term average bankroll growth.

Put another way, because the individual cannot go back in time and the bankruptcy option is always actual, it is not possible to realise the small chance of making the tail-end upside of the positive expectation value of a game/investment without taking on the significant risk of non-survival/bankruptcy. In other words, the individual lives in one universe, on one time path, and so is faced with the reality of time averaging as opposed to an ensemble average in which one can call upon the gains of parallel investors/game players on parallel timelines in essentially parallel worlds.

To summarise, the difference between 100 people going to a casino and one person going to the casino 100 times is the difference between understanding probability in conventional terms and through the lens of path dependency.

4.10.1 Exercise

In the case of ensemble averages, what eventually removes the randomness from the sample? In the case of time averages, what removes randomness?

4.10.2 Reading and Links

Syl, L.P. 2016. What is ergodicity? 23 November. https://larspsyll.wordpress.com/201 6/11/23/what-is-ergodicity-2/

Syl, L.P. 2012. Non-ergodic economics, expected utility and the Kelly criterion. 21 April. https://larspsyll.wordpress.com/2012/04/21/non-ergodic-economics-expected-utility-and-the-kelly-criterion/

Taleb, N. 2017. Medium: The logic of risk taking. 25 August. http://nassimtaleb.org/tag/ergodicity/

Time and Chance. TEDx Talks. Peters, O. 28 April 2011.. YouTube. https://www.youtube.com/watch?v=LGqOH3sYmQA

Time for a Change: Introducing Irreversible Time in Economics. Peters, O. 18 December. YouTube. https://www.youtube.com/watch?v=f1vXAHGIpfc

4.11 Options, Spreads, and Wagers

Thales of Miletus, the sixth-century BC Greek philosopher, is perhaps best known for his idea that scientific method can be used to reach general conclusions about the universe. It is for his option trading, however, that Thales should also be celebrated, as it is the first use of a derivative transaction in recorded history. Aristotle tells the story in Book 1 of "The Politics".

According to Aristotle's account, Thales puts a deposit during the winter on all the olive presses in Chios and Miletus, which would allow him exclusive use of the presses after the harvest. Because the harvest was in the future, and nobody could be sure whether the harvest would be plentiful or not, he was able to secure the contracts for a very low price. In fact, we are informed that there was not one bid against him. From the olive press owners' point of view, they were protecting themselves against a poor harvest by earning at least some money up front regardless of how things turned out.

The harvest was excellent and there was heavy demand for the presses. Thales held the monopoly and was able to rent them out at a huge profit. Either he was an expert weather forecaster or he had calculated that a bad harvest would not lose him much in terms of lost deposits, whereas the upside of a good harvest was enormous. "He made a lot of money, and so demonstrated that it is easy for philosophers to become rich, if they want to; but that is not their object in life" (p.90), wrote Aristotle.

In effect, Thales had exercised an options contract, more than 2,500 years ago. Today we would term it a "call option", i.e. an option to buy something at some designated price at some future date for a fixed fee (or "premium"). Stated another way, it is an agreement that gives the purchaser the right (but not the obligation) to buy a commodity, stock, bond, or other instrument at a specified price (the "strike price") at the end of or within a specified time period. When the price exceeds the strike price, the option is said to be "in the money".

Properly used, options can be an excellent vehicle for risk management. In the example of Thales and the olive presses, the owners of the presses were ensuring that they didn't lose their entire earnings in the event of a bad harvest. From Thales' point of view, he was taking some risk that he'd lose all the deposits he'd paid, but was confident that the expected return outweighed this risk. Today we'd say that Thales was risking not being able to exercise his call options.

Modern options can be traced back to the "tulip bulb mania" in seventeenth-century Holland. Tulip dealers bought options known as "Calls" when they wanted the assurance that they could increase their inventories when prices were rising. These options gave the dealer the right, but not the obligation, to call on the other side to deliver tulips at a pre-arranged price. Growers seeking protection against falling prices would buy options known as "Puts" that gave them the right to put, or sell, to the other side at a pre-arranged price. The other side of these options – the sellers – assumed these risks in return for premiums paid by the buyers of the options, designed to compensate sellers of calls for taking the risk that prices would rise and to compensate sellers of puts for taking the risk that prices would fall.

In simple terms, there are, therefore, two types of option – call and put options. They offer you the right but not the obligation to buy or sell a given market at a given price, known as the "strike" price, on or before a fixed date in the future. That option is itself traded. Buying a call option gives you the right to buy at a certain level if you think the market will go up. Buying a put option gives you the right to sell at a certain level, so you want the market to go down. You can both buy and sell call options and put options. If you buy an option, the most you can lose is the amount you paid for the option, and your gain can be substantial. However, you don't have the same protection if you choose to sell an option. Selling an option highlights a potentially risky area of trading. Your losses can be very substantial, whereas the most you can gain is the premium you sell the option for.

The advantage of buying options over exposure to the full market swings is that your downside is limited in advance.

If you expose yourself to the market without an option and buy the FTSE 100 at 6100, say, and it falls to 5000, you have lost 1100 times your unit stake. Similarly, if you sell at 6100 and it rises to 6500, you have lost 400 times your unit stake. Options are a way of limiting that downside.

Buying a Call Option

When buying a Call option, you hope that the market will rise. To take an example, in May, with the FTSE 100 standing at 6100, you are offered a quote for a June Call option with a strike price of 6250 at 40-48. This means you can buy the option for a premium of 48 (the offer, or ask, price) or sell the option for a premium of 40 (the bid price). So, this gives you the right to buy the index at 6250, for a premium of 48, which you accept for, say, £5 a point.

The index now rises to 6395. You exercise your right to buy, for £5 a point. The difference between the market price and the strike price = 6395 − 6250 = 145. However, the premium is 48. So profit = 145 − 48 = 97 × £5 = £485.

In contrast, take the example where you buy a 6250 Call at 48 for a stake of £5 per point. The index then falls sharply. You do not exercise your option to buy, since your Call is worthless. Irrespective of how far the market drops, however, your loss once the index falls below 6250 is your stake multiplied by the premium (in this case 48):

So, your maximum loss = 48 × £5 = £240.

What determines the price of a call option?

a. **The period of time until the option is due to expire.**

When the time to expiration is long, the option will be worth more than when the time is short. The longer the time to maturity, the longer the period in which the call can potentially move into-the-money and be exercised at a profit by the holder. A call option is known as "in-the-money" if the asset price exceeds the strike price. If the asset price equals the strike price, the option is known as "at-the-money".

If the asset price is less than the strike price, the option is known as "out-of-the-money".

b. **The spread between the current price of the stock and the price specified in the option contract at which the owner can buy or sell the stock, i.e. the strike price.**

Clearly, an increase in the strike (exercise) price will reduce the payoff from a call and hence its price. Another way of looking at this is that the higher the strike price the more difficult it will be for the call to be in-the-money.

c. **The interest the buyer can earn while waiting to exercise the option.**

An increase in interest rates will increase call premiums. Why? Because it is cheaper to buy a call option than to own the stock, and so it is possible to earn more by investing the difference as rates increase.

d. **The expected volatility of the underlying asset.**

What matters is how far the stock price might move, not the direction in which it moves. This is because of the asymmetric nature of the option itself, i.e. the investor's potential loss is limited to the premium, while the potential profit is unlimited.

Buying a Put Option

Now let's turn to Put options. Take an example, in May again, with the FTSE 100 standing at 7050, and you are offered a quote for a June Put option with a strike price of 6900 at 84-94. This gives you the right to sell the index at 6900 on or before the expiry date, having paid a premium of 94, which is the offer price. Say you buy the option for £5 a point.

The index now falls to 6710. Since your Put strike price was 6900, you exercise your option. The difference between the strike price and the market price = 6900 − 6710 = 190. However, the premium is 94.

So your profit = 190 − 94 = 96 × £5 = £480.

In an alternative scenario, you buy a 6900 Put with a stake of £5 per point. The index then rises. You do not exercise your option to sell, since your Put is worthless. Irrespective of how far it rises, your loss once the index falls below 6900 is limited to your stake multiplied by the option price (in this case 94).

Loss = 94 × £5 = £470.

To illustrate the difference between using options and directly buying and selling at the bid and offer prices in the market, let's look at a mythical World Cup football market.

World Cup "Total Goals" Market

Let's take the example of "Total Goals in the World Cup". Let's say the underlying market price is 166-170, i.e. this is the spread on offer.

So if you believe that the number of goals will be higher than the spread, you may buy at 170. This is an example of what is known as "spread betting" or "index betting".

If goals finish at 200, you make 30 (200-170) times your stake. If goals finish at 140, you lose 30 times your stake (140-170).

If you don't want to risk losing 30 times your stake, let's say you are offered the chance to *Buy* a call option instead. Let's consider a call option with a level (or Strike at 170.

If you *Buy* this, you now have the right or choice or option (but not the obligation) to *Buy* goals at a level of 170 at the end of the tournament. This "right" is obviously worth something to the buyer as there is a fair chance that goals will be higher than 170. Now let's assume that the cost (or premium) to buy this option is six goals. In this case, you pay six times your stake upfront. You now own this option with complete control of any decisions that are made, and once this is done, the *most* you can ever lose on this bet is six goals times your stake per goal. Then at the end of the tournament, you have a decision to make.

Do you or do you not want to *Buy* goals at 170?

Let's look at some possible outcomes. If goals finished at 200, you will buy at 170, for a gross profit of 30 times your stake (200-170). From this is deducted the premium (in this example, six times your stake) for a *net profit* of 24 times your stake. This is not as profitable as buying the actual market at 170 would have been, but your maximum risk throughout the bet has been limited to the premium paid. Now, what happens if goals at the end of tournament finished up at 120?

If you'd bought goals at 170, you'd be looking at a loss of 120-170 = 50 times your stake.

But what would happen if you'd bought the *call option* instead at 170? In this case, you'd have the right to exercise the option to *Buy* at 170. In practice, though, you wouldn't exercise the option, so your loss is limited to the premium of six times your stake.

So far we've been talking about Call options. But very similar principles apply to *Put Options*, the big difference being that you *Buy* a *Put Option* if you think the market is going to end up down, not up.

Normally, if you think goals are going to be lower you would *Sell* at 166. However, if goals go up to 200, you'd lose 34 times your stake. Instead you could *Buy* a *Put Option* at 160 for, say, four goals, paid upfront. As you are buying the Put option, your maximum loss is what you paid = four times your stake. If goals go to 200, you obviously decide *not* to sell goals and lose the premium of four times your stake. If goals scored turn out to be 120, you exercise the option to win a gross profit of 40 (160 − 120). Your *net profit* is your gross profit minus the upfront premium of 4, i.e. 40 − 4 = 36 times your stake.

Selling an Option

Buying an option limits your potential loss to the upfront premium. In contrast, selling an option can potentially lead to significant losses. It is, therefore, a high-risk way of trading. Think back to the previous example, where you bought the 170 call for six goals. Goals go to 200 and you make 200 − 170 − 6 = 24 times your stake. The seller of the call option, of course, makes an equivalent size loss. The seller hoped that goals would be less than 170 and that they would keep the six you paid them for a six goal profit. But that is the most the seller of the option could make. Imagine a scenario in which goals rose to 300. In this case, the seller of the 170 call option would be looking at a net loss of 124 (130 minus the premium of 6), while their maximum profit could only ever have been 6.

Summarising options

If you BUY an option, you decide what you want to do, i.e. whether to exercise the option.

If you BUY a CALL OPTION you want the market to go up.

If you BUY a PUT OPTION you want the market to go down.

If you BUY an option, the most you can lose is the amount you pay for that option, and your gain could be very substantial.

If you SELL an option your losses could be very substantial, and the most you can gain is the premium you sell the option for.

4.11.1 Appendix

Arbs and Quarbs

What is an arbitrage (or "arb")?

Say three market-makers offer the following spreads in a World Cup "Total Goals" market.

Market-maker 1: 140-150

Market-maker 2: 140-150

Market-maker 3: 129-139

Traders can *buy* with Market-maker 3 (at the offer price of 139) and *sell* with Market-maker 1 and/or 2 and take a riskless profit of a unit stake.

Arbs are uncommon, however, and fleeting.

Quasi-arbitrages ("Quarbs") are much more common and often persist. These exist when a market-maker is sufficiently out of line with the market average to make the price offered apparent value. They rely on the market "consensus" being a better indicator of the true price than prices offered by an outlier ("maverick").

Quarbs take different forms, but here is an example.

Quarbs:

Say three market-makers offer the following spreads in a World Cup "Total Goals" market.

Market-maker 1: 131-141

Market-maker 2: 140-150

Market-maker 3: 140-150

Mid-point of Market-maker 1 = 136

Mid-point of Market-maker 2 = 145

Mid-point of Market-maker 3 = 145

Market mid-point = (136 + 145 + 145) / 3 = 142

Market mid-point (142) lies outside the spread offered by Market-maker 1. We term this a quasi-arbitrage or "Quarb". Quarb prices can, in principle, also appear in fixed-odds markets in cases where an odds-setter is out of line with the market average.

It is an empirical issue whether, by taking advantage of the "maverick" market-maker, it is possible to earn a significant and consistent profit over time.

4.11.2 Exercise

A. Buy Call Option

World Cup "Total Goals" market.

Based on a market of 166-170.

175 CALL OPTION

Bid price = 3.0

Offer price = 5.5

This means you can BUY the 175 Call option at 5.5 or SELL the 175 Call option at 3.0.

You BUY the 175 CALL OPTION for £10 per goal.

1. Outcome A

   Total Goals = 200. What is your net profit?

2. Outcome B

   Total Goals = 140. What is your net profit?

3. Outcome C

   Total Goals = 180. What is your net profit?

B. Buy Put Option

Based on a market of 166-170.

175 PUT OPTION

Bid price = 8.5

Offer price = 11.5

This means you can BUY the 175 Put option at 11.5 or SELL the 175 Call option at 8.5.

You BUY the 175 PUT OPTION for £10 per goal.

1. Outcome A

   Total Goals = 200. What is your net profit/loss?

2. Outcome B

   Total Goals = 140. What is your net profit/loss?

3. Outcome C

   Total Goals = 170. What is you net profit/loss?

C. Sell Call Option

Based on a market of 166-170.

175 CALL OPTION

Bid price = 3.0

Offer price = 5.5

This means you can BUY the 175 Call option at 5.5 or SELL the 175 Call option at 3.0.

You SELL the 175 CALL OPTION for £10 per goal.

1. Outcome A

   Total Goals = 200. What is your net profit/loss?

2. Outcome B

Total Goals = 140. What is your net profit/loss?

3. Outcome C

Total Goals = 180. What is your net profit/loss?

D. Sell Put Option

Based on a market of 166-170.

175 PUT OPTION

Bid price = 8.5

Offer price = 11.5

This means you can BUY the 175 Put option at 11.5 or SELL the 175 Call option at 8.5.

You SELL the 175 PUT OPTION for £10 per goal.

1. Outcome A

Total Goals = 200. What is your net profit/loss?

2. Outcome B

Total Goals = 140. What is your net profit/loss?

3. Outcome C

Total Goals = 170. What is your net profit/loss?

Summary Puzzle

World Cup – Total Goals
Based on a market of 166-170.

Strike	Option	Price Bid	Offer	Option	Price Bid	Offer
165	Call	6.5	9.5	Put	3.0	6.0

Stake = £10 per goal.

1. Assuming the market makes up at 190:

   What is your net profit if you BUY a CALL OPTION?

   What is your net profit if you BUY a PUT OPTION?

   What is your net profit if you SELL a CALL OPTION?

   What is your net profit if you SELL a PUT OPTION?

2. Repeat if the market makes up at 130.

4.11.3 Reading and Links

Aristotle. *The Politics.* Book 1. Translated by T. A. Sinclair. Revised and Re-presented by T.J. Saunders. 1981. London: Penguin Books Ltd.

Haigh, J. and Vaughan Williams, L. 2008. Index betting for sports and stock indices. In *Handbook of Sports and Lottery Markets, and* Ed. D. Hausch and W. Ziemba, pp. 357–383. North Holland.

Paton, D. and Vaughan Williams, L. 2005. Forecasting outcomes in spread betting markets: Can bettors use 'quarbs' to beat the book? *Journal of Forecasting,* 24(2), 139–154. https://onlinelibrary.wiley.com/doi/pdf/10.1002/for.949?casa_toke n=RQOGbeXjg-sAAAAA:9WmWDWn0ewsdpCdWyeR7CUgt4oNck-4AoU RwV3AUBn2ekLu6tcSsov4K0t59IUag_hsLXKS2u2v1XpNI

Smith, M.A., Paton, D. and Vaughan Williams, L. 2005. An assessment of quasi-arbitrage opportunities in two fixed-odds horse-race betting markets. In *Information Efficiency in Financial and Betting Markets*, Ed. Vaughan Williams, L. Chapter 4, Cambridge: Cambridge University Press.

5

Probability, Truth, and Reason

In this chapter we explore some mind-bending problems, paradoxes, and ideas drawn from the worlds of philosophy and science. We consider Hempel's paradox (also called the Raven paradox), which is the counter-intuitive proposition that observing something which is neither black nor a raven (such as a red apple) increases the likelihood that all ravens are black! From there we jump to considering whether anything really exists, or whether this is all a simulation. We delve into the quantum world, where we consider the competing claims of the "Copenhagen Interpretation" and the "many-worlds" interpretation (MWI) of quantum mechanics. More basically, we ask how we are able to consider these matters at all when existence itself seems to be vanishingly improbable. Along the way, we take a speculative trip into the idea of a multiverse. Finally, we learn about Occam's Razor, the idea that the simplest explanation of the evidence is the most likely.

5.1 Does Seeing a Blue Tennis Shoe Increase the Likelihood That All Flamingos Are Pink?

You spot a pink flamingo and wonder to yourself whether all flamingos are pink. What would it take to confirm or disprove the hypothesis? The nice thing about this sort of hypothesis is that it's testable and potentially falsifiable. All it takes is to find a flamingo that is not pink, and I can conclude that not all flamingos are pink. Just one observation can change my flamingo worldview. It doesn't matter how many pink flamingos I witness, there is no number that can prove the hypothesis short of the number of flamingos that potentially exist. Still, the more you see that are pink, the more probable it becomes that all flamingos are actually pink. How probable you consider that is at any given time is related to how probable you thought it was before you saw the latest one. While considering this, you see someone wearing blue tennis shoes. Does this make it more likely that all flamingos are pink? This is one example of a broader paradox first formally identified by Carl Gustav Hempel, sometimes known as Hempel's paradox or else the Raven paradox.

The Raven paradox arises from asking whether observing a red apple increases the likelihood that all ravens are black, assuming that you don't know the answer. Intuitively, it would seem not. Why should seeing a red apple tell you anything about the colour of ravens? The way to answer this is to re-state "All ravens are black" as "If something is a raven, then it is black", and therefore "If something is not black, then it is not a raven". In fact, these statements are logically equivalent. As an example, assume there are just two ravens and two tennis shoes (one right-foot, one left-foot) in the whole world. Now you identify the colour of each of these objects. You observe that both tennis shoes are blue and the other two objects are black. So you announce that everything that is not black (each of the tennis shoes) is not a raven. This is identical to saying that all ravens are black. The logic universalises to any number of objects and colours.

Assume now we see just one of the tennis shoes and it turns out to be blue. You can now announce that one possible thing that is not black is not a raven. If you see the other tennis shoe and it is blue, that means that there are now two things that are not black that are not ravens. Each time you see something, it is possible that you would not be able to say this, i.e. you would say instead that you have seen something not black and it is a raven. It is like being dealt a playing card from a deck of four which contains only blue or black cards. You are dealt a black card, and it shows a raven. You know that at least one of the other cards is a raven, and it could be a black card or a blue card. You receive a blue card. Now, before you turn it over, what is the chance it is a raven? You don't know, but whatever it is, the chance that only black cards show ravens improves if you turn the blue card over and it shows a tennis shoe. Each time you turn a blue card over it could show a raven. Each time that it doesn't makes it more likely that no blue card shows a raven.

Substitute all non-ravens for tennis shoes and all colours other than black for the blue cards, and the result universalises. Every time you see an object that is not black and is not a raven, it makes it just that tiny bit more likely that everything that is not black is not a raven, i.e. that all ravens are black. How much more likely? This depends on how observable non-black ravens would be if they exist. If there is no chance that they would be seen even if they exist, because non-black ravens never emerge from the nest, say, it is difficult to falsify the proposition that all ravens are black. So when you observe a blue tennis shoe it offers less evidence for the "all ravens are black" hypothesis than when it is just possible that the blue thing you saw would have been a raven and not a tennis shoe. More generally, the more likely a non-black raven is to be observed if it exists, the more evidence that the observation of a non-black object that is not a raven offers for the hypothesis that all ravens are black.

So how can we test the hypothesis that all ravens are black? We could go out, find some ravens, and see if they are black. On the other hand, we could simply take the logically equivalent contrapositive of the hypothesis, i.e. that all non-black things are non-ravens. This suggests that we can conduct meaningful research on the colour of ravens from our home or office without observing a single raven, but by simply looking at random objects, noting that they are not black, and checking if they are ravens. As we proceed, we collect data that increasingly lend support to the hypothesis that all non-black things are non-ravens, i.e. that all ravens are black.

Is there a problem with this approach?

There is no logical flaw in the approach, but the reality is that there are many more non-black things than there are ravens, so if there exists a pair (raven, non-black), then we would be much more likely to find it by randomly sampling a raven than by sampling a non-black thing. Therefore, if we sample ravens and fail to find a non-black raven, then we're much more confident in the truth of our hypothesis that all ravens are black, simply because the hypothesis had a much higher chance of being falsified by sampling ravens than by sampling random non-black things.

The same goes for pink flamingos.

To summarise, let's take the propositions in the thought experiment in turn. Proposition 1: all flamingos are pink. Proposition 2 (logically equivalent to Proposition 1): everything that is not pink is not a flamingo. Proposition 3 (advanced here as what I term the "Possibility Theorem"): if something might or might not exist, but is unobservable, it is more likely to exist than something which can be observed, with any positive probability, but is not observed. Similarly, if something might or might not exist, and is not observed, it is more likely to exist if it is less observable than something else which is more observable.

Following from these propositions, when I see two blue tennis shoes I am ever so slightly more confident that all flamingos are pink than before, and especially so if any non-pink flamingos that might be out there would be easy to spot. And I'd still be wrong, but for all the right reasons, until I saw an orange flamingo, and then I'd know for sure that not all flamingos are pink.

5.1.1 Exercise

Statement 1: All flamingos are pink.

Statement 2: If something is not pink, then it is not a flamingo".

Are statements 1 and 2 logically distinct or equivalent? Is either statement true?

5.1.2 Reading and Links

Hempel's Ravens Paradox. Platonic Realms. http://platonicrealms.com/encyclopedia/Hempels-Ravens-Paradox

The Raven Paradox – A Hiccup in the Scientific Method. Up and Atom. 10 May 2019. YouTube. https://youtu.be/Ca_sxDTPo60

The Raven Paradox (Cark Hempel and the Paradox of Confirmation). Carneades.org. 13 March 2016. YouTube. https://youtu.be/7_dbh6RbdCM

The Paradox of the Ravens. Wireless Philosophy. 24 July 2015. YouTube. https://youtu.be/_SKmqh5Eu4Y

5.2 The Simulated World Question

Do we live in a simulation created by an advanced civilisation, in which we are part of some sophisticated virtual reality experience? For this to be a possibility we can make the obvious assumption that sufficiently advanced civilisations will possess the requisite computing and programming power to create what philosopher Nick Bostrom termed "ancestor simulations". These simulations would be complex enough for the minds that are simulated to be conscious and to be able to experience the type of experiences that we do. The creators of these simulations could exist at any stage in the development of the universe, even billions of years into the future.

The argument around simulation goes like this. At least one of the following three statements must be true.

1. That civilisations at our level of development always or almost always disappear before becoming technologically advanced enough to create ancestor simulations. Otherwise stated, the human species is very likely to go extinct before reaching a "posthuman" stage.

2. That the proportion of these technologically advanced civilisations that wish to create these simulations is zero or almost zero. Otherwise stated, any "posthuman" civilisation is extremely unlikely to run a significant number of simulations of their evolutionary history (or variations thereof).

3. That we are almost sure to be living in a simulation.

To see this, let's examine each proposition in turn.

For the first proposition to be untrue, civilisations must be able to go through the phase of being able to wipe themselves out, either deliberately or by accident, carelessness or neglect, and never or almost never do so. This might seem unpromising based on our experience of this world but becomes more likely if we consider all other possible worlds.

For the second proposition to be untrue, we would have to assume that virtually all civilisations that were able to create these simulations would decide not to do so. This might seem unlikely.

If we consider both propositions, and we think it is unlikely that no civilisation would survive to create "ancestor simulations", then anyone considering the question is left with a stark conclusion. They really are living in a simulation.

To summarise. An advanced "technologically mature" civilisation would have the capability of creating simulated minds. Based on this, at least one of three propositions must be true.

1. The proportion of civilisations that survive to "technological maturity" is close to zero or zero.

2. The proportion of these advanced civilisations that wish to run these simulations is close to zero or zero.
3. The proportion of those who consciously consider this question who are living in a simulation is close to one.

If the first of these propositions is true, our civilisation will almost certainly not survive to become "technologically mature". If the second proposition is true, virtually no advanced civilisations are interested in using their power to create such simulations. If the third proposition is true, then we are almost certainly living in a simulation.

Through the veil of our ignorance, it might seem reasonable to assign equal credence to all three, and to conclude that, unless we are currently living in a simulation, descendants of this civilisation will almost certainly never be in a position to run these simulations.

What if we are the only civilisation at our stage, however, that there will ever have been? What are the implications of this for the argument? In e-mail correspondence with me, Nick Bostrom answers the question in clear fashion:

> If we are the only civilization at our stage there will ever have been, then the equation remains true, although some of the possible implications become less striking: notably, (1) and (2) would not be so surprising if there is only one civilization. For example, if there is one civilization, and it decides never to run ancestor simulations after reaching technological maturity, then it would be true that "at most a negligible fraction of technologically mature civilizations remain interested in creating significant numbers of ancestor simulations", but this would now boil down to "zero out of one technologically mature civilization remained interested in running ancestor simulations" – which seems a lot less difficult to believe than that; e.g. "only 3 out of 32,995,242,342,834,800 technologically mature civilizations remained interested in running ancestor simulations". So the probability that we are not in a simulation is increased if ours is the only civilization that will have ever existed throughout the multiverse.

(Nick Bostrom e-mail, 10 February 2021)

Strangely indeed, the probability that we are living in a simulation increases as we draw closer to the point at which we are able and willing to do so. At the point that we would be ready to create our own simulations, we would paradoxically be at the very point when we were almost sure that we ourselves were simulations. Only by refraining from doing so could we in a certain sense make it less likely that we were simulated, as it would show that at least one civilisation that was able to create simulations refrained from doing so. Once we took the plunge, we would know that we were almost certainly doing so as simulated beings. And yet there must have been someone or something that created the first simulation. Could that be us, we would be asking ourselves? In our simulated hearts and minds, we would already know the answer!

5.2.1 Exercise

With reference to Bostrom's "simulation" reasoning, generate an estimate as to the probability that we are living in a simulated world. No solution is provided. This exercise is for personal or group consideration.

5.2.2 Reading and Links

Ananthaswamy, A. 2020. Do we Live in a Simulation? Chances Are about 50-50. 13 October. Scientific American. https://www.scientificamerican.com/article/do-we-live-in-a-simulation-chances-are-about-50-50/

Bostrom, N. 2006. Do We Live in a Computer Simulation? *New Scientist*. 00 Month, 8–9. https://www.simulation-argument.com/computer.pdfhe Simulation Argument. https://www.simulation-argument.com/

Bostrom, N. 2003. Are You Living in a Computer Simulation? *Philosophical Quarterly*, 53 (211): 243–255. https://www.simulation-argument.com/simulation.pdf

Colagrossi, M. 2019. 3 superb arguments for why we live in a matrix – and 3 arguments that refute them. 17 January. *Big Think*. https://bigthink.com/arguments-live-matrix--arguments-against?rebelltitem=2#rebelltitem2

Eggleston, B. Review of Bostrom's Simulation Argument. Stanford University. https://web.stanford.edu/class/symbsys205/BostromReview.html

Kipping, D. 2021. Can we know if our universe is a simulation? 27 May. *Polytechnique Insights*. https://www.polytechnique-insights.com/en/columns/science/can-we-know-if-our-universe-is-a-simulation/

Kipping, D. 2020. A Bayesian Approach to the Simulation Argument. *Universe*, 6 (8), 109. https://www.researchgate.net/publication/343401093_A_Bayesian_Approach_to_the_Simulation_Argument

Moskowitz, C. 2016. Are We Living in a Computer Simulation? *Scientific American*. 7 April. https://www.scientificamerican.com/article/are-we-living-in-a-computer-simulation

The Simulation Argument. https://www.simulation-argument.com/

Kurzgesagt – In a nutshell. Is Reality Real? The Simulation Argument. 21 September 2017. YouTube. https://www.youtube.com/watch?v=tlTKTTt47WE

Nick Bostrom – The Simulation Argument. Science, Technology & the Future. 21 February 2013. YouTube. https://youtu.be/nnl6nY8YKHs

Nick Bostrom – Could Our Universe Be a Fake? Closer to Truth. 31 July 2015. YouTube. https://youtu.be/GV1B33rjh5A

5.3 Quantum World Thought Experiments

Is it possible to be both alive and dead at the same time? This is the question central to the famous Schrödinger's Cat thought experiment. In the version posed by Erwin Schrödinger, a cat is placed in an opaque box for an hour with a small piece of radioactive material which has an equal probability of decaying or not in that time period. If some radioactivity is detected by a Geiger counter also placed in the box, a relay releases a hammer which breaks a flask of hydrocyanic acid, killing the cat. If no radioactivity is detected, the cat lives. Before we open the box at the end of the hour, we estimate the chance that the radioactive material will decay and the cat will be dead at 50/50, the same as that it will be alive. Before we open the box, however, is the cat alive (and we don't know it yet), dead (and we don't know it yet), or both alive and dead (until we open the box and find out)?

Common sense would seem to indicate that it is either alive or dead, but we don't know which until we open the box. Traditional quantum theory suggests otherwise. The cat is both alive, with a certain probability, and dead, with a certain probability, until we open the box and find out, when it must become one or the other with a probability of 100%. In quantum terminology, the cat is in a superposition (two states at the same time) of being alive and dead, which only collapses into one state (dead or alive) when the cat is observed. This might seem absurd when applied to a cat. After all, surely it was either alive or dead before we opened the box and found out. It was simply that we didn't know which. That may be true, when applied to cats. But when applied to the quantum world, such common sense goes out the window as a description of reality. For example, photons (the smallest measure of light) can exist simultaneously in both wave and particle states and travel in both clockwise and anti-clockwise directions at the same time. Each state exists in the same moment. As soon as the photon is observed, however, it must settle on one unique state. In other words, the common sense that we can apply to cats cannot be applied to photons or other particles at the quantum level.

So, what is going on? The traditional explanation as to why the same quantum particle can exist in different states simultaneously is known as the Copenhagen Interpretation. First proposed by Niels Bohr in the early twentieth century, the Copenhagen Interpretation states that a quantum particle does not exist in any one state but in all possible states at the same time, with various probabilities. It is only when we observe it that it must in effect choose which of these states it exists as. At the sub-atomic level, then, particles seem to exist in a state of what is called "coherent superposition", in which they can be two things at the same time and only become one when they are forced to do so by the act of being observed. The total of all possible states is known as the "wave function". When the quantum particle is observed, the superposition "collapses" and the object is forced into one of the states that make up its wave function.

The problem with this explanation is that all these different states exist. By observing the object, it might be that it reduces down to one of these states, but what has happened to the others? Where have they disappeared to?

This question lies at the heart of the so-called "Quantum Suicide" thought experiment.

It goes like this. A man (not a cat) sits down in front of a gun which is linked to a machine that measures the spin of a quantum particle (a quark). If it is measured as spinning clockwise, the gun will fire and kill the man. If it is measured as spinning anti-clockwise, it will not fire and the man will survive to undergo the same experiment again.

The question is: Will the man survive? And if the experiment is repeated indefinitely, for how long will he survive? This thought experiment, proposed by Max Tegmark, has been answered in different ways by quantum theorists depending on whether they adhere to the Copenhagen Interpretation. In that interpretation, the gun will go off with a certain probability, depending on which way the quark is spinning. Eventually, by the laws of chance, the man will be killed, probably sooner rather than later. A number of theorists believe something else, however. They see both states (the particle is spinning clockwise and spinning anti-clockwise) as equally real, so there are two real outcomes. In one world, the man dies and in the other he lives. The experiment repeats, and the same split occurs. In one world there will exist a man who survives an indefinite number of rounds. In other worlds, he is dead.

The difference between these alternative approaches is critical. The Copenhagen approach is to propose that the simultaneously existing states (for example, the quark that is spinning both clockwise and anti-clockwise simultaneously) exist in one world and collapse into one of these states when observed. Meanwhile, the other states mysteriously disappear. The other approach is to posit that these simultaneously existing states are real states, and neither magically disappears, but branch off into different realities when observed. What is happening is that in one world the particle is observed spinning clockwise (in the Quantum Suicide thought experiment, the man dies) and in the other world the particle is observed spinning the other way (and the man lives). Crucially, according to this interpretation, both worlds are real. In other words, they are not notional states of one world but alternative realities. This is the so-called "many-worlds" interpretation (MWI).

Where is the burden of proof in trying to determine which interpretation of reality is correct? This depends on whether we take the one world that we can observe as the default position or the wave function of all possible states as represented in the mathematics of the wave function as the reality. Adherents of the "many-worlds" position argue that the default is to go with what is described in the mathematics that underpins quantum theory – that the wave function represents all of reality. According to this argument, the minimal mathematical structure needed to make sense of quantum mechanics is the existence of many worlds which branch off, each of which contains an alternative reality. Moreover, these worlds are real. To say that our world, the one that we are observing, is the only real one, despite all the other possible worlds or measurement outcomes, has been likened to when we believed that the earth was at the centre of the universe. There is no real justification, according to this interpretation, for saying that our branch of all possible states is the only real one and that all other branches are non-existent or are "disappeared worlds". Put another way, the mathematics of quantum mechanics describes these different worlds. Nothing in the mathematics says that this world that we observe is more real than another world. The burden of proof is on those who say it is. The viewpoint of the Copenhagen school is diametrically opposite. They argue that the hard evidence is of the world we are in, and the burden of proof is on those positing other worlds containing other branches of reality.

The default position we choose to adopt here is critical to our world view.

To side with the "many-worlds" interpretation does seem kind of crazy, though, and totally counter-intuitive. In another world, of course, I'm probably saying the exact opposite.

5.3.1 Exercise

Do you favour the Copenhagen or "many-worlds" school of thought? No solution is provided. This exercise is for personal or group consideration.

5.3.2 Reading and Links

Clark, J. Do Parallel Universes Really Exist? HowStuffWorks. https://science.hows tuffworks.com/science-vs-myth/everyday-myths/parallel-universe.htm

Clark, J. How Quantum Suicide Works. HowStuffWorks. https://science.howstuffw orks.com/innovation/science-questions/quantum-suicide.htm

Rationally Speaking Podcast. 2015. Sean Carroll on the "Many Worlds Interpretation Is Probably Correct." 3 May. RS133. http://rationallyspeakingpodcast.org/133-the-many-worlds-interpretation-is-probably-correct-sean-carroll/

Transcript of Rationally Speaking Podcast. 2015. Sean Carroll on the "Many Worlds Interpretation Is Probably Correct." 3 May. RS133. http://rationallyspeaking podcast.org/wp-content/uploads/2020/11/rs133transcript.pdf

BBC Radio 4. 2001. In Our Time. Podcast. 22 February. Quantum Gravity. https://ww w.bbc.co.uk/programmes/p00547c4

Schrödinger's Cat: A Thought Experiment in Quantum Mechanics. Chad Orzel. 14 October 2014. YouTube. https://youtu.be/UjaAxUO6-Uw

The Double Slit Experiment Explained. Jim Al-Khalili. 1 February 2013. YouTube. https://youtu.be/A9tKncAdlHQ

5.4 The Fine-Tuned Universe Puzzle

It shouldn't be possible for us to exist. But we do. That's counter-intuitive. Take, for example, the "cosmological constant". What it represents is a sort of unobserved "energy" in the vacuum of space which possesses density and pressure, which prevents a static universe from collapsing in upon itself. We know how much unobserved energy there is because we know how it affects the universe's expansion. We also know that the positive and negative contributions to the cosmological constant cancel to 120-digit accuracy yet fail to cancel beginning at the 121st digit. In fact, the cosmological constant must be as it is to within one part in roughly 10^{120} (and yet be non-zero), or else the universe would have either dispersed too fast for stars and galaxies to have formed or collapsed upon itself.

How likely is this by chance? Essentially, it is the equivalent to tossing a fair coin and needing to get heads 400 times in a row and achieving it. Put another way, the defining characteristic of this "dark energy", as it has been termed, is negative pressure, which provides a type of "dark push" to balance the pull of gravity. Change the "dark push" by an almost vanishingly small amount and none of the galaxies and stars in the sky, including the sun, would exist. We would not exist. The probability that dark energy should take on its observed value is about the same probability as setting out to toss a fair coin in a pre-announced sequence 400 times in a row. It is like failing on one of the 400 tosses and all the galaxies, stars, and planets disappear, along with us and all other complex potential lifeforms. The universe is simply too "right" for galaxies, stars, planets, and life to exist purely by chance.

Now, that's only one constant that needs to be just right for all these things to exist. There are quite a few, independent of this, which must be equally just right, such as the strength of gravity and of the strong nuclear force relative to electromagnetism and the observed strength of the weak nuclear force. Others include the difference between the masses of the two lightest quarks and the mass of the electron relative to the quark masses, the value of the global cosmic energy density in the very early universe, and the relative amplitude of density fluctuations in the early universe. If any of these constants had been slightly different, stars and galaxies could not have formed.

There is also the symmetry/asymmetry paradox. When symmetry is required of the universe, such as a perfect balance of positive and negative charge, conservation of electric charge is critically ensured. If there were an equal number of protons and antiprotons, of matter and antimatter produced in the Big Bang, however, they would have annihilated each other, leaving a universe empty of its atomic building blocks. Fortuitously for the existence of a live universe, protons outnumbered antiprotons by a factor of just 1 in 1 billion. If the perfect symmetry of the charge and almost vanishingly tiny asymmetry of matter and antimatter were reversed, if protons and antiprotons had not differed in number by that one part in a billion, we would not be here to consider the question.

In summary, then, if the conditions in the Big Bang which started our universe had been even a tiniest of a tiniest of a tiny bit different, with regard to a diverse range of independent physical constants, our galaxies, stars, and planets would not have been able to exist, let alone lead to the existence of living, thinking, feeling things.

So why are they so right?

Let us first tackle those who say that if they hadn't been so right we would not have been able to even ask the question. This sounds a clever point but in fact it is not. For example, it would be absolutely bewildering how I could have survived a fall out of an aeroplane from 39,000 feet onto the tarmac without a parachute, but it would still be a question very much in need of an answer. To say that I couldn't have posed the question if I hadn't survived the fall is no answer at all.

Others propose the argument that since there must be some initial conditions, those conditions which gave rise to the universe and made life within it possible were just as likely to prevail as any others, so there is no puzzle to be explained.

But this is like saying that there are two people, Jack and Jill, who are arguing over whether Jill can control that a fair coin lands heads or tails. Jack challenges Jill to toss the coin 400 times. He says he will be convinced of Jill's amazing skill if she can toss heads followed by tails 200 times in a row, and she proceeds to do so. Jack could now argue that a head was equally likely as a tail on every single toss of the coin, so this sequence of heads and tails was, in retrospect, just as likely as any other outcome. But clearly that would be a very poor explanation of the pattern that just occurred. That particular pattern was clearly not produced by chance. Yet it's the same argument as saying that it is just as likely that the initial conditions were just right to produce the universe and life to exist as that any of the other patterns of billions of initial conditions that would have failed to do so. There may be a reason for the pattern that was produced, but it needs a much more profound explanation than proposing that it was just coincidence.

There are those who might argue that the existence of the universe is not equivalent to the Jack and Jill challenge because there was no pre-defined outcome, so any sequence of the 400 flips is as likely as any other unless we pre-define in the challenge that there should be, say, 400 heads in a row. That is technically so, but if we observe a sequence of 400 heads in a row, one can reasonably conclude that there is something unusual about the coin or suspect about the coin tossing process. Just one tail in 400 tosses and bang goes the universe and all that's in it.

A second example. There is one lottery draw, devised by an alien civilisation. The lottery balls, numbered from 1 to 59, are to be drawn, and the only way that the human race will escape destruction is if the first 59 balls out of the drum emerge as 1–59 in sequence. The numbers duly come out in that exact sequence. Now that outcome is no less likely than any other particular sequence, so if it came out that way, a sceptic could claim that we were just lucky. That would clearly be nonsensical. A much more reasonable and sensible conclusion is that the aliens had rigged the draw to allow us to survive!

As a final example, take the case of the blindfolded man standing before a trained firing squad, all firing live bullets, yet every bullet manages to miss, and he is reprieved. Rather than wondering whether the astonishing turn of events was deliberately planned, he dismisses his luck as wholly unremarkable. After all, he reasons, he would not be there to ponder his good fortune if they had in fact not all missed. That argument clearly doesn't make much sense. It's hugely more likely that they all missed on purpose. As Brierley (2017, p. 42) puts it, "We can't wave off the fine tuning as inconsequential simply because we are here to observe it – our existence as observers is the very thing that needs explaining."

So the fact that the initial conditions are so fine-tuned deserves an explanation, and a very good one at that. It cannot be simply dismissed as a coincidence or a non-question.

An explanation that has been proposed that does deserve serious scrutiny is that there have been many Big Bangs, with many different initial conditions. Assuming that there were billions upon billions of these, eventually one would produce initial conditions that are right for the universe to have a shot at existing.

In this theory, we are essentially proposing a process statistically along the lines of aliens drawing lottery balls over and over again, countless times, until the numbers come out in the sequence 1–59.

On this basis, a viable universe could arise out of re-generating the initial conditions at the Big Bang until one of the lottery numbers eventually comes up. Is this a simpler explanation of why our universe and life exists than an explanation based on a primal cause? And in any case, does simplicity in this context matter as a criterion of truth?

Of course, the simplest state would be a situation in which nothing had ever existed. This would also be the least arbitrary and certainly the easiest to understand. Indeed, if nothing had ever existed, there would have been nothing to be explained. Most critically, it would solve the mystery of how things could exist without their existence having some cause. This is not helpful to us, though, as we know that at least one universe does exist.

Take the opposite extreme, where every possible universe exists, underpinned by every possible set of initial conditions. In such a situation, most of these might be subject to different fundamental laws, governed by different equations, and composed of different elemental matter. There is no reason in principle, in this version of reality, to believe that each different type of universe should not exist over and over again, up to an infinite number of times, so even our own type of universe could exist billions of billions of times, or more, so that in the limit everything that could happen has happened and will happen, over and over again.

Even so, our sole source of understanding about the make-up of a universe is a study of our own universe. On what basis do we scientifically propose that the other speculative universes are governed by totally different equations and fundamental physical laws?

Perhaps the laws are the same, but the constants that determine the relative masses of the elementary particles, the relative strength of the physical forces, and many other fundamentals differ. If so, what is the law governing how these constants vary from universe to universe, and where do these fundamental laws come from? From nothing? It has been argued that absolutely no evidence exists that any other universe exists but our own, and that the reason for proposing it is simply to explain the otherwise baffling problem of how our universe and life within it can exist. That may well be so, but we can park that for now as it is at least possible that they do exist.

So let's step away from requiring any evidence and move on to at least considering the possibility that there are a lot of universes, but not every conceivable universe. One version of this is that the other universes have the same fundamental laws, subject to the same fundamental equations, and composed of the same elemental matter as ours, but differ in the initial conditions and the constants. But this leaves us with the question as to why there should be only just so many universes and no more. A hundred, a thousand, a hundred thousand, whatever number we may choose requires an explanation of why just that number. This is again very puzzling. If we didn't know better, our best "a priori" guess is that there would be no universes, no life. We happen to know that's wrong, so that leaves our universe; or else a limitless number of universes where anything that could happen has or will, over and over again; or else a limited number of universes, which begs the question, why just that number?

Is it because certain special features have to obtain in the initial conditions before a universe can be born and that these are limited in number? Let us assume for the sake of argument that this is so. This only begs the question, however, of why these limited features cannot occur more than a limited number of times. If they could, there is no reason to believe the number of universes containing these special features would be less than limitless in number. So, in this view, our universe exists because it contains the special features which allow a universe to exist. If so, we are back with the problem arising in the conception of all possible worlds, but in this case it is only our own type of universe (i.e. obeying the equations and laws that underpin this universe) that could exist limitless times.

The alternative is to adopt an assumption that there is some limiting parameter to the whole process of creating universes, such as some 10^{500} different types of universe, different versions of space-time, that can in string theory make up the so-called "landscape" of reality. For some sort of context, though, note that there are only (relatively speaking) about 10^{80} atoms in the known universe.

Before summarising where we have got to, a quick aside on the "Great Filter" idea, relating to the question of how life of any form could arise out of inanimate matter, and ultimately to human consciousness. Observable civilisations don't seem to happen much based on what we currently know, and possibly have only occurred once. Indeed, even in a universe that manages to exist, the mind-numbingly small improbability of getting from inanimate matter to conscious humans seems to require a series of steps of apparently astonishing improbability. The Filter refers to the causal path from simple inanimate matter to a viable civilisation. The underpinning logic is that almost everything that starts along this path is blocked along the way, which might be by means of one extremely hard step or many very hard steps. Indeed, it's commonly supposed that it has only once ever happened here on earth. Just exactly once, traceable to LUCA (our Last Universal Common Ancestor), from which all organisms now living on earth have a common descent. If so, it may be why the universe out there seems for the most part to be quite dead. The biggest filter, so the argument goes, is that the origin of life from inanimate matter is itself very, very, very hard. It's a sort of resurrection from death but, in a sense, an order of magnitude harder because the "dead stuff" had never been alive, nor had anything else! And that's just the first giant leap along the way.

So let's summarise. If we didn't know better, our best guess, the simplest description of all possible realities, is that nothing exists. But we do know better, because we are alive and conscious, and considering the question. But our universe is far, far, far too fine-tuned to exist by chance if it is the only universe. So there must be more, if our universe is caused by the roll of the dice, a lot more. But how many more? If there is some mechanism for generating experimental universe upon universe, why should there be a limit to this process? If there is not, that means there will be limitless universes, including limitless identical universes, in which in principle everything possible has happened, and will happen, over and over again.

Even if we accept there is some limiter, we have to ask what causes this limiter to exist, and even if we don't accept there is a limiter, we still need to ask what governs the equations representing the initial conditions to be as they are, to create one universe or many. What puts life into the equations and makes a universe or universes at all? Why should the mechanism generating life into these equations have infused them with the physical laws that allow the production of any universe at all?

Some have speculated that we can create a universe or universes out of nothing. A particle and an anti-particle could, for example, in theory be generated spontaneously out of what is described as a "quantum vacuum". According to this theoretical conjecture, the universe "tunnelled" into existence out of nothing.

This would be a helpful handle for proposing some rational explanation of the origin of the universe and of space-time if a "quantum vacuum" was in fact nothingness. But that's the problem with this theoretical foray into the quantum world. In fact, a quantum vacuum is not empty or nothing in any real sense at all. It has a complex mathematical structure; it is saturated with energy fields and virtual-particle activity. In other words, it is a thing with structure and things happening in it. As such, the equations that would form the quantum basis for generating particles, anti-particles, fluctuations, a universe, actually exist and possess structure. They are not nothingness, not a void.

To be more specific, according to relativistic quantum field theories, particles can be understood in terms of specific arrangements of quantum fields. So, one arrangement could correspond to there being 28 particles, another to 240, another to no particles at all, and another to an infinite number. The arrangement which corresponds to no particles is known as a "vacuum" state. But these relativistic quantum field theoretical vacuum states are indeed particular arrangements of elementary physical stuff, no less so than our planet or solar system. The only case in which there would be no physical stuff would be if the quantum fields didn't exist. But they do exist. There is no something from nothing. And this something, and the equations which infuse it, has somehow had the shape and form to give rise to protons, neutrons, galaxies, planets, and us.

The key question is what gives life to this structure, because without that structure, no amount of "quantum fiddling" can create anything, can produce something out of nothing. Even an empty space is something with structure and potential. More basically, how and why should such a thing as a "quantum vacuum" ever have existed, let alone be infused with the potential to create a universe or universes? All of this is quite apart from the question of how conscious life emerged out of non-conscious somethingness. Put another way, why was there something rather than nothing? Why is there something rather than nothing?

It is certainly a puzzle, and perhaps the most amazing thing about it is that we can ask the question at all.

5.4.1 Exercise

Do we live in a universe or a multiverse? How relevant is the answer for the "Fine-Tuned Universe" problem, as well as for the "Why is there something rather than nothing" question? This exercise is for personal or group consideration. No solution is provided.

5.4.2 Reading and Links

Albert, D. 2012. "On the Origin of Everything," Sunday Book Review, *The New York Times*, 23 March. https://www.nytimes.com/2012/03/25/books/review/a-universe-from-nothing-by-lawrence-m-krauss.html?ref=oembed

Bailey, D. 2017. What Is the Cosmological Constant Paradox, and What Is Its Significance? 1 January . http://www.sciencemeetsreligion.org/physics/cosmo-constant.php

Bostrom, N. Was the Universe Made for Us? http://www.anthropic-principle.com/?q=book/chapter_2#2a

Brierley, J. 2017. Unbelievable? London: SPCK.

Hanson, R. 1998. The Great Filter – Are We Almost Past It? 15 September . http://mason.gmu.edu/~rhanson/greatfilter.html

Horgan, J. 2012. "Science Will Never Explain Why There's Something Rather than Nothing," *Scientific American*, 23 April. https://blogs.scientificamerican.com/cross-check/science-will-never-explain-why-theres-something-rather-than-nothing/

Kuhn, R.L. 2015. Confronting the Multiverse: What Infinite Universes Would Mean. 23 December. https://www.space.com/31465-is-our-universe-just-one-of-many-in-a-multiverse.html

Lewis, G.F. 2016. A Universe Made for Me? Physics, Fine-Tuning and Life. Cosmos. 18 December. https://cosmosmagazine.com/physics/a-universe-made-for-me-ph ysics-fine-tuning-and-life

Parfit, D. 1998. "Why Anything? Why this? Part 1". London Review of Books, 20, 2, 22 January, pp. 24–27. https://www.lrb.co.uk/v20/n02/derek-parfit/why-anything-why-this

Parfit, D. 1998. "Why Anything? Why This? Part 2". London Review of Books, 20, 3, 5 February, pp. 22–25. https://www.lrb.co.uk/v20/n03/derek-parfit/why-any thing-why-this

Piippo, J. 2012. Giving Up on Derek Parfit, 22 July. http://www.johnpiippo.com/2012/ 07/giving-up-on-derek-parfit.html

Last Common Universal Ancestor (LUCA). Wikipedia. https://en.wikipedia.org/ wiki/Last_universal_common_ancestor

BBC Radio 4. In Our Time. 2008. Podcast. 21 February. The Multiverse. https://www. bbc.co.uk/programmes/b008z744

5.5 Occam's Razor

William of Occam (also spelled William of Ockham) was a fourteenth-century English philosopher. At the heart of Occam's philosophy is the principle of simplicity, and Occam's Razor has come to embody the method of eliminating unnecessary hypotheses. Essentially, Occam's Razor holds that the theory which explains all (or the most) while assuming the least is the most likely to be correct. Put another way, when presented with competing hypotheses about the same prediction, we should prefer the solution which makes the fewest assumptions. This is the principle of parsimony. Put more elegantly, it is the principle of "pluritas non est ponenda sine necessitate" (plurality must never be posited beyond necessity).

Empirical support for the Razor can be drawn from the principle of "overfitting". In statistics, "overfitting" occurs when a statistical model describes random error or noise instead of the underlying relationship. Overfitting generally occurs when a model is excessively complex, such as having too many parameters relative to the number of observations. Critically, a model that has been overfit will generally have poor predictive performance, as it can exaggerate minor fluctuations in the data.

We can also look at it through the lens of what is known as Solomonoff Induction. Whether it is a detective trying to solve a crime, a physicist trying to discover a new universal law, or an entrepreneur seeking to interpret some latest sales figures, all are involved in collecting information and trying to infer the underlying causes. The problem of induction is this: we have a set of observations (or data) and want to find the underlying causes of those observations, i.e. to find hypotheses that explain our data. We'd like to know which hypothesis is correct, so that we can use that knowledge to predict future events. In doing so, we need to create a set of defined steps to arrive at the truth, a so-called algorithm for truth.

If all of the hypotheses are possible but some are more likely than others, how do you weight the various hypotheses? This is where Occam's Razor comes in.

Consider, for example, the two 32 character sequences:

abababababababababababababababab

4c1j5b2p0cv4w1x8rx2y39umgw5q85s7

The first can be written "ab 16 times". The second probably cannot be simplified further.

Now consider the following problem. A computer program outputs the following sequence of numbers: 1, 3, 5, 7. What rule do you think gave rise to the number sequence 1, 3, 5, 7? If we know this, it will help us to predict what the next number in the sequence is likely to be if there is one. Two hypotheses spring instantly to mind. It could be $2n - 1$, where n is the step in the sequence. So the third step, for example, gives $2 \times 3 - 1 = 5$. The fourth step is $2 \times 4 - 1 = 7$. If this is the correct rule generating the observations, the fifth step in the sequence will be 9 ($2 \times 5 - 1$).

But it's possible that the rule generating the number sequence is: $2n - 1 + (n - 1)(n - 2)(n - 3)(n - 4)$. The third step, for example, gives $2 \times 3 - 1 + (3 - 1)(3 - 2)(3 - 3)(3 - 4) = 5$. The fourth step is $2 \times 4 - 1 + (4 - 1)(4 - 2)(4 - 3)(4 - 4) = 7$. In this case, however, the fifth step in the sequence will be $2 \times 5 - 1 + (5 - 1)(5 - 2)(5 - 3)(5 - 4) = 9 + 24 = 33$.

But doesn't the first hypothesis seem more likely? Occam's Razor is the principle behind this intuition. "Among all hypotheses consistent with the observations, the simplest is the most likely".

More generally, say we have two different hypotheses about the rule generating the data. How do we decide which is more likely to be true? To start, is there a language in which we can express all problems, all data, all hypotheses? Let's look at binary data. This is the name for representing information using only the characters "0" and "1". In a sense, binary is the simplest possible alphabet. With these two characters we can encode information. Each 0 or 1 in a binary sequence (e.g. 01001011) can be considered the answer to a yes-or-no question. In principle, all information can be represented in binary sequences. Being able to do everything in the language of binary sequences simplifies things greatly and gives us great power. We can treat everything contained in the data in the same way.

Now that we have a simple way to deal with all types of data, we need to look at the hypotheses, in particular how to assign prior probabilities to the hypotheses. When we encounter new data, we can then use Bayes' Theorem to update these probabilities.

To be complete, to guarantee we find the real explanation for our data, we have to consider all possible hypotheses. But how could we ever find all possible explanations for our data?

By using the language of binary, we can do so.

Here we look to the concept of Solomonoff Induction, in which the assumption we make about our data is that it was generated by some algorithm, i.e. the hypothesis that explains the data is an algorithm. Now we can find all the hypotheses that would predict the data we have observed. Given our data, we find potential hypotheses to explain it by running every hypothesis, one at a time. If the output matches our data, we keep it. Otherwise, we discard it. We now have a methodology, at least in theory, to examine the whole list of hypotheses that might be the true cause behind our observations.

The first thing is to imagine that for each bit of the hypothesis, we toss a coin. Heads will be 0, and tails will be 1. Take as an example 010101010, so the coin landed heads, tails, heads, tails, and so on. Because each toss of the coin has a 50% probability of landing heads, and similarly tails, each bit contributes a half to the final probability. Therefore, an algorithm that is one bit longer is half as likely to be the true algorithm. This intuitively fits with Occam's Razor: a hypothesis that is 8 bits long is much more likely than a hypothesis that is 34 bits long. Why bother with extra bits? We'd need evidence to show that they were necessary. So why not take the shortest hypothesis and call that the truth? Because all of the hypotheses predict the data we have so far, and in the future we might get data to rule out the shortest one. The more data we obtain, the easier it is likely to become to pare down the number of competing hypotheses which fit the data.

Let's turn now to "ad hoc" hypotheses and the Razor. An "ad hoc hypothesis" is a hypothesis added to a theory in order to save it from being falsified. Ad hoc hypothesising compensates for anomalies not anticipated by the theory in its original form. For example, you say that there is a leprechaun in your garden shed. A visitor to the shed sees no leprechaun. This is because he is invisible, you say. He spreads flour on the ground to see the footprints. He floats, you declare. He wants you to ask him to speak. He has no voice, you say. More generally, for each accepted explanation of a phenomenon, there is any number of possible, more complex alternatives. Each true explanation may therefore have had many alternatives that were simpler and false, but any number of additional alternatives that are more complex and also potentially false.

This leads us to the idea of what I term "Occam's Leprechaun". Any new and more complex theory can always be possibly true. For example, if an individual claims that leprechauns were responsible for breaking a vase that he is suspected of breaking, the simpler explanation is that he is not telling the truth, but ongoing ad hoc explanations (e.g. "That's not me on the CCTV, it's a leprechaun disguised as me") counter outright falsification. An endless supply of elaborate competing explanations, called "saving hypotheses", prevent ultimate falsification of the leprechaun hypothesis, but appeal to Occam's Razor helps steer us towards the probable truth. Another way of looking at this is that simpler theories are more easily falsifiable and hence possess more empirical content.

All assumptions introduce possibilities for error; if an assumption does not improve the accuracy of a theory, its only effect is to increase the probability that the overall theory is wrong.

It can also be looked at this way. The prior probability that a theory based on n + 1 assumptions is true must be less than a theory based on n assumptions, unless the additional assumption is a consequence of the previous assumptions. For example, the prior probability that Jack is a train driver must be less than the prior probability that Jack is a train driver *and* that he owns a Mini Cooper, unless all train drivers own Mini Coopers, in which case the prior probabilities are identical.

Again, the prior probability that Jack is a train driver and a Mini Cooper owner and a ballet dancer is less than the prior probability that he is just the first two, unless all train drivers are not only Mini Cooper owners but also ballet dancers. In the latter case, the prior probabilities of the n and n + 1 assumptions are the same.

From Bayes' Theorem, we know that reducing the prior probability will generally reduce the posterior probability, i.e. the probability that a proposition is true after new evidence arises.

Occam's Razor thus points to the simplest explanation that is consistent with the data available at a given time, but even so the simplest explanation may be ruled out as new data become available. This does not invalidate the Razor, which does not state that simpler theories are necessarily more true than more complex theories, but that when more than one theory explains the same data, the simpler should be accorded more probabilistic weight. The theory which explains all (or more) and assumes less is more likely to be true. So Occam's Razor advises us to keep explanations simple. But it is also consistent with multiplying entities which are in fact necessary to explain a phenomenon. A simpler explanation that fails to explain as much as a more complex explanation is not necessarily the better one.

Another good way of looking at this seems to be to abandon certainties and think probabilistically using the idea of entropy. Entropy is a measure of randomness or disorder in a system. High entropy means high disorder and low energy. Energy would need to be put into the system to increase order. The tendency of isolated systems is to move towards disorder, from low entropy to high entropy. For example, assembling a deck of cards in a defined order requires introducing some energy to the system. If you drop the deck, they become disorganised and won't re-organise themselves automatically. This is central to the Second Law of Thermodynamics, which implies that time is asymmetrical with respect to the amount of order: as the system advances through time, it will become more disordered. By "order" and "disorder" we mean how compressed the information is that is describing the system. So if all your papers are in one neat pile, then the description is "All paper in one neat pile". If you drop them, the description becomes "One paper to the right, another to the left, one above, one below, etc. etc." The longer the description, the higher the entropy. According to Occam's Razor, we want a theory with low entropy, i.e. low disorder and high simplicity. The lower the entropy, the more likely it is that the theory is the true explanation of the data, and hence that the theory should be assigned a higher probability.

In summary, Occam's Razor stipulates that among all hypotheses we should prefer the simplest of those that are consistent with the known evidence. In terms of a prior distribution over hypotheses, this is the same as giving simpler hypotheses a higher a priori probability, and more complex ones a lower a priori probability.

5.5.1 Exercise

"The simplest explanation is the most probable explanation". This proposition is for personal or group consideration.

5.5.2 Reading and Links

Altair, A. 2012. An Intuitive Explanation of Solomonoff Induction. LESSWRONG. 11 July. https://www.lesswrong.com/posts/Kyc5dFDzBg4WccrbK/an-intuitive-explanation-of-solomonoff-induction

Heylighen, F. 1997. Occam's Razor. Principia Cybernetica Web. 7 July. http://pespmc1.vub.ac.be/OCCAMRAZ.html

Hiroshi, S. 1997. What Is Occam's Razor? http://math.ucr.edu/home/baez/physics/General/occam.html

Lumen. The Second Law of Thermodynamics. https://courses.lumenlearning.com/wm-biology1/chapter/reading-the-second-law-of-thermodynamics/#:~:text=Entropy%20is%20a%20measure%20of,have%20higher%20entropy%20than%20solids.&text=High%20entropy%20means%20high%20disorder,think%20of%20a%20student's%20bedroom.

Occam's Razor. Simple English Wikipedia. https://simple.wikipedia.org/wiki/Occam%27s_razor

Occam's Razor. Wikipedia. https://en.wikipedia.org/wiki/Occam%27s_razor

BBC Radio 4. 2007. In Our Time. Podcast. 31 May. Ockham's Razor. https://www.bbc.co.uk/programmes/b007m0w4

BBC Radio 4. 2007. In Our Time. Podcast. 8 February. Popper. https://www.bbc.co.uk/programmes/b00773y4

BBC Radio 4. 2014. In Our Time. Podcast. 18 December. Truth. https://www.bbc.co.uk/programmes/b04v59gz

Occam's Razor. C0nc0rdance. 21 December 2009. YouTube. https://youtu.be/9XEA3k_QIKo

6

Anomalies of Choice and Reason

In this chapter we examine how markets work and explore anomalies in the way that they do so. We look at the idea of information efficiency, as well as economic theories of adverse selection, signalling, and screening. We investigate how markets can be used to aggregate and process information and to provide forecasts, and we examine how modern theories of human perception and behaviour, notably prospect theory, can be used to explain some financial puzzles and other puzzles of behaviour and decision-making. We present a special focus on the so-called "wisdom of crowds" and expert forecasting, and we find an anomaly at the interface between probability, gambling, and taxation.

6.1 Efficiency and Inefficiency of Markets

The Efficient Market Hypothesis (EMH) holds, in its strictest form, that market prices always reflect all relevant information.

Prices may change when new information is released, but this new information is unpredictable. For this reason, the best estimate of the price likely to prevail at any point in the future is the price now.

This is dismal, if true, because it would mean that it is not possible to beat the market, except by chance. But this can't be true or else it creates a paradox. If the market was always efficient, traders would have no economic incentive to acquire information, since information acquisition and processing is not a costless activity and would add nothing to what can be obtained by simply looking at current market prices.

It has been demonstrated (Grossman and Stiglitz, 1980) that when information is not costless to obtain or process, asset prices can never fully reflect all the information available to traders. So, in the real world, markets are not completely efficient, in the strictest sense of information efficiency. They cannot be. This result is a relief, at least in principle, to those seeking to beat the market. The equilibrium proposed by Grossman and Stiglitz can be interpreted as one in which *some* profits are available to *some* investors.

Essentially, rational "informed" traders will seek to acquire and process new information whenever the benefits of doing so are greater than the costs. Up to the point, economists say, where the marginal costs equal the marginal

DOI: 10.1201/9781003083610-6

benefits of obtaining and processing information. But to the extent that trading is a zero-sum game, or worse, as most betting markets are, winners need losers. So, who are the winners and who are the losers?

To take the example of a poker game, good players need weak players in the game. In the financial literature these "weak players" are known as "noise (or 'uninformed') traders". Noise makes trading in financial markets possible and thus allows us to observe prices for financial assets. But noise also causes markets to be somewhat inefficient.

Imagine a world with no noise traders, no information costs, no trading costs – an efficient market. This is a market in which it would be irrational to place any trades. In such a market, the Efficient Market Hypothesis in its strictest form cannot be true. It is possible in principle, therefore, to beat the market. How might we do this? By a "technical" strategy which uses information contained in past and present prices or odds? Or by a "fundamental" strategy which uses information about real variables? Or, by some combination of these? Those practical matters can be examined separately. Here we are looking exclusively at whether markets are informationally inefficient as a matter of principle, in a world of positive information costs, or indeed transaction costs, and we can conclude that the answer, in the strictest sense of information efficiency, is yes.

More commonly, however, we call a market informationally efficient if it is not possible to earn abnormal returns by trading in that market except by chance. By an abnormal return we mean a return which is more than sufficient to compensate for all the costs and risks involved.

This is a less strict form of the Efficient Market Hypothesis than requiring that all information is at all times incorporated into prices.

In passing, let us note that for practical investigation information is commonly classified into three categories, popularised in a 1970 paper by Eugene Fama – weak form information, semi-strong form information, and strong form information.

Weak form information is information contained in the set of historical prices and returns. Thus, in a weak form efficient market, investors should not be able to beat the market, allowing for costs and risks incurred, by using trading rules that are based on information about past prices and returns. In other words, investors should not be able to make what is termed an "abnormal return" based on weak form information.

This definition of efficiency is consistent with share prices following a so-called "random walk". In a random walk, the next movement in a particular time series cannot be predicted from an examination of previous movements. The current share price equals the share price in the previous period plus a random error term. So past prices provide no systematic information about movements in future prices.

Semi-strong form information is information which is publicly available. If a market is semi-strong form efficient, investors should not be able to make abnormal returns on the basis of publicly available information, such as the information contained in the annual reports of companies. Semi-strong form information includes weak form information.

The strong form of market efficiency requires that asset prices in capital markets fully incorporate all relevant public and private information. This requires that investors who possess private, price-sensitive information could not outperform the market, net of costs and risks incurred. The market should already discount their private information. Hence strong form tests are concerned with monopolistic access to any information relevant for price formation. Strong form efficiency, if true, would imply that insider trading does not lead to excess returns!

So, semi-strong form efficiency is concerned with how well the market processes the information disclosed to it, whereas strong efficiency is concerned primarily with the adequacy of the information disclosure process. In this sense, strong efficiency should not be considered as a progression from the weak and semi-strong forms of efficiency. The reason is that this confuses the ability of the market to respond to and interpret information with the failure of the market to supply information (what has been termed the "information production function"). Jack Dowie, in an article in *Economica* in 1976, makes a fundamental distinction between strong form efficiency, which tells us about access to and availability of information, and the other forms of efficiency (weak and semi-strong), which are concerned with how well the market responds to information. Since strong form inefficiency implies the existence of subsets of investors who possess monopolistic access to information (which can be exploited to earn abnormal returns), he uses the term "equitable" to describe markets which pass the strong test and the term "efficient" to describe those that pass the weak and semi-strong tests.

So far, we have looked at information efficiency, but there is also a different sense in which markets might be inefficient, and that is because of the existence of asymmetric information. A notable case of this is called "adverse selection", which refers to a situation in which the buyer or seller of a product knows something about the product quality or condition that the other party does not know, allowing them to have a better estimate of what the true price of the product should be. This can lead to the breakdown of a market in which it exists. George Akerlof's seminal article, "The Market for Lemons", published in 1970 in the *Quarterly Journal of Economics*, which examined the problem of adverse selection in the market for used cars, has important implications for any market characterised by adverse selection.

Here is the problem. If Mr. Smith wants to *sell* me his horse, do I really *want* to buy it? It's a question as old as markets and horses have existed, but it was for many, many years, one of the unspoken questions of economics. So how do we solve this paradox? For most of the history of economics, the answer was quite simple. Simply assume perfect markets and perfect information, so the horse buyer would know everything about the horse, and so would the seller, and in those cases where the horse is worth more to the buyer than the seller, both can strike a mutually beneficial deal. There's a term for this: "gains from trade".

In the real world, the person selling the horse is likely to know rather more about it than the potential purchaser. This is called "asymmetric information", and the buyer is facing what is called an "adverse selection" problem, as he has adverse information relative to the seller. Akerlof had become intrigued, however, by the way in which economists were limited by this assumption of well-functioning markets characterised by perfect information. For example, the conventional wisdom was that unemployment was simply caused by money wages adjusting too slowly to changes in the supply and demand for labour. This was the so-called "neo-classical synthesis", and it assumed classical markets, albeit they could be a bit slow to work.

At the same time, economists had come to doubt that changes in the availability of capital and labour could in themselves explain economic growth. The role of education was called upon as a sort of magic bullet to explain why an economy grew as fast as it did. But how can we distinguish the impact on productivity of education itself from the extent to which education simply helped grade people? The idea here is that more able people will tend on average to seek out more education. So how far does education contribute to growth, and how far is it simply a signal and a screen for employers? In the real world, of course, these signals could be useful because employers are like the horse buyers – they know less about the potential employees than the potential employees know about themselves, the classic adverse selection problem.

Akerlof turned to the used car market for the answer, not least because at the time a major factor in the business cycle was the big fluctuation in sales of new cars. Just as in the market for horses, the first thing a potential used car buyer is likely to ask is, "Why should I *want* to buy that used car if he wants so much to *sell* it to me?" The suspicion is that the car is what Americans call a "lemon", a sub-standard pick of the crop. Owners of better quality used cars, called "plums", are much less likely to want to sell.

Now let's say that you are willing to spend £10,000 on a plum but only £5,000 on a lemon. In such a case, the best price you are likely to be willing to pay is about £7,500, and only then if you thought there was an equal chance of a lemon and a plum. At this price, though, sellers of the plums will tend to back out, but sellers of the troublesome lemons will be very happy to accept your offer.

But as a buyer you know this, so you will not be willing to pay £7,500 for what is very likely to be a lemon. The prices that will be offered in this scenario may well spiral down to £5,000, and only the worst used cars will be bought and sold. The bad lemons have effectively driven out the good plums, and buyers will start buying new cars instead of plums. Just as with horses, asymmetric and imperfect information in the used car market has the potential to severely compromise its effective operation.

Look at it this way. We can assume that the demand for used cars depends most strongly on two variables – the price of the car and the average quality of used cars traded. Both the supply of used cars and the average quality will depend upon the price. In equilibrium, the supply equals the demand for the given average quality. As the price declines, normally the quality will also fall. And it's quite possible that no cars will be traded at any price.

This same idea applies to medical insurance. In a free market for medical insurance, people above a certain age, for example, will have great difficulty in buying medical insurance. So why doesn't the price rise to match the risk? The answer is that as the price level rises the people who insure themselves will be those who are increasingly certain that they will need the insurance. In consequence, the average medical condition of insurance applicants deteriorates as the price level rises. This is strictly analogous to the car case, where the average quality of used cars supplied fell with a corresponding fall in the price level. The principle of "adverse selection" is potentially present in all lines of insurance. Adverse selection can arise whenever those seeking insurance have freedom to buy or not to buy, to choose the insurance plan, and to continue or discontinue as a policy holder.

There are ways to counteract the effects of quality uncertainty, such as guarantees on consumer durables. Brand names perform a complementary function. Brand names not only indicate quality but also give the consumer a means of retaliation if the quality does not meet expectations. Chains – such as hotel chains or restaurant chains – perform a similar function in this respect to brand names. Licensing practices also reduce quality uncertainty, and education and labour markets themselves have their own "brand names".

So, one of the big problems that confront markets is the fact that some of the participants often don't know certain things that others in the market do know. This includes the market for most consumer durables, virtually all jobs markets, many financial markets, etc. In these cases, one of the roles of economics is to ask what system of incentives is most likely to address this problem of imperfect and asymmetric information. In economics, signalling is the idea that one party (termed the agent) credibly conveys some information about itself to another party (the principal). Signals should be distinguished from what have been called "indices" (a term coined by Robert Jervis in his 1968 PhD thesis). Indices are attributes over which one has no control. Think of these as generally unalterable attributes of something or someone. Signals are visible and in part designed to communicate. They can be thought of as alterable attributes. Employees send a signal about their ability level to the employer by acquiring certain education credentials. The informational value of the credential comes from the fact that the employer assumes it is positively correlated with having greater ability.

Education credentials can be used as a signal to the firm, indicating a certain level of ability that the individual may possess, thereby narrowing the informational gap. In a seminal article on signalling, published in 1973 by Michael Spence, he proposes the key assumption that good-type employees "pay less" for one unit of education than bad-type employees. In Spence's model it is optimal for the higher ability person to obtain the credential (the observable signal) but not for the lower ability individual. The premise for the model is that a person of higher ability incurs a lower cost for obtaining a given level of education relative to a person of lower ability. Cost can be in terms of tuition costs, or intangible costs, such as stress and time and effort in obtaining the qualification. Thus, if both individuals act rationally it is optimal for the higher ability person to obtain the qualification but not for the lower ability person so long as the employers respond to the signal correctly. This will result in the workers self-sorting into the two groups. For this to work effectively, it must be excessively costly, or impossible, to project a false image. The basic argument follows from the intuition that a behavioural attribute that costs nothing can be equally well acquired by anyone and so provides no information. It follows that perceivers should focus on behaviour which is costly to undertake. Signalling is an action by a party with good information that is confined to situations of asymmetric information. The market works best when signals are genuine and cannot, or only with difficulty, be faked.

The concept of screening should be distinguished from signalling, the latter implying that the agent moves first. When there is asymmetric information in the market, screening can involve incentives that encourage the better informed to self-select or self-reveal. Joseph Stiglitz pioneered the theory of screening, examining how a less informed party can induce the other party to reveal their information. They can provide a menu of choices in such a way that the optimal choice of the other party depends on their private information. For example, a company seeking a salesperson might offer a low basic salary, topped up with a good commission on sales made. A potential applicant who is good at selling will be attracted to this type of pay structure compared to someone else who is poor at making sales.

In summary, can markets be efficient? In the strictest informational sense, the answer is no. But there are ways in which they can be made more efficient in the broader sense of the term than they would be in their natural state.

6.1.1 Exercise

Are markets mostly efficient or mostly inefficient? This exercise is for personal or group discussion.

6.1.2 Reading and Links

Akerlof, G. 1970. The market for lemons: Quality, uncertainty and the market mechanism, *Quarterly Journal of Economics*, 84(3), 488–500. https://www.yourhome worksolutions.com/wp-content/uploads/edd/2020/04/answers-2.pdf

Akerlof, G.A. 2001. Behavioral macroeconomics and macroeconomic behavior, Nobel Prize Lecture, December 8. https://www.nobelprize.org/uploads/2018/06/ak erlof-lecture.pdf

CORE. 12.6 missing markets: Insurance and lemons. https://core-econ.org/the-econ omy/book/text/12.html#126-missing-markets-insurance-and-lemons

Dowie, J. 1976. On the efficiency and equity of betting markets, *Economica*, 43(170), 139–150. https://www.jstor.org/stable/2553203?seq=1#metadata_info_tab_contents

Fama, E.F. 1970. Efficient capital markets: A review of theory and empirical work, *Journal of Finance*, 25(2), 383–417. https://onlinelibrary.wiley.com/doi/full/10.1111/j.1540-6261.1970.tb00518.x?casa_token=gw7SP-2gbeYAAAAA:EGHnbCwdeYl2SHkaP8L2P2wEuCqqGuleOw5bFvSFiCcSRiGRmHKG5aqlhjSaJh1lhJgUb76KLSg8wS2J

Grossman, S.J. and Stiglitz, J. 1980. The impossibility of informationally efficient markets, *American Economic Review*, 70 (3), 393–408. https://www.jstor.org/stable/pdf/1805228.pdf?casa_token=dnfS4I0ols8AAAAA:6_lRGPqJi8ip9WzVzLiUgxVtrHuRqIZlAVkIES-0mL-tqtOJU5BjMfYqOSmtYGZ6EFaXxCOjV5ZBENb8Wg_sTV1gSlXnrMk-l3eqxeitn15uJDK4IpvY

Jervis, R. 2002. Signaling and perception: drawing inferences and projecting images. In Monroe, K.R. (ed.) *Political Psychology*, Mahwah, NJ: Earlbaum.

Keane, S.M. 1987. *Efficient Markets and Financial Reporting*. Edinburgh: Institute of Chartered Accountants of Scotland.

Malkiel, B. 2003. *The Efficient Market Hypothesis and Its Critics*. CEPE Working paper 91. Princeton University. http://www.vixek.com/Efficient%20Market%20Hypothesis%20and%20its%20Critics%20-%20Malkiel.pdf

Spence, A. M. 1981. Signaling in retrospect and the informational structure of markets, Nobel Prize Lecture, 8 December. https://www.nobelprize.org/uploads/2018/06/spence-lecture.pdf

Spence, M. 1973. Job market signaling, *Quarterly Journal of Economics*, 87(3), 355–374. http://bibliotecadigital-old.econ.uba.ar/download/Pe/187895.pdf

Stiglitz, J. E. 1981. Information and the change in the paradigm of economics, Nobel Prize Lecture, December 8. https://www.nobelprize.org/uploads/2018/06/stiglitz-lecture.pdf

Stiglitz, J. E. 1975. The theory of 'screening' education, and the distribution of income, *American Economic Review*, 65(3), 283–300. https://www.jstor.org/stable/pdf/1804834.pdf?casa_token=bBYiQ6rOcbQAAAAA:WguZreSvwwGQ5BWDsN9tEYBy-AX0uxqKczHsEWXSj0_-MZHFUo3JPdfz-MaC8ma89sZMsaAD9L1yW_Mpi20tHVf7S8HkB00GtYm4wNOBQkKKa7-Vj5EO

Vaughan Williams, L. Ed. 2005. *Information Efficiency in Financial and Betting Markets*. Cambridge: Cambridge University Press.

Vaughan Williams, L. 1999. Information efficiency in betting markets: A survey, *Bulletin of Economic Research*, 51(1), 1–39. https://onlinelibrary.wiley.com/doi/pdf/10.1111/1467-8586.00069?casa_token=9Ph0y2DicOgAAAAA:nWvz0aY1OXwqEbOqp8M4xCEQu04DJjDJjrCK2pfv0hMVqAmUqMcISRHQf2qo-TLAERxArVnolL86nZUH

6.2 Curious and Classic Market Anomalies

The Bank of England was founded in 1694, with the original purpose of acting as the government's banker and debt manager. It was the year that the French Enlightenment writer and philosopher, Francois-Marie Arouet, better known as Voltaire, was born.

The year 1694 is also the year from which we can trace the so-called "Halloween Effect". This is the effect seminally confirmed by Sven Bouman and Ben Jacobsen in a paper titled, "The Halloween Indicator, Sell in May and Go Away: Another Puzzle". In this 2002 paper, they test the hypothesis that stock market returns tend to be significantly higher in the November–April period than in the May–October period and find it to be true in 36 of the 37 developed and emerging markets studied in their sample between 1973 and 1998. They further find the effect is particularly pronounced in the countries of Europe and that it persists over time. The puzzle was especially noteworthy because the anomaly had been widely recognised for years, though not previously rigorously tested.

Some analysts trace this back to the practice of the landed classes of selling off stocks in May of each year as they headed to their country estates for the summer months and re-investing later in the year. Times have moved on, but summer vacations and attitudes may have not. That's a theory, at least, but perhaps an unlikely one to explain modern-day investment strategies.

In a bigger follow-up study by Jacobsen and Zhang, published in 2012, they looked at 108 countries, using over 55,000 observations. They found that the effect was confirmed for 81 of the 108 countries, with the post-Halloween returns outperforming the pre-Halloween period returns by 4.52% on average and by 6.25% over the past 50 years looked at alone. They updated this in Zhang and Jacobsen (2021), finding the effect to be "remarkably robust with returns on average 4% higher during the November-April period than during May-October" (Abstract).

The Halloween Indicator is an example of a financial market anomaly known as a "calendar effect". Market anomalies are conditions in a financial market which systematically offer the opportunity of earning above-average or abnormal returns. The literature on calendar anomalies includes identification of a January effect and a weekend effect among others. The January effect is the idea that stock performance improves or is unusually good in January. Rozeff and Kinney (1976) represent seminal work in this area. The weekend effect, traceable to findings by Cross (1973), is the proposition that large stock decreases tend to occur between the Friday close and the Monday close. Although the January and weekend effects are the best documented of the calendar effects, they are not the only ones. There is also significant evidence of a "holiday effect", which mirrors the weekend effect.

Strange but true!

Yet not quite as strange, perhaps, as the Super Bowl Indicator of stock market performance.

> Few prediction schemes have been more accurate, and at the same time more perplexing, than the Super Bowl Stock Market Predictor, which asserts that the league affiliation of the Super Bowl winner predicts stock market direction. In this study, the authors examine the record and statistical significance of this anomaly and demonstrate that an investor would have clearly outperformed the market by reacting to Super Bowl game outcomes.

This is from the abstract to a paper published in 1990 by Thomas Krueger and William Kennedy in the *Journal of Finance*.

"If the Super Bowl is won by a team from the old National Football League (now the NFC, or National Football Conference)", they wrote,

> then the stock market is very likely to finish the year higher than it began. On the other hand, if the game is won by a team from the old American Football League (now the AFC, or American Football Conference), the market will finish lower than it began.

It should be noted, though, that some AFC teams count as NFL wins because they originated in the old NFL.

Over the 22-year history of the Super Bowl to the date of submission of their study in 1988, they documented a 91% accuracy rate for their predictor. What happened in 1989? The NFC team, San Francisco 49ers, beat the AFC's Cincinnati Bengals and the stock market rose 27%. This was further evidence in favour of an idea originally proposed by *New York Times* sportswriter Leonard Koppett in 1978 and highlighted by Robert Stovall in *Financial World* in 1989.

So, what happened in 1990? The NFC's San Francisco 49ers won a second consecutive victory, beating the AFC's Denver Broncos, by 55 points to 10. But the stock market fell in 1990, by 4.3%.

The Super Bowl Indicator soon returned to form, however, correctly predicting the direction of the stock market in 1991, 1992, 1993, 1994, 1995, 1996, and 1997. From the launch of the Super Bowl that made for 28 correct predictions out of 31 (a success rate of 90.3%).

Since then, the Super Bowl Indicator has had a much more chequered record, predicting correctly only about half the time since 1997. In 2009, Robert Stovall, a strategist for Wood Asset Management in Sarasota, Florida, and an early champion of the Stock Market Indicator wrote:

> Nothing seems to be working anymore [in the stock market] ... Used to be, I was only happy when it was over 90% performance, and when it was still above 80% I was pleased. But certainly 79% is still far above a failing grade.

(quoted on 12 January 2009, in WSJ.com)

Prior to Super Bowl 2017, the Indicator had called it right since then five times (2010, 2011, 2012, 2014, and 2015) and wrong twice (2013 and 2016). Over the whole run of Super Bowls, the indicator had been right a total of 40 times out of 50, as measured by the S&P 500 index. That year the AFC's New England Patriots stormed from 25 points behind at one point in the game to beat the Atlanta Falcons by 34 points to 28 in overtime. It should have presaged a bad year for the stock markets, but in fact the markets climbed. They should also have climbed following the 2018 victory of the NFC's Philadelphia Eagles over the Patriots, but the reverse happened.

For those still retaining some faith in the Indicator and wanting to see a good year ahead for the stock market, the team to cheer for in 2019 was the LA Rams, of the NFC. Having said that, their opponents, the AFC's New England Patriots, won the 2017 Super Bowl, a good year on the markets. In the betting, the Patriots were the marginal favourites to win in 2019 and triumphed by 13 points to 3, yet the stock market performed very well. So, the Indicator, as of Super Bowl 2020, had been right 40 times out of 53, with a failing record for each of the previous four years.

In 2020, the Kansas City Chiefs of the AFC beat the San Francisco 49ers by 31 points to 20, predicting a poor year for the stock market. Then came the pandemic!

So is the Super Bowl Indicator a real forecasting tool, or is it simply descriptive of what has happened rather than containing any predictive value? You decide, but it's always wise to consider chance and to consider data mining whenever an anomaly comes up.

Fischer Black, quoted in the *Economist* (5 December 1992, pp. 23–26), said about a very different apparent anomaly:

> it sounds like people searched over thousands of rules till they found one that worked in the past. Then they reported it, as if past performance were indicative of future performance. As we might expect, in real life the rule did not work anymore.

Another curious anomaly involves a quick look to the heavens.

"The weather in New York City has a long history of significant correlation with major stock indexes ... Investor psychology influences asset prices ... [these findings] cast doubt on the hypothesis that security markets are entirely rational". Not a piece of blind speculation but the conclusions of a study published by Edward Saunders in the *American Economic Review* in 1993.

Using a different data set, examining the relation between morning sunshine and stock returns at 26 stock exchanges, David Hirshleifer and Tyler Shumway, in a 2003 paper, find that "Sunshine is strongly correlated with daily stock returns. There were positive net-of-transaction costs profits to be made from substantial use of weather-based strategies".

A series of later papers, though not all, have found similar evidence.

Could it be related to a 2008 study published in the *Proceedings of the National Academy of Sciences of the United States of America*? The authors, John Coates and Joe Herbert, found that a trader's morning testosterone level predicts his day's profitability. They also found that a trader's cortisol rises with both the variance of his trading results and the volatility of the market. Their results suggest that higher testosterone may contribute to economic return, whereas cortisol is increased by risk. Coates and Herbert argue that since testosterone and cortisol have cognitive and behavioural effects, it is possible that high market volatility may shift risk preferences and even affect a trader's ability to engage in rational choice.

Sometimes there is a clear reason to question whether an apparent anomaly is an actual anomaly at all. For example, the small-firm effect refers to the tendency displayed by smaller firms to outperform larger firms. Early academic evidence of this was reported by Banz (1981), who identified a negative correlation between the average return to stocks and the market value of the stocks. Fortune (1991) compared the accumulated values of two investments notionally made in January 1926, the first in a portfolio represented by the Standard and Poor 500 (S&P 500) and the second in a portfolio of small-firm stocks. He reported that the latter portfolio significantly outperformed the former.

An explanation may lie in the risk profile of small-firm stocks compared to large-firm stocks. If they are more volatile, and investors are risk-averse, we can expect from the normal forces of supply and demand that small-firm stocks will earn a higher expected return than large-firm stocks, to compensate for their higher risk profile.

More generally, the idea of market anomalies gives rise to an obvious question. If we can identify an apparent market inefficiency that we believe we can exploit to our financial advantage, how might we go about doing so? In thinking about this, a key question to ask is why other traders (if an anomaly is genuine) haven't exploited it to the degree that the inefficiency is used up and disappears. We might believe that it's because we are cleverer than anyone else, or more industrious, or have a better algorithm or system. These explanations are possible but unlikely except in perhaps a very few isolated cases. More likely, perhaps, is that we misunderstood something, or that the costs of turning that financial advantage are prohibitive, in terms perhaps of risk or negative skew or the incurrence of unforeseen downside. Possibly we are taking on risk by, say, providing something which others value, like insurance or liquidity, in which case both sides might potentially gain, or maybe we have not fully understood the terms or extent of the exposure we have opened ourselves up to. It's important, therefore, to try and understand in what ways a trade that we are considering has unattractive features. We might also consider whether it is worth trading in the size that is available to trade – is it worth the time and effort? Having said all this, if an anomaly presents itself that really is easy to exploit and offers no discernible downside or cost, it is just possible that it might look too good to be true and yet still be true. Think about the £20 note on the pavement that the financial economist ignored because it must be fake or someone would have already picked it up. Sometimes it might just pay to take a quick look!

6.2.1 Exercise

What is your favourite market anomaly? Do any of the "anomalies" mentioned in this section convincingly contradict the Efficient Market Hypothesis? No solution is provided. This exercise is for personal or group consideration.

6.2.2 Reading and Links

Amoateng, K.A. 2019. Did Tom Brady Save the US stock market? Market Anomaly or Market Efficiency? *International Journal of Economics and Finance*, Canadian Center of Science and Education, 11(5), May, 128. https://econpapers.repec.org/article/ibnijefaa/v_3a11_3ay_3a2019_3ai_3a5_3ap_3a128.htm

Banz, R.W. 1981. The relationship between return and market value of common stocks, *Journal of Financial Economics*, 9(1), 3–18. http://citeseerx.ist.psu.edu/viewdoc/download?doi=10.1.1.554.8285&rep=rep1&type=pdf

Black, F. 1992. Beating the market: Yes, it can be done. *Economist*, December 5, pp. 23–26. Published in Economist Online. https://www.economist.com/free-exchange/2013/10/16/beating-the-market-yes-it-can-be-done

Boumen, S. and Jacobsen, B. 2002. The Halloween indicator: "Sell in May and go away": Another puzzle, *American Economic Review*, 92(5), 1618–1635. https://www.jstor.org/stable/pdf/3083268.pdf?casa_token=Bxq8njYrL_UAAAAA:gOi2oU36ri53nUmxMpHagSRMRC9qRsYJWlyBdlUwP3Dnpto7BlcPCp2yfe1Lj-kZNdWjKsbHNs7IKk7MNLNq8wkwQO5TQZdbjHMVHhb6YmroWB-SJgfj

Brumfiel, G. 2008. The testosterone of trading. *Nature*. 14 April. https://www.nature.com/articles/news.2008.753

Carr, M. 2018. The S&P 500 taps this team to win the Super Bowl. Banyan Hill, January 17. https://banyanhill.com/super-bowl-indicator-2/

Carrazedo, T., Curto, J.D. and Oliveira, L. 2016. The Halloween effect in European sectors, *Research in International Business and Finance*, 37, 489–500. https://www.sciencedirect.com/science/article/pii/S0275531916300034

Coates, J.M. and Herbert, J. 2008. Endogenous steroids and financial risk taking on a London trading floor, *Proceedings of the National Academy of Sciences of the United States of America*, 15(16), 6167–6172. https://www.pnas.org/content/pnas/105/16/6167.full.pdf?sid=17

Cross, F. 1973. The behavior of stock prices on Fridays and Mondays, *Financial Analysts Journal*, 29(6), 67–69. https://www.jstor.org/stable/pdf/4529641.pdf?casa_token=1rpkgze4VFEAAAAA:BFNJ-yjytHhhMaey6KsB_JI3dTHU7u2LifSzMxu2N7WiK6sipjRRWEpOcAxyeeJNXTMYLjfh2zMvlAI4sZMrEOBR5_BITY4DhZ2sJiTbxNIHQxWntB4

Fortune, P. 1991. Stock market efficiency: An autopsy? *New England Economic Review*, Federal Reserve Bank of Boston, March/April, 17–40.

Haggard, K.S. and Witte, H.D. 2010. The Halloween effect: Trick or treat, *International Review of Financial Analysis*, 19(5), 379–387. https://www.sciencedirect.com/science/article/pii/S1057521910000608

Hirshleifer, D. and Shumway, T. 2003. Good day sunshine: Stock returns and the weather, *Journal of Finance*, 58(3), June, 1009–1062. https://onlinelibrary.wiley.com/doi/pdf/10.1111/1540-6261.00556?casa_token=XLpSblDo3hYAAAAA:DQt16P_2s9HDPead7ZWQFIAQAY-qE8vEKZ62ryLOhpfkpLGd-OsjicGuUYitooAr51WO2RKOM9anaGLs

Jacobsen, B. and Zhang, C.Y. 2012. The Halloween indicator: Everywhere and all the time. Working Paper. Massey University. http://www.trend-friends.be/sites/default/files/9/SSRN-id2154873.pdf

Krueger, T.M. and Kennedy, W.F. 1990. An examination of the super bowl stock market predictor, *Journal of Finance*, 45(2), 691–697. https://onlinelibrary.wiley.com/doi/abs/10.1111/j.1540-6261.1990.tb03712.x

Prelec, D., Seung, H.S. and McCoy, J. 2017. A solution to the single-question crowd wisdom problem. *Nature*, 541, 532-535. https://www.nature.com/articles/nature21054

Rozeff, M.S. and Kinney, W.R. 1976. Capital market seasonality: The case of stock returns, *Journal of Financial Economics*, 3, October, 379–402.

Saunders, E. 1993. Stock prices and wall street weather, *American Economic Review*, 83(5), 1337–1345. https://www.jstor.org/stable/pdf/2117565.pdf?casa_token=u3VT2RKIwXsAAAAA:Cg4xTl3g5gIS7dyOh11jQ4jmBEvMcwaVYTQWgNg8G8FCncMCQIjSDd1BloirEtz3qetcjChtZD4rdU13ufGhmpUV1KbgwEcl4c_BcHj4y_56GQZY58Gl

Schmidt, B. and Clayton, R. 2017. Super bowl indicator and equity markets: Correlation not Causation. *Journal of Business Inquiry*, 17(2), 97–103.https://journals.uvu.edu/index.php/jbi/article/view/235

Stovall, R. 1989. The super bowl predictor. *Financial World*. 158, 24 January: 72.

Thompson, T.H. & Sen,K.C. 2017. Exploring a market curiosity: an examination of the Super Bowl Indicator, *Managerial Finance*, 43 (2), February, 167-177. https://ideas.repec.org/a/eme/mfipps/mf-05-2016-0131.html

WSJ.com (2019). William Power. Checking the Super-Bowl Indicator, January 12.

Zhang, C.Y. and Jacobsen, B. 2021. The Halloween indicator, "Sell in May and go away": Everywhere and all the time, *Journal of International Money and Finance*, 110, February, 1–49. https://www.sciencedirect.com/science/article/pii/S0261560620302242?casa_token=jA_-3vykvxwAAAAA:EI0TfiJ2V7mLmd9EaQ2h83zHIEhLLPJlD_3pAp5AYhEgOocRRoIRS6IfRF_oc96JS0hpOlB01A

Calendar Effects. Wikipedia. https://en.wikipedia.org/wiki/Calendar_effect

Market anomaly. Wikipedia. https://en.wikipedia.org/wiki/Market_anomaly

Asymmetric information and life insurance. Marginal Revolution University. 23 September 2015. YouTube. https://youtu.be/pUkRo9COd38

Asymmetric information and used cars. Marginal Revolution University. 8 January 2015. YouTube. https://youtu.be/sXPXpJ5vMnU

Market failure: Asymmetric information (Akerlof's Lemons). Mark Seccombe. 20 February 2016. YouTube. https://youtu.be/UzZO0l6A5uU

Super bowl winner can predict the markets. *Wall Street Journal*. 12 January 2016. YouTube. https://youtu.be/WAE2lO2riEg

Surprising stock market indicators. ISM university. 30 July 2013. YouTube. https://youtu.be/RbH78AhvuU8

The stock market can predict the super bowl winner – Michael Carr. Banyan Hill Publishing. 29 January 2019. YouTube. https://youtu.be/j_y8124aYc4

6.3 Ketchup Anomalies, Financial Puzzles, and Prospect Theory

Traditional finance is more concerned with checking that the price of two 8-ounce bottles of ketchup is close to the price of one 16-ounce bottle than it is in understanding the price of the 16-ounce bottle. Such is the view of Lawrence ("Larry") Summers, former Director of the White House's

National Economic Council, writing in the *Journal of Finance* in 1985. "They have shown", he went on,

> that two quart bottles of ketchup invariably sell for twice as much as one quart bottle of ketchup except for deviations traceable to transactions costs ... Indeed, most ketchup economists regard the efficiency of the ketchup market as the best-established fact in empirical economics.

If so, this represents an example of the LOOP ("Law of One Price") principle in economics, i.e. identical goods should have identical prices.

But are they right? To find out, check out the prices on offer at any supermarket chain of a smaller and larger bottle of tomato ketchup. You will find that a much bigger bottle will not be a commensurately much bigger price. Does this defy the LOOP principle? Does this indicate something anomalous about the market? Well, there is nothing obviously wrong with the market, since there's no clear way to exploit the "mispricing", short of tipping the contents of the bigger bottle into the smaller bottles and selling them yourself. Summers would call this a deviation due to transactions costs. More fundamentally, the smaller bottle offers advantages that the larger bottle doesn't have. For one thing, it's easier to store. Perhaps it also looks nicer on the table.

Trading financial assets, on the other hand, is a different issue altogether. Transactions costs are relatively small and assets trading in different markets are often identical, so in these cases one would expect the LOOP principle to more clearly apply. What's the evidence? One apparent violation is the case of Royal Dutch Shell. Royal Dutch and Shell merged their interests in 1907 on a 60/40 basis but continued as separate legal entities. On this basis, the Royal Dutch shares should automatically have been priced at 50% more than Shell shares. However, they diverged from this considerably, from 30% too low in 1981 to more than 15% too high in 1996. (Lamont and Thaler, 2003a).

When the company 3Com spun off shares of its mobile phone subsidiary Palm into a separate stock offering, 3Com kept most of Palm's shares for itself (Lamont and Thaler, 2003b). So, a trader could invest in Palm simply by buying 3Com stock. 3Com stockholders were guaranteed to receive three shares in Palm for every two shares in 3Com that they held. This seemed to imply that Palm shares could trade at an absolute maximum of 2/3 of the value of 3Com shares. Rather than being worth less than 3Com shares, however, Palm shares instead traded at a higher price for a period of several months. This should have allowed an investor to make a guaranteed profit by buying 3Com shares and shorting Palm – a virtual no-risk arbitrage opportunity, the equivalent of exchanging, say, $1,000 for £800 in the UK and almost simultaneously exchanging the £800 for $1,250 in the US.

How about prediction markets (speculative markets used for making predictions)? Is it possible to buy low and sell high across different prediction markets? Seems so! For an example, we need only point to the 2008 and 2012 US Presidential elections when it was for several days possible at some times to back John McCain and Mitt Romney, respectively, on the Betfair betting exchange at a healthy shade of odds against and simultaneously to do likewise with Barack Obama on the Intrade exchange.

An informationally efficient market might be defined as one in which deviations from the efficiency hypothesis are within information and transactions costs. On this basis, there would appear to be at least some evidence that markets (in particular respect of the "Law of One Price") are not always informationally efficient.

Before concluding this section, it's worth turning to the application of prospect theory for our understanding of some classic financial puzzles.

The field of prospect theory is a body of work pioneered by Daniel Kahneman and Amos Tversky (see, for example, Kahneman and Tversky, 1984). In prospect theory, people perceive outcomes in terms of gains and losses measured relative to some reference point. This is in line with how our perceptual system works in other contexts. We are more attuned to changes in brightness, volume, and temperature, for example, than to their absolute magnitudes. In other words, people think about life in terms of changes, not levels, and derive feelings of well-being in this way. These can be changes from the status quo or changes from what was expected. See, for example, McGraw et al. (2005). Linked to this so-called "reference weighting" is the concept of loss aversion, that we are more sensitive to losses than to similar sized gains. It is not always clear, however, how to define what a gain or loss is. Say, for example, we wish to identify investment gains and losses. Does a perceived stock market gain mean that the return was positive, or does it mean that the return exceeded what the investor expected to earn? Are we focused on annual gains and losses, or is our focus more short term, on monthly, weekly, or even shorter fluctuations?

Koszegi and Rabin (2006, 2007, 2009) propose a framework for applying prospect theory based on the idea that the reference point people use to compute gains and losses is their expectations, or beliefs held in the recent past about outcomes. In this way, people derive utility from the difference between consumption and expected consumption, where their utility function exhibits loss aversion and diminishing sensitivity. Koszegi and Rabin (2006) argue that we should not abandon models which link utility to absolute levels of consumption, but that we should also factor in the utility derived from gains and losses in consumption.

While it is widely agreed that prospect theory offers an accurate description of risk attitudes in experimental settings, some have questioned whether its predictions will retain their accuracy in the real world, where the stakes can be much higher and where people may have significant experience making the relevant decisions. What is the evidence?

Post et al. (2008) look at evidence from "Deal or No Deal", a game show with large potential payouts. Kachelmeier and Shehata (1992) examine evidence from poor country environments where the US researcher's budget represented a large amount of money. Both studies find that prospect theory continues to provide a good description of behaviour under strong financial incentives.

In the context of prospect theory, Barberis and Huang (2008) study asset prices in an economy populated by investors who derive prospect theory utility from the change in the value of their portfolios. The model yields the prediction that a security's skewness in the distribution of its returns is valued for itself, and so a security whose return distribution has a right tail longer than its left tail will be relatively overpriced and earn a lower average return than would be predicted by a traditional financial model.

The intuition is that by taking a significant position in a positively skewed stock, investors give themselves the chance, however small, of becoming very wealthy. In effect, they over-weight the unlikely state of the world in which they make a lot of money by investing in the positively skewed stock. As a result, they are willing to pay a relatively high price for the stock, even when it means earning a low average return on it.

This is an element of prospect theory called probability weighting. In prospect theory, people do not weight outcomes by their objective probabilities but by transformed probabilities or decision weights. The weighting function over-weights low probabilities and under-weights high probabilities. People also tend to over-weight the tails of any distribution, i.e. to over-weight unlikely extreme outcomes.

Several papers, using a variety of techniques to measure skewness, have confirmed the basic prediction that more positively skewed stocks will have lower average returns. Conrad et al. (2013) is an example of the literature.

A related, well-established, puzzle is that the long-term average return of stocks that conduct an IPO (initial public offering) is below that of similar stocks that do not conduct an IPO. One explanation is that returns on stocks that offer an IPO are highly skewed; most don't perform particularly well, but some do extremely well. As such, prospect theory indicates that stocks that do issue an offering may be valued more than similar stocks, even though the long-run average returns of the former are much lower than that of the latter. Consistent with this hypothesis, Green and Hwang (2012) find that the higher the predicted skewness of an initial public offering, the lower is its long-term average return.

This brings us to the Equity Premium Puzzle, which refers to the fact that the average return of the US stock market has historically been very much higher than the average return to Treasury bills, the difference in magnitude (the "equity risk premium") being far too great to be plausibly explained by simple investor risk aversion.

Benartzi and Thaler (1995) explain the disparity to be a consequence of a combination of loss aversion and narrow framing, i.e. investors consider the distribution of one specific component of their wealth (stock returns) in annual terms, so the high dispersion of stock returns considered this way is very unappealing. To compensate, the stock market needs to have a high average return.

Benartzi and Thaler's explanation relies on loss aversion and an assumption of narrow framing, which occurs when an individual evaluates a risk separately from other concurrent risks. Narrow framing has been linked to many empirical findings. Barberis et al. (2006), for example, argue that aversion to an even probability bet to win $110 or lose $100 is evidence not only of loss aversion but also of narrow framing, i.e. the decision is taken without proper consideration of the wider context of all other risky choices undertaken. These other risks are diversifying the risk of the 50/50 bet. Without these explanations, risk aversion (they argue) would have to be implausibly high. A combination of loss aversion and narrow framing might also explain the "non-participation puzzle", i.e. the fact that most households have never participated in the stock market.

Probability weighting might also contribute to the high equity premium predicted by prospect theory. The context here is that the aggregate stock market is negatively skewed: it is subject to occasional large crashes. If investors over-weight these rare events, they will require compensation in terms of a higher equity premium than that predicted by loss aversion alone (De Giorgi and Legg, 2012). Probability weighting can therefore generate both the high average return on the overall stock market and the low average return on, for example, IPOs. In each case, the skewness of the asset, positive or negative, plays a key role.

Sydnor (2010) studied the insurance decisions of 50,000 customers of a home insurance company. The choice facing these households is the level of "excess" (deductible) from a menu of four possibilities: $100, $250, $500, and $1,000.

Sydnor found that the households that chose a $500 "excess" paid an average premium of $715 per year. In choosing this policy, these households all turned down a policy with a $1,000 "deductible" whose average premium was just $615. Given that the annual claim rate is approx. 5%, these households agreed to pay $100 a year to insure against a 5% chance of paying an additional $500 in the event of a claim.

In an expected utility framework, this can only be rationalised by unreasonably high levels of risk aversion.

Sydnor favours a probability weighting explanation, i.e. households over-weight the low probability that they will have to pay the excess (deductible) and so over-value policies with a lower excess.

The other factor he proposes is the reference point. If this is not the household's wealth, but rather expectations about future outcomes (Koszegi and Rabin, 2007), then we can see the premium as a payment that the household *expects* to make, but the excess as a payment that arises only in the unlikely event of a claim. Seen like this, the household doesn't experience as much loss aversion when it pays the premium as when it pays the excess. As a result, it is willing to pay a higher premium.

A related puzzle is the "annuitisation puzzle", the proposition that, at the point of retirement, people allocate a much smaller fraction of their wealth to annuity products than they should. Hu and Scott (2007) argue that loss aversion is the key factor. The annuity is unappealing because the individual is more sensitive to the potential loss on the annuity (early death) than to the potential gain (extended life).

See Barberis (2013) for a review and assessment of the literature on prospect theory.

6.3.1 Exercise

Is there enough evidence of genuine deviations from the LOOP (Law of One Price) principle to convincingly reject the Efficient Market Hypothesis? To what extent can prospect theory explain some classic financial puzzles? This exercise is for personal or group consideration.

6.3.2 Reading and Links

Barberis, N.C. 2013. Thirty years of prospect theory in economics: a review and assessment, *Journal of Economic Perspectives*, 27(1), 173-196.

Barberis, N. and Huang, M. 2008. Stocks as lotteries: The implications of probability theory for security prices, *American Economic Review*, 98(5), 2066–2100. https://www.nber.org/system/files/working_papers/w12936/w12936.pdf

Barberis, N., Huang, M. and Thaler, R. 2006. Individual preferences, monetary gambles, and stock market participation: A case for narrow framing, *American Economic Review*, 96(4), 1069–1090. https://www.nber.org/system/files/workin g_papers/w12936/w12936.pdf

Benartzi, S. and Thaler, R.H. 1995. Myopic loss aversion and the equity premium puzzle, *Quarterly Journal of Economics*, 110(1), 73–92. https://www.jstor.org/sta ble/pdf/2118511.pdf?casa_token=PXBT-deVbygAAAAA:aydShRH4l2gT7K Huaw-UAh3ig-80CPy_QPZTUXMqu1PQNKV-qhtfIBqmDmR_lWmfnXeuW uIaS1jLj9otIEfo2A38bHtJXuuUj5sKZdaApmustTvxk4o

Conrad, J., Dittmar, R.F. and Ghysels, E. 2013. Ex ante skewness and expected stock returns, *Journal of Finance*, 68(1), 85–124. https://onlinelibrary.wiley.com/doi/ pdf/10.1111/j.1540-6261.2012.01795.x?casa_token=kNNsO_jtFTsAAAAA:bmQz fn5eLkSMJ7g68PqVUuV0yUSux8owwYQbMYCpUeaBMaxVuJD177_auS2 6L3mCju-IlOx5msMHhk8

DeGiorgi, E.G. and Legg, S. 2012. Dynamic portfolio choice and asset pricing with narrow framing and probability weighting, *Journal of Economic Dynamics and Control*, 36(7), 951–972. https://www.sciencedirect.com/science/article/pii/ S0165188912000255?casa_token=G1eoq0jsaz8AAAAA:kO7GW9OzzNMIgz ZPfgX5J4p2UeJYsz7UrI03s9bV8vDa8ddLJwS6ugt4xhDdVesSkHLpBe_Ykw

Green, T.C. and Hwang, B.-H. 2012. Initial public offerings as lotteries: Skewness preferences and first-day returns, *Management Science*, 58(2), 432–444. https:// www.jstor.org/stable/pdf/41406398.pdf?casa_token=rYmPfEwxIDUAAAAA: WYvEH65YuUxSOnC5bR0ZMs1pnMe3zuSqdDgRXXX1e6jlAo8zqDnp2GoZkhf FHNnqErDREZWOhYRaZalg8JVmVz_Seztof-dW8sgQMnmHHOkAdLBpXGQ

Hu, W.-Y. and Scott, J.S. 2007. Behavioral obstacles in the annuity market, *Financial Analysts Journal*, 63(6), 71–82. https://www.corp.financialengines.com/employe rs/FE-AnnuityMarketBehavior-FAJ-07.pdf

Kachelmeier, S.J. and Shehata, M. 1992. Examining risk preferences under high monetary incentives: Experimental evidence from the People's Republic of China, *American Economic Review*, 82(5), 1120–1141. http://www.its.caltech.edu/~pbs/ expfinance/Readings/KachelmeierShehata1992.pdf

Kahneman, D. and Tversky, A. 1984. Choices, values, and frames, *American Psychologist*, April, 341–350. http://www.columbia.edu/itc/hs/medinfo/g6080/ misc/articles/kahneman.pdf

Koszegi, B. and Rabin, M. 2006. A model of reference-dependent preferences, *Quarterly Journal of Economics*, 121(4), 1133–1165. http://citeseerx.ist.psu.edu/v iewdoc/download?doi=10.1.1.197.5740&rep=rep1&type=pdf

Koszegi, B. and Rabin, M. 2007. Reference-dependent risk attitudes, *American Economic Review*, 97(4), 1047–1073. https://www.jstor.org/stable/pdf/3003408 4.pdf?casa_token=131GKhFai28AAAAA:X3WQ-Lpr-Gc9f1lD7Z461bTUTjB1_ B9q01S6HnsDqrr1E7-yrseQ_5Hv3s7cEzldPknkK02UqxJmLcKkv7c-fnL8S C7Xgb_WZAjOhoHm53-_PxPkkKY

Koszegi, B. and Rabin, M. 2009. Reference-dependent consumption plans, *American Economic Review*, 99(3), 909–936. https://www.jstor.org/stable/pdf/25592487 .pdf?casa_token=R1OqWcWE5DEAAAAA:OcuhH9YVZX6IAu0WrAgXCRbZ PiPOR9VsOy3NUqryPa9-1hBHAhG0hyoOs1t0aloMvo8hQC1wQoJaxbYbMq-e XTzKFyxs3oWXm1ycCl0XOPBis1dF9Ak

Lamont, O.A. and Thaler, R.A. 2003a. Anomalies: The law of one price in financial markets, *Journal of Economic Perspectives*, 17(4), 191–202. https://pubs.aeaweb.org/ doi/pdfplus/10.1257/089533003772034952

Lamont, O.A. and Thaler, R.A. 2003b. Can the market add and subtract? Mispricing in tech stock carve-outs? *Journal of Political Economy*, 111(2), 227–268. https:// www.journals.uchicago.edu/doi/full/10.1086/367683?casa_token=1b1MDD d_cOoAAAAA%3APe5Ky_eGvANL308JDYFidVJzZC2CuTBhzxPh47J6rX1efcs 9tA325-7bg1P_PM4pqRDLPuGs3XYX&

Mankiw, N.G. and Zeldes, S.P. 1991. The consumption of stockholders and nonstockholders, *Journal of Financial Economics*, 29(1), 97–112. https://www0.gsb.colum bia.edu/mygsb/faculty/research/pubfiles/440/440.pdf

McGraw, A.P., Mellers, B.A. and Tetlock, P.E. 2005. Expectations and emotions of Olympic athletes, *Journal of Experimental Social Psychology*, 41(4), 438–446. https://www.sciencedirect.com/science/article/pii/S0022103104001118?casa_ token=fkDUJBi56OQAAAAA:9bOlf1kz-fMbFgYOxOADzCRUawr8u2D- dtRaPMraXXw94d137bU8pTX9OcVllo-Nj5GPTHkTZieg

Post, T., Van den Assem, M.J., Baltussen, G. and Thaler, R.H. 2008. Deal or no deal? Decision making under risk in a large-payoff game show, *American Economic Review*, 98(1), 38–71. https://www.jstor.org/stable/pdf/29729963.pdf?casa_toke n=49O9q5wgLesAAAAA:ufEY3T_ijSYJkmkeDVayT-wr-lfUN7zPUmPaHzq nev8Ab6KecCGrIsfFAGXfBPF3yZUBlLJlklMFjvcgmUPKtl_t7Q6NdyFAxaE H5MGVz3Pjg_DLAzM

Summers, L.H. 1985. On economics and finance, *Journal of Finance*, 40(3), 633–635. http://m.blog.hu/el/eltecon/file/summers_ketchup%5B1%5D.pdf

Sydnor, J. 2010. (Over)insuring modest risks, *American Economic Journal: Applied Economics*, 2(4), 177–199. https://www.jstor.org/stable/pdf/25760237.pdf?casa _token=RuOj5di8N88AAAAA:YMrMxUrfo96sUEfVitL-1t1j_d8qxacTqbqTf1n KLpHEUCqi2tZfL1sRTZDoecsYkPJK42_EGF5Kpr3RJmlPNA2A3yRHGkmqlpy TQXCR2F7DSMg4fPk

Law of one price. *Wikipedia*. https://en.wikipedia.org/wiki/Law_of_one_price

6.4 The Wisdom of Crowds

The original idea of group wisdom is well illustrated by the "Galton's ox" experiment. In 1906, Sir Francis Galton, the English anthropologist and scientist, visited the West of England Fat Stock and Poultry Exhibition. He came across a competition in which visitors could, for 6 pence, guess the weight of an ox. Those who guessed closest would receive prizes. Eight hundred people entered the competition. Galton added the contestants' estimates and calculated the average of the estimates, publishing the results in 1907 in *Nature*. Using the mean, the crowd had guessed that the ox would weigh 1,197 pounds. In fact, the ox weighed 1,198 pounds, almost exactly reflecting the wisdom of the crowd. The median estimate was 1,207 pounds, still within 1% of the correct weight.

Other classic studies include Jack Treynor's 1987 "Bean Jar" experiment in which he asked his class of 56 students to guess the number of jellybeans in a jar. The mean guess was 871. The actual number was 850, but notably only one student guessed closer than the group average. Kate Gordon's seminal study in 1921 involved 200 students estimating the weights of items. The group (average) result was 94.5% correct: only five students performed better than this.

James Surowiecki, in his book, *The Wisdom of Crowds*, gives an example of the use of real-world aggregated "crowd wisdom". The example dates to May 1968, when the submarine USS *Scorpion* was declared missing with all 99 men aboard. It was known that the vessel must be lost at some point below the surface of the Atlantic Ocean within a circle 20 miles wide. This information was of some help, but not enough to determine even five months later where she could be found. Their top deep-water scientist came up with a plan which pre-dated the explosion of interest in markets to aggregate crowd-based information, asking a group of submarine and salvage experts to bet on the probabilities of what could have happened. Aggregating their responses, he correctly identified the location of the missing vessel to within 200 metres of where it was found.

So, what is the best way to effectively aggregate disparate pieces of information that are spread among many different individuals? In other words, how can we best access the "wisdom of the crowd"? A key idea within this context is to use prediction markets, which are speculative markets (like betting markets) created or employed for the purpose of aggregating information and making predictions.

Their theoretical underpinning derives from the Efficient Market Hypothesis, the idea that markets incorporate all relevant information, and stems from the view that relevant information concerning the likelihood of future events is dispersed among the knowledge, opinions, and intuitions of many people.

While the mechanisms underlying prediction markets vary, they all offer a means of aggregating this information. Market prices represent probabilities. Many of these markets are open to the public, while others are open to defined groups.

The markets have already been used to forecast uncertain outcomes ranging from influenza outbreaks to the spread of infectious diseases, the demand for hospital services, the box office success of movies, climate change, vote shares, and election outcomes, to the probability of meeting project deadlines at Google. Prediction markets may also be used as a mechanism to help market participants hedge their exposure to risk. In some cases, they have even "predicted" the "news" before it was "news". An example was the capture of Saddam Hussein in December 2003. One well-traded prediction market was flooded with bets that he had been captured several hours before this was rumoured and later announced publicly. This is a case of being able to use the markets to find "tomorrow's news today".

The effective use of prediction markets has the potential not only to help track and forecast events at a national and international level, but also to assist companies. For example, they can provide improved estimates of the potential market size for a new product idea or the launch date of new products and services. Examples of companies that have used internal prediction markets for a range of business forecasts include Hewlett-Packard, Google, and General Electric.

Insights gained have many potentially valuable policy applications, not least when accurate forecasts are required in relation to quantifiable targets. Moreover, the information provided by prediction markets may have value in the advance warning that managers may be given of weak performance in identifiable areas. This can help improve resource allocation. Not least, they process and provide information in real time.

A key aspect of well-designed prediction markets, therefore, is that they offer substantial promise as a tool of information aggregation as well as forecasting. They can also be applied at a macroeconomic and microeconomic level to yield information that is valuable for government and commercial policymakers.

A seminal paper on prediction markets, published by Charles Plott and Kay-Yut Chen in 2002, puts it this way:

> Many business examples share the following characteristic: small bits and pieces of relevant information exist in the opinions and intuition of individuals who are close to an activity ... In these cases very little is known by any single individual but the aggregation of the bits and pieces of information might be considerable.

Important research questions include the impact of prediction market design on performance and the impact of the nature of rewards on the level of accuracy of prediction markets. Indeed, the design of the incentive programme may be critical to optimising performance, insofar as people may invest more thought and energy into expressing their opinion when there's a meaningful incentive to do so.

It is also important that the questions posed are in a form which is unambiguous and which can be quantified, and need to be closed in nature. This requires an assessment of who should be involved in responding and ensuring that each of these contributors has an equivalent understanding of the meaning of what is being asked and that these answers can usefully be pooled. The set-up will vary depending on the diversity of contributors, both geographically and functionally.

Specific questions include: Who would be asked to make predictions? How many people? What medium of exchange would the participants use in the market – their own money, company money, or points? Does every participant have the same purchasing power? If not, how is it distributed? What information would be made available to the participants? Who would have access to the results? There is also the issue of the number of markets to run, as well as the length of these markets and how often new markets should be introduced.

In summary, simple rules applied to prediction markets are to ask the right questions, involve the right people, and provide the appropriate incentives.

Properly designed prediction markets offer a potentially valuable tool that may be used to synthesise the specific knowledge of those directly involved with implementing policy at a lower level. The specific nature of targets relating to, for example, waiting list times, educational outcomes, are both specific and quantifiable and hence ideal candidates for operating a trading market. Taking the example of health care targets, the number of people involved from nurses and doctors to administrators further suggests that the operation of markets in this context is feasible. Additional value will be provided by the advance warning that politicians and managers will be given of weak performance in particular areas. This has the potential to improve resource allocation to make it more likely that key targets are met.

Let's look at an example of such a market in operation. The question is whether policy A or policy B should be undertaken to reduce waiting lists.

Let's say the current waiting list for an appointment at the hospital clinic is 30 days. The contract is designed to pay £1 for the length of the waiting list in days. It currently trades at £30. Participants in the market can *buy* the contract at £30 if they think the waiting list will increase and *sell* if they think it will decrease.

For example, if they *sell* at £30 and the waiting list decreases to 25 days, they will win £5 (30-25). But if the waiting list increases to 35 days, they lose £5 (30-35). By comparing the "Waiting list with policy A" contract with the "Waiting list with policy B" contract, the policymaker has gained information about what the "market" thinks about the relative impacts of introducing policy A and policy B on the length of the waiting list. If a policy is not implemented, the contract relating to that event is simply voided, and the market is settled on the policy that is implemented.

Prediction markets can, therefore, be used to aggregate information about anything from a hospital clinic's waiting times to a major national election. In forecasting elections, it has been shown (Vaughan Williams and Reade, 2015; Reade and Vaughan Williams, 2019) that they are particularly effective in terms of accuracy and precision. Any biases tend to be systematic and therefore easy to adjust for. For reference, a very simple measure of accuracy is the proportion of correct forecasts, i.e. how often a forecast correctly predicts the outcome. Precision relates to the spread of the forecasts. Higher precision means a smaller dispersion in the range of forecasts. See Kahneman et al. (2021) for the associated concept of *Noise*.

A related but distinctly different concept to accuracy is unbiasedness. An unbiased probability forecast is also, on average, equal to the probability that the candidate wins the election.

Forecasts that are accurate can also be biased, provided the bias is in the correct direction. To explain, if polls are consistently upward biased for candidates that eventually win, then despite being biased they will be very accurate in predicting the outcome, whereas polls that are consistently downward biased for candidates that eventually win will be inaccurate as well as biased.

An important aside about election forecasting, and event forecasting more generally, relates to the concept of probability. Candidates who are considered by an efficient prediction market to have a 75% chance of winning an election (1 to 3 in betting terms) can still be expected to lose one time in four.

More general examples of the applications of prediction markets range from traditional finance forecasting, such as sales and costs, to product development support. They can also be applied to forecast on-time project delivery times or the likelihood of regulatory approval for new drugs. Other uses are for innovative decision support, such as evaluating the impact of switching advertising agencies or forecasting the market receptivity of new software releases. They can also be used for disease surveillance and projection.

Although there are many strategies for gathering opinions about the future trends of infectious diseases, the resulting data are often behind the curve and not amendable to standard statistical and epidemiological methods. Prediction markets, on the other hand, are well known for their ability to quickly collect, summarise, and deliver information in an accessible way, in real time.

By combining the strengths of prediction markets with the knowledge of a well-defined community, markets can be used to report and project infectious disease activity and related developments quickly and effectively, allowing planning and allocation of resources for treatment and prevention. Prediction markets can also be used to estimate the likelihood of new drug or vaccine developments and the likely success of alternative interventions.

Still, there are limits to the wisdom of crowds; however, this wisdom is aggregated. Can the crowd, for example, predict the lottery numbers? If not, why? Assuming the lottery is not rigged or biased in some way, the answer is no. Why not? Because if the lottery numbers are drawn randomly, no model or individual or crowd or other means of aggregating information can predict them because random numbers are unpredictable. If the lottery numbers were, for whatever reason, not drawn randomly, we might have a different story to tell.

6.4.1 Exercise

How wise are crowds? This question is for personal or group consideration.

6.4.2 Reading and Links

Brown, A., Reade, J. and Vaughan Williams, L. 2019. When are prediction market prices most informative? *International Journal of Forecasting*, 35(1), 420–428. https://www.sciencedirect.com/science/article/pii/S0169207018300852?casa _token=xp1rlyl9coMAAAAA:llJ-6bZj46C2nyQ1hArBLabh9jT0t-GtXBXfkIM rPfGOT0ONrLprbAaZwvJccFWrBUzjxUnNoqU

Dai, M., Jia, Y., and Kou, S. 2021. The wisdom of the crowd and prediction markets. *Journal of Econometrics*, 222 (1), Part B, May, 561-578.

Galton, F. 1907. Vox Populi, *Nature*, 75, 7 March, 450–451. https://www.all-about-psychology.com/support-files/the-wisdom-of-crowds.pdf

Gordon, K.H. 1921. Group judgments in the field of lifted weights, *Psychological Review*, 28(6), November, 398–424. https://psycnet.apa.org/record/1926-0819 8-001

Hanson, R. and Oprea, R. 2009. A manipulator can aid prediction market accuracy, *Economica*, 76(302), 304–314. https://onlinelibrary.wiley.com/doi/pdf/10.1111/ j.1468-0335.2008.00734.x?casa_token=9ftoqlti8uIAAAAA:7yzrQxHFzgxbHU iclNIvRCm4sgw5IA_yHON-uF1xCeGtvT7mtbAIotpPee4zu67ze7bQdz9Mtt M93lf8

Paton, D., Siegel, D. and Vaughan Williams, L. 2010. Gambling, prediction markets and public policy, *Southern Economic Journal*, 76(4), 878–883. https://onlinelibrar y.wiley.com/doi/pdf/10.4284/sej.2010.76.4.878?casa_token=L_b68xlYohcAA AAA:VNuZFz5rHkQ-mpDtGMzIELipZEDMErfgydWYPR7TjFrV5POZ8Z_6La pOc2Z4j5D2ZY_RqaYrP00gQ6La

Paton, D., Siegel, D. and Vaughan Williams, L. 2009. The growth of gambling and prediction markets: Economic and financial implications, *Economica*, 76(302), 219–224. https://onlinelibrary.wiley.com/doi/pdf/10.1111/j.1468-0335.2008. 00753.x

Plott, C.R. and Chen, K-Y. 2002. *Information Aggregation Mechanisms: Concept, Design and Implementation for a Sales Forecasting Problem.* Working papers 1131, California Institute of Technology. https://authors.library.caltech.edu/44358/1/ wp1131.pdf

Reade, J. and Vaughan Williams, L. 2019. Polls to probabilities: Comparing prediction markets and opinion polls, *International Journal of Forecasting*, 35(1), 336–350. https://www.sciencedirect.com/science/article/pii/S0169207018300633?ca sa_token=eV--m4IDTq0AAAAA:2jMfGq24DTAajW5CRzcp0n-e8iX7W0CJaC 5YD2AvmMYP5Depx58Gzix2XMrZdZL1jGPNshGR0Q4

Surowiecki, J. 2004. The Wisdom of Crowds. Why the Many Are Smarter Than the Few. London: Little, Brown.

Tabarrok, A. 2008. Manipulation of prediction markets. 18 October. *Marginal Revolution.* https://marginalrevolution.com/marginalrevolution/2008/10/ma nipulation-of.html

Tabarrok, A. 2012. Intrade manipulation fail. 23 October. *Marginal Revolution.* https:// marginalrevolution.com/marginalrevolution/2012/10/intrade-manipulation-f ail.html

Treynor, J. 1987. Market efficiency and the bean jar experiment, *Financial Analysts Journal*, 43, 50–53. https://www.jstor.org/stable/pdf/4479031.pdf?casa_token =gRcNVGoGpsYAAAAA:1IY2pBRfUKeMNzFE4oib41gb9jyGPPV4ada4yP_ jXBDXcFpz4iNi0wl0lntHW3Gn6ohtr9W0BPXq6ePrL6QPYEO8j55p03omsQY spjD63tGk0jQHShXJ

Vaughan Williams, L. 2020. Joe Biden: How betting markets foresaw the result of the 2020 US election. 13 November. https://theconversation.com/joe-biden-how-betting-markets-foresaw-the-result-of-the-2020-us-election-150095

Vaughan Williams, L. 2018. Written evidence (PPD0024). House of Lords Political Polling and Digital Media Committee. 16 January. http://data.parliament. uk/writtenevidence/committeeevidence.svc/evidencedocument/political-polling-and-digital-media-committee/political-polling-and-digital-media/written/72373.pdf

Vaughan Williams, L. 2016. How the wisdom of crowds could solve the mystery of Shakespeare's 'lost plays.' 14 April. https://theconversation.com/how-the-wisdom-of-crowds-could-solve-the-mystery-of-shakespeares-lost-plays-57705

Vaughan Williams, L. 2015. What happened to MH370? Prediction markets might give us the answer. 4 August. https://theconversation.com/what-happened-to-mh370-prediction-markets-might-give-us-the-answer-45528

Vaughan Williams, L. 2015. The Nobel Prize prediction industry: Far from perfect, but pretty impressive. 12 October. https://theconversation.com/the-nobel-prize-prediction-industry-far-from-perfect-but-pretty-impressive-48916

Vaughan Williams, L. 2015. Forecasting the decisions of the US Supreme Court: Lessons from the 'Affordable Care Act' judgment, *The Journal of Prediction Markets*, 9(2), 64–78. http://irep.ntu.ac.uk/id/eprint/29729/1/Pubsub7245_Vaughan-Williams.pdf

Vaughan Williams, L. Ed. 2011. *Prediction Markets: Theory and applications.* Routledge International Studies in Money and Banking. Abingdon: Routledge.

Vaughan Williams, L. and Paton, D. 2015. Forecasting the outcome of closed-door decisions: Evidence from 500 Years of betting on Papal Conclaves, *Journal of Forecasting*, 34(5), 391–404. https://onlinelibrary.wiley.com/doi/pdf/10.1002/for.2339?casa_token=eQf-fBGGnSAAAAAA:-mPj4JfEPbdJSi34YWVuSmtmgdQreg54TrroAh_-2EAMddc_H19HbxZjDWhqx4-ck6YllhvfjP3se4Or

Vaughan Williams, L. and Reade, J. 2016. Prediction markets, social media and information efficiency, *Kyklos*, 69(3), 518–556. https://onlinelibrary.wiley.com/doi/pdf/10.1111/kykl.12119?casa_token=J2tBTnYoaHoAAAAA:TWmNaXjXzGB-UengtsudDZamVLjczIAPw_ZARloitHFM29UHZ3QgdIyuQkPP9aEaoZOoO9dnEoLVh79h

Vaughan Williams, L. and Reade, J. 2015. Forecasting elections, *Journal of Forecasting*, 35(4), 308–328. https://onlinelibrary.wiley.com/doi/pdf/10.1002/for.2377?casa_token=SgoM-QRVglwAAAAAA:lQ3O8Gtx9Ni3jnW6jNv79o0StWzDfew_9ixiwGB4zsSfg3OoicGE_oYDyGZLC0xwkiP9sr7BwAp6wa7d

Vaughan Williams, L., Sung, M. and Johnson, J.E.V. 2019. Prediction markets: Theory, evidence and applications, *International Journal of Forecasting*, 35(1), 266–270. http://irep.ntu.ac.uk/id/eprint/35265/1/12800_Vaughan-Williams.pdf

Vaughan Williams, L. and Vaughan Williams, J. 2009. The cleverness of crowds, *The Journal of Prediction Markets*, 3(3), 45–47.

6.5 Superforecasting

"Superforecasting" is a term popularised from insights gained as part of an idea known as the "Good Judgment Project", which consists of running tournaments where entrants compete to forecast the outcome of national and international events.

The key conclusion of this project is that an identifiable element of those taking part (the so-called Superforecasters) were able to consistently and significantly out-predict their peers. This idea of expert forecasting in a sense, however, runs contrary to the idea of the "wisdom of crowds".

So, what is special about these expert Superforecasters? A key distinguishing feature of these wizards of prediction is that they tend to update their estimates more frequently than regular forecasters, and they do so in smaller increments. Moreover, they tend to break big intractable problems down into smaller tractable ones.

They are also much better than regular forecasters at avoiding the trap of under-weighting new information or over-weighting it. In particular, they are good at evaluating probabilities dispassionately using a Bayesian approach, i.e. establishing a prior (or baseline) probability that an event will occur and then constantly updating that probability as new information emerges, incrementally updating in proportion to the weight of the new evidence.

Superforecasters in the field of sports betting can benefit from betting in-running, while the event is taking place. Their evaluations are likely to be data-driven and are updated as frequently as possible. They will be aware of players who tend to struggle to close the deal, whether in golf, tennis, snooker, or whatever, or shaky starters, like batsmen in cricket whose average belies their likely performance once they get into double figures. This information is only valuable, however, if the market doesn't already incorporate it.

Superforecasters also gain an edge by access to and dispassionate analysis of large data sets. Moreover, they are very aware that patterns and conclusions derived from small data sets can be dangerous.

Trading contracts on the future is, however, very much about striking the right balance between under- and over-confidence and between prudence and decisiveness. The problem is that confidence is not necessarily positively correlated with accuracy. Indeed, it's been argued that "even experienced risk takers often bet more when they're wrong than when they're right; and the most confident people are generally the least reliable" (Brown, 2015, p.14). The "Dunning–Kruger effect" (the claimed tendency for those of low ability to over-estimate their ability relative to others) is a related cognitive bias.

So how might these expert forecasters go about constructing a sports forecasting model?

Let's say they want to construct a model to forecast the outcome of a football match or a golf tournament. In the former, they might focus on assessing the likely team line-up before its announcement and draw on a hopefully extensive data set to eke out an edge. The football market is very liquid and likely to be quite efficient to known information, so any forecasting edge in terms of estimating future information, like team shape, can be critical. The same might apply to rugby, cricket, and other team games.

In terms of golf, they could include statistics on the average length of drive of the players, their tee to green percentages, their putting performance, the weather, the type of course, and so on. But where is the edge over the market?

They could try to develop a better model than others, including using new, state-of-the-art econometric techniques. In trying to improve the model, they could also seek to identify additional explanatory variables.

They might also turn to the field of prospect theory. The central idea in prospect theory is that people derive utility from "gains" and "losses" measured relative to a reference point.

An excellent seminal paper on this effect in golf (Pope and Schweitzer, 2011) is an example of how a study of the economic literature can improve sports modelling. The key contribution of the Pope and Schweitzer paper is that it shows how prospect theory can play a role even in the behaviour of highly experienced and well-incentivised professionals when a lot of real money is at stake. In particular, they demonstrate, using a database of millions of putts, that professional golfers are significantly more likely to make a putt for par (the professional standard) than a putt for birdie (one stroke better than the professional standard), even when all other factors, such as distance to the pin and break, are allowed for. But why? And how does prospect theory explain it?

To find the explanation, they examine several possible explanations and reject them one by one until they determine the true explanation. They find it is because golfers see par as the "reference" score, and so a missed par is viewed (subconsciously or otherwise) by these very human golfers as a significantly greater loss than a missed birdie. They perform differently in response. The researchers show that equivalent birdie putts are more likely to be laid up short of the hole than par putts. They are also significantly more likely to miss the hole to the left or right. This is valuable information for even the casual bettor, but more likely to be picked up by Superforecasters. It is also valuable information for a sports psychologist. If only someone could stand close to a professional golfer every time they stand over a birdie putt and whisper in their ear "This is for par", it would over time make a significant difference to their performance and prize money.

So Superforecasters will improve their model by increments, taking account of factors which more conventional thinkers might not even consider, and will apply due weight to updating their forecasts as new information emerges.

In conclusion, how might we sum up the difference between a Superforecaster and an ordinary mortal? Watch them as they view the final holes of the Masters golf tournament. What's the chance that the 10-footer on the 17th will find the hole? The ordinary mortal will just see the putt, the distance to the hole, and the potential break of the ball on the green. The Superforecaster is going one step further and also asking whether the 10-footer is for par or birdie. It really does make a difference, and it's why she has earned the privilege of watching from the members' area at the Augusta National Golf Club.

6.5.1 Exercise

Why do professional golfers hole par putts more often than birdie putts of equal difficulty, according to Pope and Schweitzer (2011)?

6.5.2 Reading and Links

Brown, A. 2015. Superforecasting, *Wilmott*, 78, 12–15. https://onlinelibrary.wiley.com /doi/pdf/10.1002/wilm.10429?casa_token=DmoSp1-sZ3oAAAAA:zCTwVfv1J YJZyxcGC5FeZisE8seJhQynR8ivJ2CtFtmu09KVBy37fyvSt6xpmabIHMNY_ lP4OuPvTw8

Dunning, D. 2011. Chapter Five – The Dunning-Kruger effect: On being ignorant of one's ignorance, *Advance in Experimental Social Psychology*, 44, 247–296. Available online 11 June. https://www.sciencedirect.com/science/article/pii/B978012385 5220000056

Pope, D.G. and Schweitzer, M.E. 2011. Is Tiger Woods loss averse? Persistent bias in the face of experience, competition and high stakes, *American Economic Review*, 101(1), 129–157. https://repository.upenn.edu/cgi/viewcontent.cgi?article=1215 &context=mgmt_papers

Smith, M., Paton, D. and Vaughan Williams, L. 2009. Do bookmakers possess superior skills to bettors in predicting outcomes? *Journal of Economic Behavior and Organization*, 71(2), 539–549. https://www.sciencedirect.com/science/article/ pii/S0167268109000833?casa_token=Sgp56tUyToYAAAAA:jD5RckvjnWrOHp VBLGjef2rhKVePFdg3S15MFvR35ZQGfbk_3mu89gypp0hCcfepYDSF_Q8fN1w

Stringfellow, W. 2017. Superforecasting: The art and science of predicting. Review and Summary. 24 January. https://medium.com/west-stringfellow/superforecast ing-the-art-and-science-of-prediction-review-and-summary-e075be35a936

Tetlock, P. and Gardner, D. 2016. *Superforecasting: The Art and Science of Prediction*. London: Random House.

Vaughan Williams, L. 2000. Can forecasters forecast successfully? Evidence from UK betting markets, *Journal of Forecasting*, 19(6), 505–513. https://onlinelibrary. wiley.com/doi/pdf/10.1002/1099-131X(200011)19:6%3C505::AID-FOR756%3E3. 0.CO;2-2?casa_token=IPhf6-KAACgAAAAA:nlXBGaikVmHvDe5n2JM2pnPut pasHu0GlBw8DVe8pL3QyDIFKoKC5Tm43i5ltYUV9Uf91mMTIkUhtpRX

Rationally Speaking Podcast. 2015 Phil Tetlock on "Superforecasting: The Art of Science and Prediction." 20 October. RS145. http://rationallyspeakingpodcast. org/145-superforecasting-the-art-and-science-of-prediction-philip-tetlock/

Transcript of Rationally Speaking Podcast. 2015. Phil Tetlock on "Superforecasting: The Art of Science and Prediction." 20 October. RS145. http://rationallyspe akingpodcast.org/wp-content/uploads/2020/11/rs145transcript.pdf

Rationally Speaking Podcast. 2012. Crowdsourcing and the wisdom of Crowds. 16 December. RS76. http://rationallyspeakingpodcast.org/76-crowdsourcing-and- the-wisdom-of-crowds/

Rationally Speaking Podcast. 2012. Intuition. 8 April. RS58. http://rationallyspe akingpodcast.org/58-intuition/

#3 Aaron Brown – Risk management, gambling and wall street. Speaking to legends. 21 May 2020. YouTube. https://www.youtube.com/watch?v=svunbY5OOgo

Predicting the future: A lecture by Philip Tetlock. American Enterprise Institute. 2015, 19 October. YouTube. https://www.youtube.com/watch?v=xBXDTQdmNyw

Superforecasting: How to predict the future – Philip Tetlock – Animated Book Review. Practical Psychology. 2016, 16 May. YouTube. https://youtu.be/rV5Gicb66WA

Why incompetent people think they're amazing – David Dunning. Ted-Ed. 2017, November 9. YouTube. https://www.youtube.com/watch?v=pOLmD_WVY-E

Why an open mind is key to making better decisions. KnowledgeAtWharton. 2 October 2015. YouTube. https://youtu .be/ -07DJ7xVBis

6.6 Anomalies of Taxation

We can distinguish explicitly between a *commodity* (or unit or specific) system of taxation, which is levied on quantity, and an *ad valorem* system of taxation, which is based on price. The intuitive case for an ad valorem tax (tax on price) is that it should provide an incentive for firms that have any price-setting ability to adopt a low-price/high-turnover strategy. This is in contrast to a low-turnover/high-price strategy, which might intuitively be expected to be incentivised by the implementation of a commodity tax (tax on quantity, or turnover). In other words, the underlying intuition is that taxing price ("ad valorem" taxation) will tend to reduce price, while taxing quantity ("commodity" taxation) will tend to reduce quantity.

Price and quantity are usually easy to distinguish, and therefore it should be easy to distinguish between an ad valorem tax and a commodity tax. There is one context, however, in which intuition can lead us astray. This context is the world of gambling, in which a wager of a particular size is placed on a predicted outcome, such as the result of a horse race. If we tax the "gross profit" of the market-maker or game operator, we tax the amount they win from the player (stake received from, minus winnings returned to, the player). Alternatively, we can tax turnover, i.e. the total amount staked.

Is this a commodity tax or an ad valorem tax?

In fact, a turnover tax (tax on stakes) is not an ad valorem tax, even though casual observation might suggest it is similar in operation and effect to a value added tax. This is because there is a significant economic difference.

To explain, the quantity of output in betting markets may be specified as the number of unit bets placed. Using £1, for example, as the standard unit, a £20 stake is equivalent to 20 unit bets and the quantity of output may, therefore, be defined more generally as equivalent to the total amount staked.

Critically, however, the price of each unit bet can be defined not as the full unit bet staked but as the part of it that is retained by the market-maker or game operator. Based on this approach, the gross profit of bookmakers is the amount they retain after paying out winnings, rather than the total amount of money received in stakes. This retained amount is essentially, therefore, the price paid by the bettor.

Now, a tax that is levied as a proportion of the price charged to bettors is equivalent to an ad valorem tax. It is clear, therefore, that a tax on gross profits is effectively a tax on what may be termed "net revenue" and not on betting stakes. In contrast, a tax on betting stakes, levied as a percentage of stakes, is not a tax on price or value, but is a tax on quantity. The value of a stake of £100 on a horse is not £100, except in limiting circumstances. The intuition lies in the definition of the price of a bet. Say, for example, someone places 100 £1 bets with each of two bookmakers. One returns £80, the other just £60. The value can be interpreted in terms of the total price paid for these 100 bets. The price of the bets is not £100 in each case but £20 and £40, respectively. A 10% turnover tax would yield £10 in both cases. This is only proportional to price if the price of these two bets is the same. So, a tax on gross profits is an ad valorem tax and a tax on turnover is a commodity (or unit or specific) tax.

As a related matter, say that you plan to place a £10 bet on a horse at odds of 5 to 1 against (£1 staked to win a net £5). Before the introduction of gross profits tax on betting in the UK in 2001, bettors were given the option of paying the tax up front on the bet stake or on any winnings from that bet. Say for simplicity the tax is levied at 10%. Now assume, for illustration, that the bettor wishes to place a bet designed to return a net £45 if the horse wins. By staking £10 and choosing to pay the tax on any winnings, the bettor wins £45 (net) if the horse wins (£50 minus 10% of £50). By staking £9 on the same horse, however, and choosing to pay the tax up front on stake, the bettor is risking just £9.90 (£9 plus 10% of the stake) to win £45 if the horse wins. If the horse loses, therefore, the bet of £9 plus the tax on stake loses 10 pence less than the £10 bet (choosing tax on winnings), with both winning the same amount (£45) if the horse wins. This example generalises to any bet, such that paying the tax on stake always dominates, from the point of view of the bettor, paying the tax on winnings.

6.6.1 Exercise

Is a tax on betting stakes a commodity tax or an ad valorem tax? Is a tax on losses by bettors (operator gross profits) a commodity tax or an ad valorem tax?

6.6.2 Reading and Links

Garrett, T., Paton, D. and Vaughan Williams, L. 2020. Taxing gambling machines to enhance public and private revenue, *Kyklos*, 73(4), 500–523. https://onlinelibrar y.wiley.com/doi/pdf/10.1111/kykl.12247

Paton, D., Siegel, D. and Vaughan Williams, L. 2002. A policy response to the E-Commerce revolution: The case of betting taxation in the UK, *Economic Journal*, 12(480), F296–F314. https://www.researchgate.net/profile/Donald-S iegel-2/publication/4890566_A_Policy_Response_to_the_E-Commerce_Revolu tion_The_Case_of_Betting_Taxation_in_the_UK/links/5b2103a1458515270fc6 3b5c/A-Policy-Response-to-the-E-Commerce-Revolution-The-Case-of-Betting-Taxation-in-the-UK.pdf

Paton, D., Siegel, D. and Vaughan Williams, L. 2004. Taxation and the demand for gambling: New evidence from the United Kingdom, *National Tax Journal*, 57(4), 847–861. https://www.jstor.org/stable/pdf/41790262.pdf?casa_token=CL8fc _JzTlgAAAAA:BrqxWszCH-Wo2-NM84Lk6qGSOu2NK_uG2aCFUUSdFa-ar0 BA2oZNkDMOPojMsMawLOygVgufk8381n7caEUUmoytgnlxKIR1HvpqqIq _zcLCI5hoZZCy

Paton, D., Siegel, D. and Vaughan Williams, L. 2002. Gambling taxation: A comment, *Australian Economic Review*, 34(4), 437–440. https://onlinelibrary.wiley.com/doi/ pdf/10.1111/1467-8462.00211?casa_token=MRt_7AOSSpoAAAAA:A1USeFVb bF6AyvGocHul5p-v5rNDkvr6ewlmAyGK_ezV1aPJ2BI4O4Ok2OjS-HVEdO_B xnmq0ycEKOEL

Vaughan Williams, L., Garrett, T. and Paton, D. 2020. Taxing gambling machines to enhance tourism, *The Journal of Gambling Business and Economica*, 13(2), 83–90. http://irep.ntu.ac.uk/id/eprint/42037/1/1400750_Vaughan_Williams.pdf

Vaughan Williams, L. 2014. The Churchill betting tax, 1926-30: A historical and economic perspective, *Economic Issues*, 19(2), 21–38. http://irep.ntu.ac.uk/id/eprint/ 29654/1/PubSub7244_Vaughan_Williams.pdf

Vaughan Williams, L. and Paton, D. 2013. The taxation of gambling machines: A theoretical perspective. In *The Oxford Handbook of the Economics of Gambling*, Ed. L. Vaughan Williams and D. Siegel. New York: Oxford University Press.

7

Game Theory, Probability, and Practice

In this chapter we look at game theory, introducing the reader to the Nash equilibrium and the dominant strategy equilibrium, using a range of examples to illustrate these. We introduce the Prisoner's Dilemma and consider how the game-theoretic outcome can be sub-optimal for those involved and how this might be addressed in a real-world setting. We look at the theory and practice of game-theoretic strategies in a game show setting. We move on from one-shot games to consider multi-stage games, of defined and of undefined duration, and consider optimal strategy in these settings, and whether there is a dominant strategy in each. Finally, we consider the use of mixed strategies, considering when it is optimal for participants to randomise strategies with a degree of probability attached to each. We use the example of a penalty-taker and a goalkeeper to derive a solution to the optimal mixed strategy problem.

7.1 Game Theory: Nash Equilibrium

What is game theory? Game theory is the study of models of conflict, cooperation, and interaction between rational decision-makers. A key idea in the study of game theory is the Nash equilibrium (named after John Nash), which is a solution to a game involving two or more players who want the best outcome for themselves and must take account of the actions of others.

Specifically, if there is a set of "game" strategies with the property that no "player" can benefit by changing strategy while the other players keep their strategies unchanged, then that set of strategies and the corresponding pay-offs constitute the Nash equilibrium.

To summarise, a Nash equilibrium is a situation where nobody can improve their own payoff given what others are doing. So each person is making the best possible choice that they can given the choices everyone else is making.

Assume, for example, there is a simple two-player game in which each player (Bill and Ben) can adopt a "Friendly" (smiles) or a "Hostile" (scowls) approach. Now, depending on their respective actions, let's say the game organiser awards monetary payoffs to each player.

DOI: 10.1201/9781003083610-7

An example of a payoff structure is shown in the next table and is known to each player.

	Ben "Friendly"	Ben "Hostile"
Bill "Friendly"	750 to A; 1,000 to B	25 to A; 2,000 to B
Bill "Hostile"	1,000 to A; 50 to B	30 to A; 51 to B

Now, what is Bill's best response to each of Ben's actions?

If Ben acts "Friendly", Bill's best payoff is to act "Hostile". This yields a payoff of 1,000. If he had acted "Friendly" he would have earned a payoff of only 750.

If Ben acts "Hostile", Bill's best response is if he acts "Hostile". He earns 30 instead of a payoff of 25 if he acted "Friendly".

In both cases his best response is to act "Hostile".

Now, what is Ben's best response to each of Bill's actions?

If Bill acts "Friendly", Ben's best payoff is if he acts "Hostile". This yields a payoff of 2,000. If he had acted "Friendly" he would have earned a payoff of only 1,000.

If Bill acts "Hostile", Ben's best response is if he acts "Hostile". He earns 51 instead of a payoff of 50 if he acted "Friendly".

In both cases his best response is to act "Hostile".

A Nash equilibrium exists when Ben's best response is the same as Bill's best response.

Bill and Ben have the same best response to either action of his opponent. Both should act "Hostile", in which case Bill wins 30 and Ben wins 51.

But if both had been able to communicate and reach a joint, enforceable decision, they would both have gained hugely by acting "Friendly".

So, in conclusion, they would have been better off by smiling. Instead, they both scowled, which was the rational thing for them both to do, even though it was the less satisfactory outcome for both. A case of the best strategy being the worst strategy.

Let's turn now to the world of espionage in seeking out a Nash equilibrium. Let's assume that there are two possible codes, and Agent Anna can select either of them and so can Agent Barbara. The payoff to selecting non-matching codes is zero. An example of a payoff structure is shown in the next slide and is known to each agent.

	Barbara uses Code "A"	Barbara uses Code "B".
Anna uses Code "A"	1,000 to Anna; 500 to Barbara	0 to Anna; 0 to Barbara
Anna uses Code "B"	0 to Anna; 0 to Barbara	500 to Anna; 1,000 to Barbara

So, where is the Nash equilibrium?

Let's look at the top-left box. Here neither Agent Anna nor Agent Barbara can increase their payoff by choosing a different action to the current one. So there is no incentive for either agent to switch given the strategy of the other agent. So this is a Nash equilibrium.

How about the bottom-right box? This is the same. Again, neither Agent Anna nor Agent Barbara can increase their payoff by choosing a different action to the current one. So there is no incentive for either agent to switch given the strategy of the other agent. So this is also a Nash equilibrium.

How about the top-right box? By choosing to use Code B instead of Code A, Agent Anna obtains a payoff of 500, given Agent Barbara's actions. Similarly for Agent Barbara, who would gain by switching to Code A, given Agent Anna's strategy. So this box (Agent Anna uses Code A, and Agent Barbara uses Code B) is *not* a Nash equilibrium, as both agents have an incentive to switch given what the other agent is doing.

How about the bottom-left box? This is the same as the top right. There are again incentives to switch given what the other agent is doing. So it is *not* a Nash equilibrium.

In conclusion, this game has two Nash equilibria – top left (both agents use Code A) and bottom right (both agents use Code B).

Let's turn now to the classic "Live or Die" problem. In this problem, there are two drivers, Peter and Paul. If both Peter and Paul drive on the left of the road, they will be safe, while they will crash if one decides to adhere to one side of the road and the other to the opposite.

	Paul drives on the left	Paul drives on the right
Peter drives on the left	Safe, Safe	Crash, Crash
Peter drives on the right	Crash, Crash	Safe, Safe

At the top left and at the bottom right, there is no incentive for either driver to switch to the other side of the road given the driving strategy of the other driver. They will both be safe if they adopt this strategy. So both the top left and the bottom right are Nash equilibria.

In both other scenarios (top right and bottom left), there is a powerful incentive to switch to the other side, given the driving strategy of the other driver. So neither the top right nor the bottom left is a Nash equilibrium.

In summary, there are two Nash equilibria in the "Live or Die" problem.

Now let's consider the case of two companies, Alligator PLC and Crocodile PLC, both of which have the option of using one of two emblems. Let's call the first the Blue Badger emblem and the other the Black Bull emblem.

	Crocodile uses Black Bull emblem	Crocodile uses Blue Badger emblem
Alligator uses Black Bull emblem	1,000 to Alligator; 500 to Crocodile	500 to Alligator; 1,000 to Crocodile
Alligator uses Blue Badger emblem	500 to Alligator; 1,000 to Crocodile	1,000 to Alligator; 500 to Crocodile

Top left: Crocodile gains by switching from Black Bull to Blue Badger.

Top right: Alligator gains by switching from Black Bull to Blue Badger.

Bottom left: Alligator gains by switching from Blue Badger to Black Bull.

Bottom right: Crocodile gains by switching Blue Badger to Black Bull.

So this game has no Nash equilibrium. There is always an incentive to switch.

So how many Nash equilibria can there be in these sorts of games? Let us recall that if there is a set of "game" strategies with the property that no "player" can benefit by changing their strategy while the other players keep unchanged their strategies, then that set of strategies and the corresponding payoffs constitute what is known as the "Nash equilibrium".

There may be one (e.g. the Friendly/Hostile game). There may be more than one (e.g. Spy problem, "Live or Die" problem). There may be none (e.g. company emblems problem).

An example of a Nash equilibrium drawn from classic economic theory is the case of the Bertrand Duopoly model, where two firms compete solely on price, and customers respond solely to price. The resolution of this model of competition is a situation where both firms are selling at a low price floor, knowing that if either increased their price, they would lose all their customers to the other firm. Although this outcome is not to the benefit of either firm, it is another example of a Nash equilibrium. By communicating and co-operating, both firms would gain.

This leads us to the classic "Prisoner's Dilemma" problem. In this scenario, two prisoners, linked to the same crime, are offered a discount on their prison terms for confessing if the other prisoner continues to deny it, in which case the other prisoner will receive a stiffer sentence. However, they will both be better off if both deny the crime than if both confess to it. The problem each faces is that they can't communicate and strike an enforceable deal. The box diagram below shows an example of the Prisoner's Dilemma in action.

	Prisoner 2 Confesses	Prisoner 2 Denies
Prisoner 1 Confesses	Two years each	Freedom for P1; eight years for P2
Prisoner 1 Denies	Eight years for P1; Freedom for P2	One year each

The Nash equilibrium is for both to confess, in which case they will both receive two years. But this is not the outcome they would presumably have chosen if they could have agreed in advance to a mutually enforceable deal. In that case, they would have chosen a scenario where both denied the crime and received one year each.

Note that the action that gave each of the prisoners the least jail time did not depend on what the other prisoner did. There was what is called a "dominant strategy" for each player, and hence a single dominant strategy equilibrium. That's the definition of a dominant strategy. It is the strategy that will give the highest payoff whatever the other person does.

Often there is no dominant strategy. We have already looked at such a situation, which is whether to drive on the right or on the left. If other drivers choose the right, the best response is to drive on the right too. If others drive on the left, the best response is also to drive on the left. In the US, where people drive on the right, there is an equilibrium, as no one would want to alter their strategy given what others are doing. If everyone is responding optimally to the strategies of everyone else, there is a Nash equilibrium. In the UK, driving on the left is the Nash equilibrium. So the Live or Die "game" has two Nash equilibria but no dominant strategy equilibrium.

Many games have no dominant strategy equilibria, but if we can find a Nash equilibrium, it gives us a prediction of what we might observe. So a Nash equilibrium is a stable state that involves interacting participants in which none can gain by a change of strategy as long as the other participants remain unchanged. It is not necessarily the best outcome for the parties involved, but it is the outcome we would most likely predict. We find that the best strategy in this world of rational self-interested people may not be the one that is actually in their self-interest.

Another example of the Prisoner's Dilemma in action is at the poker table. The casino sets a prize money sum for a tournament and an entry fee to the game, say £100 for 5,000 chips. Players can continue in the tournament until all their chips are gone or they have won all the chips of the other players. After entering the competition, they are now offered an opportunity to purchase an extra 2,000 chips for £8. This purchase is optionaland does not contribute to the prize money, which is set. The additional chips give a significant advantage to anyone purchasing them, but not if the other players also purchase them. In a real-world setting, purchasing the chips is likely to be the dominant strategy for each player, in order not to suffer serious disadvantage against the other players. The outcome is that the casino benefits from all the additional chip buys but all the players face each other on effectively the same terms as before. This is a clear Prisoner's Dilemma situation in action, unless the players are able and willing to act co-operatively.

Perhaps the best example of an attempted real-life resolution to the Prisoner's Dilemma was demonstrated in the TV "Golden Balls" game show. In the game, two players must select a ball which, unknown to the other player, is either a "Split" or "Steal" Ball. Before doing so, they can talk to their opponent. If both choose "Split", they share the prize money. If both choose "Steal" they both go away with nothing. If one chooses "Steal" and one chooses "Split", the contestant who chooses "Steal" wins all the money and the contestant who chooses "Split" gets nothing. In this game, the Nash equilibrium among self-interested players is Steal–Steal as Steal dominates Split (wins all the money compared to sharing the money if choosing Split) but loses nothing to Steal compared to choosing Split (wins nothing either way). Steal in the Golden Balls game is thus equivalent to confess in the traditional Prisoner's Dilemma game. The difference between the Prisoner's Dilemma scenario and the Golden Balls TV show is that there is no communication between the players in the Prisoner's Dilemma. In Golden Balls they both gain half the prize money if they both can agree to choose Split, while they both lose the entire prize money if they both try to Steal. There are many examples on the show where both agree to Split before one or both cheat and choose to Steal. Indeed, the more credible is the promise to Split, the more tempting it may be for the opponent to Steal. This tells us that not even communication and agreement can resolve the Prisoner's Dilemma when there is no way to enforce the agreement.

The YouTube video shown linked below (Golden Balls, 3 February 2012) is a classic demonstration of an attempt to resolve this dilemma, in which one player tells the other that he will definitely "Steal" and then offers to share the money after the show. If the other player believes this threat to "Steal" and thinks there is a chance the money will be shared, the optimal strategy is to "Split". If not, both go away with nothing.

Finally, let's consider the "Dollar Auction" paradox. In this problem, Mr. Moneymaker holds an auction in which he sells dollar notes. Each dollar will be sold to the highest bidder, with the condition that the second highest bidder must also pay the amount they bid and will get nothing in return. How might this work out in practice? Say the first bid is 1 cent, offering a profit of 99 cents, so a second bidder offers 2 cents, and so on a cent at a time up to 99 cents. At this point, the under-bidder stands to lose 98 cents and so bids a dollar, at which point the new under-bidder stands to lose 99 cents, so bids a dollar and a cent, and so on. Without co-operation between bidders, the only winner is the auctioneer.

The Dollar Auction is a very good example of how conflict escalation can proceed. To examine one way that this could happen, consider a first bid of 99 cents. There is no longer any incentive for someone to bid a dollar. Someone might bid a dollar, however, perhaps out of spite or perhaps just for the fun of playing the game. If this happens, the first bidder now stands to lose 99 cents, and may wish to retaliate, potentially leading to a cycle of mutually destructive escalation, perhaps even where there was no ill intent in the first place.

Having highlighted some successes and failures of communication and co-ordination, let's conclude with a nod to focal points (also referred to as Schelling points, after Thomas Schelling). A Schelling point is a strategy that is natural or special in some way, because it allows people to co-ordinate without communication. An example offered by Schelling is when people were asked to meet a stranger in New York City on a particular day, with no instructions on when or where to do so and no communication allowed between the participants before meeting. A focal point emerged quite quickly of 12 noon at Grand Central Station.

7.1.1 Exercise

1. Is every Nash equilibrium a dominant strategy equilibrium? Is every dominant strategy equilibrium a Nash equilibrium? Illustrate using an example.

2. In the Golden Balls game, with no communication allowed outside the game format, is there a dominant strategy for each player? Is there a dominant strategy equilibrium? Is there are a Nash equilibrium? If so, what is it?

3. You are to meet a stranger in London on Midsummer's Day, with no way to communicate between yourselves. Where and when would you show up? What about Paris? Is a consensus focal point likely to emerge? No solution is provided. This is for personal or group consideration.

7.1.2 Reading and Links

Altruistic Preferences in the Prisoners' Dilemma. CORE. https://core-econ.org/the-economy/book/text/04.html#45-altruistic-preferences-in-the-prisoners-dilemma

Equilibrium in the Invisible Hand Game. CORE. https://core-econ.org/the-economy/book/text/04.html#42-equilibrium-in-the-invisible-hand-game

Interaction: Game Theory. CORE. https://core-econ.org/the-economy/book/text/04.html#41-social-interactions-game-theory

Noll, D. 2000. The Dollar Auction Game: A Lesson in Conflict Escalation. https://www.mediate.com/articles/noll1.cfm

Social Preferences: Altruism. CORE. https://core-econ.org/the-economy/book/text/04.html#44-social-preferences-altruism

Social Interactions: Conclusion. CORE. https://core-econ.org/the-economy/book/text/04.html#414-conclusion

Social Interactions: Conflicts in the Choice among Nash Equilibria. CORE. https://core-econ.org/the-economy/book/text/04.html#413-social-interactions-conflicts-in-the-choice-among-nash-equilibria

The Prisoners' Dilemma. CORE. https://core-econ.org/the-economy/book/text/04.html#43-the-prisoners-dilemma

Networks. 2015. Strategies in Playing the Dollar Auction. Cornell University. 21 September. https://blogs.cornell.edu/info2040/2015/09/21/strategies-in-playing-the-dollar-auction/Social

Waniek, M., Niescieruk, A., Michalak, T. and Rahwan, T. 2015. Spiteful Bidding in the Dollar Auction. *Proceedings of the Twenty-Fourth International Joint Conference on Artificial Intelligence.* https://www.ijcai.org/Proceedings/15/Papers/100.pdf

Dollar Auction – C2 wiki https://wiki.c2.com/?DollarAuction

BBC Radio 4. In Our Time. Podcast. *Game Theory.* https://www.bbc.co.uk/programmes/b01h75xp

A Game You Can Never Win. Vsauce2. 17 September 2018. YouTube. https://www.youtube.com/watch?v=1IAsV31ru4Y&feature=youtu.be

Golden Balls. The weirdest split or steal ever. spinout3. 3 February 2012. YouTube. https://www.youtube.com/watch?v=S0qjK3TWZE8

The (Ir)Rationality of the Dollar Auction. 22 May 2019. YouTube. https://www.youtube.com/watch?v=A07LCQ5Pq88&feature=youtu.be

The Prisoner's Dilemma. This Place. 4 October 2014. YouTube. https://youtu.be/t9Lo2fgxWHw

7.2 Game Theory: Repeated Game Strategies

The Prisoner's Dilemma, which we examined in the previous section, is an example of what is called a one-stage game. There is no follow-up to the decision to confess or deny.

But what happens in games with more than one round, where players can learn from the previous moves of the other players?

Take the case of a two-round game. In this game, the payoff from the game will equal the sum of payoffs from both moves. How does optimal strategy change in a two-round game? Actually, it makes no difference. In this scenario, the second round is the final round, and so there are no future rounds to realise the benefit of any goodwill earned in the first round. Your opponent knows this, so you can assume your opponent, who is self-interested and wishes to maximise personal total payoff, will be Hostile on the second move. Your opponent will assume the same about you. Since you will both do the totally self-interested thing on the second and final move, why should either do anything in the first round to earn goodwill? So, the optimal strategy in the first round of a two-stage game is to act as if it is the only round. What if there are three rounds? The same applies. You know that your opponent will act only with regard to his or her self-interest in the final round and therefore might as well do so in the penultimate round as well. Your optimal strategy is, therefore, to do the self-interested thing in the first round, the second round, and the final round. The same goes for your opponent. And so on. In any finite, pre-determined number of rounds, the optimal strategy in any round is to act as if there is no tomorrow.

But what if the game involves an indeterminate number of moves? Suppose that after each move, you roll two dice. If you get a double-6, the game ends. For any other combination of numbers, you play another round. Keep playing until you get a double-6. Your score for the game is the sum of your payoffs. This sort of game mirrors many real-world situations. In the real world, we often don't know when the game will end.

What is the best strategy in repeated plays of the Friendly/Hostile game?

There are seven proposed strategies here.

The first is to be Friendly every time.

The second is always to be Hostile.

The third is a strategy of retaliation. Be Friendly as long as your opponent is Friendly but if your opponent is ever Hostile, you should be Hostile from that point on.

The fourth is a "Tit for Tat" strategy. Be Friendly on the first move. Thereafter, copy what your opponent did on the previous move.

The fifth is to play a random approach. On each move, toss a coin. If it lands heads, be Friendly. If it lands tails, be Hostile.

The sixth is an "Alternate" strategy. Be Friendly on even-numbered moves, and Hostile on odd-numbered moves, or vice versa.

Finally, the "Fraction" strategy. Be Friendly on the first move. Thereafter, be Friendly if the fraction of times your opponent has been Friendly until that point is more than a half. Be Hostile if it is less than or equal to a half.

Which of these is the dominant strategy in this game of repeated rounds with no defined end point? There is no dominant strategy in such games, but is there a strategy that wins if every strategy plays every other strategy in a tournament?

This has been simulated many times. And the winner is Tit for Tat.

It's true that Tit for Tat can never get a higher score than a particular opponent, but it wins tournaments where each strategy plays every other strategy. It does well against Friendly strategies, while Hostile strategies do not exploit it. Essentially, you can trust Tit for Tat. It won't take advantage of another strategy. Tit for Tat and its opponents both do best when both are Friendly.

Look at it this way. There are two reasons for a player to be unilaterally Hostile – to take advantage of an opponent or to avoid being taken advantage of by an opponent. Tit for Tat eliminates the reasons for being Hostile. What accounts for Tit for Tat's success, therefore, is its combination of being nice, retaliatory, forgiving, and clear.

In other words, success in an evolutionary "game" is correlated with the following characteristics. Be willing to be nice: co-operate and never be the first to defect. Don't be played for a sucker: return defection for defection, co-operation for co-operation. Don't be envious: focus on how well you are doing, as opposed to ensuring you are doing better than everyone else. Be forgiving if someone is willing to change their ways and co-operate with you. Don't bear grudges for old actions. Don't be too clever or too tricky. For others to co-operate, clarity is essential.

As Robert Axelrod wrote in his book, *The Evolution of Cooperation*: Tit for Tat's

> niceness prevents it from getting into unnecessary trouble. Its retaliation discourages the other side from persisting whenever defection is tried. Its forgiveness helps restore mutual cooperation. And its clarity makes it intelligible to the other player, thereby eliciting long-term cooperation.

How about the bigger picture? Can Tit for Tat perhaps teach us a lesson in how to play the game of life? Yes, in my view it probably can.

7.2.1 Exercise

How does the optimal strategy in a two-round "Prisoner's Dilemma" game differ from that in a one-round game?

7.2.2 Reading and Links

Axelrod, R. 2006. *The Evolution of Cooperation* (Revised ed.), Perseus Books Group.

Axelrod, R.1984. *The Evolution of Cooperation*, Basic Books.

Axelrod, R. and Hamilton, W.D. 1981. The Evolution of Cooperation, *Science*, 211: 1390–1396, http://www-personal.umich.edu/~axe/research/Axelrod%20and %20Hamilton%20EC%201981.pdf

The Evolution of Cooperation. Wikipedia. https://en.wikipedia.org/wiki/The_E volution_of_Cooperation

Game Theory 101: Repeated Prisoner's Dilemma (Finite). 30 March 2016. YouTube. https://youtu.be/CO3-796fGv8

The Iterated Prisoner's Dilemma and the Evolution of Cooperation. This Place. 2 July 2016. YouTube. https://youtu.be/BOvAbjfJ0x0

7.3 Game Theory: Mixed Strategies

The El Clasico game between Real Madrid and Barcelona is in the 23rd minute at the Santiago Bernabeu when Barca's star striker is brought down in the penalty box and rewarded with a spot-kick against the custodian of the Los Blancos net.

The penalty-taker knows from the team statistician that if he aims straight and the goalkeeper stands still, his chance of scoring is just 30%. But if he aims straight and the goalkeeper dives to a corner, his chance of converting the penalty rises to 90%.

On the other hand, if the striker aims at a corner and the goalkeeper stands still, his chance of scoring is a solid 80%, while it falls to 50% if the goalkeeper dives to a corner.

We are here simplifying the choices to two distinct options, for the sake of simplicity and clarity.

So this is the simple payoff matrix facing the striker as he weighs up his decision.

	Goalkeeper – stands still (%)	Goalkeeper – dive to a corner (%)
Striker – aims straight	30	90
Striker – aims at corner	80	50

The goalkeeper also knows from his team statistician that if he dives to a corner and the striker aims straight, his chance of saving is just 10%. But if he dives to a corner and the striker aims at a corner, his chance of saving the penalty rises to 50%.

On the other hand, if the goalkeeper stands still and the penalty kick is aimed at a corner, his chance of making the save is just 20%, while it rises to 70% if the striker aims straight.

Here is the payoff matrix facing the goalkeeper.

	Striker – aims straight (%)	Striker – aims at a corner (%)
Goalkeeper – stands still	70	20
Goalkeeper – dives to a corner	10	50

So what should the penalty-taker do? Aim straight or to a corner. And what should the goalkeeper do? Stand still or dive?

Game theory can help here.

Neither player has what is called a dominant strategy in game-theoretic terms, i.e. a strategy that is better than the other, no matter what the opponent does. The optimal strategy will depend on what the opponent's strategy is.

In such a situation, game theory indicates that both players should mix their strategies, with the penalty-taker aiming for a corner with a two-thirds probability, while the goalkeeper should dive with a 5/9 probability.

These figures are derived by finding the ratio where the chance of scoring (or saving) is the same, whichever of the two tactics the other player uses.

The Proof

Suppose the goalkeeper opts to stand still, then the striker's chance (if he aims for the corner with a 2/3 probability) = 1/3 × 30% + 2/3 × 80% = 10% + 53.3% = 63.3%.

If the goalkeeper opts to dive, the penalty-taker's chance = 1/3 × 90% + 2/3 × 50% = 30% + 33.3% = 63.3%.

Adopting this mixed strategy (aim for the corner with a 2/3 probability and shoot straight with a 1/3 probability), the chance of scoring is, therefore, the same. This is the ideal mixed strategy, according to standard game theory.

From the point of view of the goalkeeper, on the other hand, if the penalty shot is aimed straight, his chance of saving the penalty kick (if he dives with a 5/9 probability) = 5/9 × 10% + 4/9 × 70% = 5.6% + 31.1% = 36.7%.

If the shot is aimed at a corner, the goalkeeper's chance = 5/9 × 50% + 4/9 × 20% = 27.8% + 8.9% = 36.7%.

Adopting this mixed strategy (dive with a 5/9 probability and stand still with a 4/9 probability), the chance of scoring is therefore the same. This is the ideal mixed strategy, according to standard game theory.

The chances of scoring and making the save in each case add up to 100%, which cross-checks the calculations.

Of course, if either the striker or the goalkeeper gives away real new information about what they will do, then each of them can adjust tactics and increase their chance of scoring or saving.

To properly make operational a mixed strategy requires one extra element, and that is the ability to effectively randomise the choices. The idea is to randomise the choices so that the penalty-taker does have exactly a 2/3 chance of aiming for a corner, and the goalkeeper does have a 5/9 chance of diving for the corner. There are different ways of achieving this. One method of achieving a 2/3 ratio is for the penalty-taker to roll a die on the penalty spot and go for the corner if it comes up 1, 2, 3, or 4 and aim straight if it comes up 5 or 6. Well – perhaps not! But you get the idea.

So, game theory suggests that a rational goalkeeping strategy involves a random selection of options in order to equalise the chance of saving the penalty regardless of the strategy the penalty-taker selects. This is basic game theory, but there is an academic literature which has sought to examine how this plays out in the real world, e.g. Chiappori, Levitt, and Groseclose (2002).

There is further debate in the academic literature about the relative advantages and disadvantages of where to shoot and where to dive. In particular, Roskes, Sligte, Shalvi and De Dreu (2011) found evidence that goalkeepers dived to their right significantly more often than to their left when their team was behind. No such difference was identified if their team was in front or tied in the game. Price and Wolfers (2014) also found in a different data set of penalty kicks that 65.6% of dives were to the goalkeeper's right when their team was behind compared to 56.0% when tied and 55.6% when ahead. These differences were not statistically significant, however, given the sample size. Non-random patterns have also been identified in tennis, including published evidence (Walker and Wooders, 2001) that even professional players tend to alternate serves too regularly, while the stage of the particular game was again something of a predictor.

Back to the El Clasico game. The shot was aimed at the left corner; the goalkeeper guessed correctly and got an outstretched hand to it, pushing it back into play, only to concede a goal on the rebound. Real Madrid got a chance to equalise from the spot eight minutes later and took it. And that's how it ended at the Bernabeu. Real Madrid 1 Barcelona 1. Honours even!

7.3.1 Appendix

Penalty-taker's strategy

	Goalkeeper – stands still (%)	Goalkeeper – dive to a corner (%)
Striker – aims straight	30	90
Striker – aims at corner	80	50

x = probability with which the penalty-taker should aim at a corner

y = probability with which the penalty-taker should aim straight

So,

80x + 30y (if goalkeeper stands still) = 50x + 90y (if goalkeeper dives)

x + y = 1

So,

30x = 60y

30x = 60 (1 − x)

90x = 60

x = 2/3

y = 1/3

Goalkeeper's strategy

	Striker – aims straight (%)	Striker – aims at a corner (%)
Goalkeeper – stands still	70	20
Goalkeeper – dives to a corner	10	50

x = probability with which the keeper should dive to corner

y = probability with which the keeper should stand still

So,

10x + 70y (if penalty shot is aimed straight) = 50x + 20y (if shot is aimed at corner)

x + y = 1

So,

10x + 70y = 50x + 20y

40x = 50y

40x = 50(1 − x)

90x = 50

x = 5/9

y = 4/9

7.3.2 Exercise

How might someone randomise a mixed strategy (say 2/3 call the other player, 1/3 raise the other player) in practice, say in a poker game? Consider how this might be achieved using a wristwatch.

7.3.3 Reading and Links

Chiappori, P., Levitt, S. and Groseclose, T. 2002. Testing Mixed-Strategy Equilibria When Players Are Heterogeneous: The Case of Penalty Kicks in Soccer. *American Economic Review*, 92: 1138–1151. Game Theory: Mixed Strategies Explained. https://www.theprofs.co.uk/library/pdf/mixed-strategy-game-the ory-examples.pdf

Gauriot, R., Page, L. and Wooders, J. 2016. Nash at Wimbledon: Evidence from Half a Million Serves. Working Paper. https://opus.lib.uts.edu.au/bitstream/10453/1 16715/1/SSRN-id2850919.pdf

Price, J. and Wolfers, J. 2014. Right-Oriented Bias: A Comment on Roskes, Sligte, Shalvi, and De Dreu (2011). *Psychological Science*, 25 (11): 2109–2111. https://jo urnals.sagepub.com/doi/full/10.1177/0956797614536738?casa_token=YrA0N W-_e-oAAAAA:Pmlheb-rxunOAQK17UZaxPLAz4Tcx_0u52xhDyaGgdMA ihyspecuNPeiY3uGRbrLp29SUtVTRuS5Zg

Roskes, M., Sligte, D., Shalvi, S., Carsten, K., and De Dreu, W. 2011. The Right Side? Under Time Pressure, Approach Motivation Leads to Right-Oriented Bias. *Psychological Science*, 22 (11): 1403–1407. https://journals.sagepub.com/doi/fu ll/10.1177/0956797611418677?casa_token=ryRgVWggpPcAAAAA:-Tn8HUo81 TBoPRESzNh_9GF71fQy5mDinpSF7LXPW0llad-4jgVFaDwDKLjmUCdLO d-LNSKWHABvkQ

Walker, M. and Wooders J. 2001. Minimax Play at Wimbledon. *American Economic Review*. 91: 1521–1538. https://www.jstor.org/stable/pdf/2677937.pdf?casa_ token=KosTU_ZStjcAAAAA:6uq7c2VNMhp_HFNBX1AqejJ-MBWNqEzAir mNLBtNUafyw29_Q4gMcFR__mBbJdYKbVsTehJVuUxbSJArEHWgeSkaA qPkpgbtYHziagTd51RdiDLEUsIS

Game Theory 101: Soccer Penalty Kicks. William Spaniel. 17 June 2010. YouTube. https://youtu.be/OTs5JX6Tut4

Mixed strategies. Game Theory – Microeconomics. Policonomics. 5 May 2016. YouTube. https://youtu.be/xTQm2mNOVQY

8

Further Ideas and Exercises

In this chapter we conclude with a number of fun problems and paradoxes which to varying extents defy intuition. The "Four Card Problem" (also known as the Wason selection task) and the Bell Boy Paradox are modern classics, while Zeno's Paradox and the Thomson's Lamp thought experiment are examples of paradoxes involving infinity. An example of how "mathemagic" can sum all positive integers to produce a negative fraction, while mathematics cannot, is offered as a modern mindbender, and the book concludes with a couple of short cool down exercises drawn from the world of numbers.

8.1 The Four Card Problem

You are presented with four cards, with each card displaying either a letter or a number. You are promised that each has a letter on one side and a number on the other. This is known as the Four Card Problem (for obvious reasons) or else the Wason selection task, after Peter Cathcart Wason.

Here is an example of the problem.

The face-up sides of the cards show:

23; 28; R; B

You are now presented with the following statement: Every card with 28 on one side has R on the other side.

Question: What is the minimum number of cards needed to determine whether this statement is true? What are these cards?

Think about it: Do you need to turn over the R Card? Do you need to turn over the 23 Card? Do you need to turn over the 28 Card? Do you need to turn over the B Card?

When given this puzzle to solve, most people get it wrong.

The clue is to note that only a card with both a 28 on one side and something other than R on the other side can invalidate this rule.

DOI: 10.1201/9781003083610-8

In fact, turning over the R Card does not help you verify or falsify the statement. Regardless of what is on the other side of the R Card, it will not help you determine whether every card with 28 on one side has R on the other.

Neither does turning over the 23 Card help you verify or falsify the statement. If the 23 Card displays R, say, on the other side, this only tells us that the statement "Every card with 23 on one side has B on the other side" is false. It tells us nothing about the statement that "Every card with a 28 on one side has R on the other side". These two statements are logically separate.

The correct solution is that you must turn over the 28 Card to see if it has R on the other side. If it does not, the statement is false. You must also turn over the B Card to see if it has 28 on the other side. If it does, the statement is false.

So the minimum number of cards needed to determine whether the statement is true is two, and they are the 28 Card and the B Card.

8.1.1 Exercise

Your cousins, Gordon and Yvonne, place between them four cards on a table, each of which has a number on one side and a patch of colour on the other side. The visible faces of the cards display 1, 4, red, and yellow. What is the minimum number of cards you need to turn over to test the truth of the proposition that a card with an even number on one face is red on the other side? What are they?

8.1.2 Reading and Links

Philosophy Experiments. The Wason selection task. https://www.philosophyexp
   eriments.com/wason/
Wason Selection Task. Wikipedia. https://en.wikipedia.org/wiki/Wason_selection
   _task
How logical are you? (Psychology of Reason). WonderWhy. 6 March 2015. YouTube.
   https://www.youtube.com/watch?v=t7NE7apn-PA&feature=youtu.be
These two riddles are the same. So why is one harder? Bit Size Psych. 1 October 2016.
   https://www.youtube.com/watch?v=6I5n2aZNoUU&feature=youtu.be

8.2 The Bell Boy Paradox

Three salesmen arrive at a hotel and are quoted £30 for the only room left. They agree to share the room and pay £10 each at the check-in. Later, the manager notices that there was a discount available for that room, and they should have paid £25, not £30, for the room. He gives the £5 to the bell boy to return to the guests. Rather than splitting the whole of the £5 between the three, however, he returns a pound each to them and keeps the remaining £2 for himself. This means that each of the guests has paid £9 for the room, which is a total of £27, and the bell boy has made £2. That makes a total of £29. But they initially paid £30.

8.2.1 Exercise

What happened to the missing pound?

8.3 Can a Number of Infinite Length Be Represented by a Line of Finite Length?

8.3.1 Exercise

Take the square root of 2. The solution is an irrational number, with no finite solution. In other words, it goes on for ever. 1.4142135623730950488 ....

So, can a line of finite length be drawn that is exactly equal to this infinitely long number?

Clue: Draw a right-angled triangle. Does Pythagoras' Theorem help?

8.4 Does the Sum of All Positive Numbers Really Add Up to a Negative Number?

An Exercise in Mathemagic

What we show here is an exercise in mathemagic not mathematics. In mathematics, it is not allowed to sum a "divergent series", as opposed to a series which converges to a number (such as $1 + 1/2 + 1/4 + 1/8 + ...$, which converges to 2).

Now that we have established that, if you add up 1 and 2, what do you get? The answer is 3. Ok. Let's go one step further. What if you add up 1 and 2 and 3? What do you get now? Now the answer is 6. Now 1 plus 2 plus 3 plus 4. That sums to 10. Now what if I do this forever, in other words add up all the natural numbers right to infinity? What do I get? Most people say it is infinity. As noted above, mathematicians say that there is no sum because technically you can't sum a "divergent series", as opposed to a series which converges to a number.

But let's apply mathemagic and see where we get.

Let's start simple and add up the following series:

$$1 - 1 + 1 - 1 + 1 - 1 + 1 - 1 + 1 - 1 + ... \text{ to infinity.}$$

What is this?

If you stop at an odd step in the series, such as the first or third or fifth step, the series sums to 1. But if you stop at an even step, say the second or fourth or sixth, the series sums to 0. Both are equally likely, so it is intuitive that we can take the average of 1 and 0, which is 0.5 as the solution of this equation.

For those who aren't convinced by the intuition, however, we can show it a little more rigorously like this:

Let $S = 1 - 1 + 1 - 1 + 1 - 1 \dots$

So, $1 - S = 1 - (1 - 1 + 1 - 1 + 1 - 1 \dots) = 1 - 1 + 1 - 1 - 1 - 1 \dots = S$.

So, $1 - S = S$.

So, $2S = 1$.

Therefore, $S = 1/2$.

So the series: $1 - 1 + 1 - 1 + 1 - 1 - \dots = 0.5$.

If we accept this, the task of calculating the solution to:

$1 + 2 + 3 + 4 + 5 \dots$ becomes quite straightforward.

First, we have established that $1 - 1 + 1 - 1 + 1 - 1 + \dots = 1/2$.

We'll call this series S1.

But what if we want to calculate S2, which is the series $1 - 2 + 3 - 4 + \dots$?

The way to do this is to add it to itself, to get 2 x S2

$1 - 2 + 3 - 4 + 5 - \dots + (1 - 2 + 3 - 4 + 5 \dots)$

The easiest way to do this is to move the second series one step along, which is fine as it is an infinite series. Start with 1 and now add up each pair of the remaining terms.

So we get:

$1 + (-2 + 1) + (3 - 2) + (-4 + 3) + (5 - 4) + \dots = 1 - 1 + 1 - 1 + 1 \dots$

But we have seen this series before. It is S1 and is equal to 1/2.

So, 2 xS2 = 1/2.

Therefore, S2 = 1/4.

Now what we are trying to sum is $1 + 2 + 3 + 4 + 5 + 6 + \dots$

Let us call this S.

So, $S - S2 = 1 + 2 + 3 + 4 + 5 + 6 + \dots - (1 - 2 + 3 - 4 + 5 - 6 \dots)$.

So, $S - S2 = 0 + 4 + 0 + 8 + 0 + 12 + \dots$

This series is identical to: $4 + 8 + 12 + 16 + 20 + 24 + \dots$

This is $4 \times (1 + 2 + 3 + 4 + 5 + 6 + \dots)$.

In other words, S − S2 = 4S.

We know already that S2 = 1/4.

Therefore, S – 1/4 = 4S.

So, 3S = –1/4.

S = –1/12

And that is the "proof" that the sum of all the natural numbers up to infinity equals –1/12.

In a way, we shouldn't be surprised. Infinity throws up many such paradoxes, and it is not for nothing that mathematicians eschew the summing of these so-called divergent series, i.e. series in which the individual terms of the series do not approach zero. As mathematician N.H. Abel put it in 1828, "Divergent series are the invention of the devil, and it is shameful to base on them any demonstration whatsoever".

8.4.1 Reading and Links

Berman, D., and Freiberger, M. 2014. Infinity or – 1/12. 18 February. https://plus.ma ths.org/content/infinity-or-just-112

Dodds, M. 2018. The Ramanujan summation. 3 September. https://medium.com/c antors-paradise/the-ramanujan-summation-1-2-3-1-12-a8cc23dea793

Haran, B. 2015. This blog probably won't help. 11 January. https://www.bradyhar anblog.com/blog/2015/1/11/this-blog-probably-wont-help

Padilla, T. What do we get if we sum all the natural numbers? https://www.nottingh am.ac.uk/~ppzap4/response.html

Wikipedia. 1 + 2 + 3 + 4 + … https://en.wikipedia.org/wiki/1_%2B_2_%2B_3_% 2B_4_%2B_%E2%8B%AF

BBC Radio 4. 2003. In Our Time. Podcast. 23 October. Infinity. https://www.bbc.co. uk/programmes/p0054927

ASTOUNDING: 1 + 2 + 3 + 4 + 5 + … = -1/12. Numberphile. 2014. 9 January 2014. YouTube. https://www.youtube.com/watch?v=w-I6XTVZXww&feature=youtu.be

Sum of natural numbers (second proof and extra footage). Numberphile. 11 January 2015. YouTube. https://www.youtube.com/watch?v=E-d9mgo8FGk&feature =youtu.be

8.5 Zeno's Paradox

Zeno of Elea was a Greek philosopher of the fifth-century BC, best known for his paradoxes of motion, described by Aristotle in his book *Physics*. Of these perhaps the best known is his paradox of the tortoise and Achilles, in its various forms. In one telling of the paradox, the tortoise starts 100 metres ahead of the hare and moves at a slower speed than the hare. Will the hare ever catch the tortoise, assuming they don't slow down?

Zeno's paradox relies on the fact that when Achilles reaches the starting position of the tortoise, the antelope will have travelled a bit further along the route. When Achilles arrives at that point, the tortoise will have travelled a bit further, and so on. Zeno argued that this was an infinite process and so does not have a final, finite step. So how can Achilles ever catch the tortoise?

As Aristotle (in *Physics*, VI:9, 239b15) recounts it, "In a race, the quickest runner can never overtake the slowest, since the pursuer must first reach the point whence the pursued started, so that the slower must always hold a lead".

There is a mathematical solution to the paradox, which goes like this:

For the sake of illustration, assume this is a very fast tortoise and a very slow Achilles, so Achilles moves twice as fast as the tortoise.

Let S be the distance Achilles runs and let 1 = 100 metres.

So $S = 1 + 1/2 + 1/4 + 1/8 + 1/16 + 1/32 \ldots..$

$1/2\,S = 1/2 + 1/4 + 1/8 + 1/16 + 1/32 \ldots..$

Therefore, $S - 1/2\,S = 1$.

Therefore, $S = 2$.

Achilles catches the tortoise in 200 metres.

In this way, an infinite process, with no final step, has a finite conclusion.

That's the mathematical solution, but does that meaningfully solve the paradox? After all, how can an infinite process, with no final step, ever come to an end?

The Thomson's Lamp thought experiment, devised by philosopher James F. Thomson in 1954, might provide a clue. Think of a lamp with a switch. You flick the switch to turn the light on. At the end of one minute exactly you flick it off. At the end of a further half minute, you turn it on again. At the end of a further quarter minute you turn it off. And so on. The time between each turning on and off of the lamp is always half the duration of the time before. Assume you have the superpower to do each turning on and turning off instantaneously.

Adding these up gives one minute plus half a minute plus a quarter of a minute ....

$$1 + 1/2 + 1/4 + 1/8 + 1/16 + 1/32 + \ldots = 2.$$

In other words, all of these infinitely many time intervals add up to exactly two minutes.

This leads to an obvious question. At the end of two minutes, is the lamp on or off? Say the lamp starts out being off and you turn it on after one minute, then off after a further half minute and so on. Does this make any difference to the answer?

8.5.1 Exercise

You are presented with a lamp that is on. After 1 minute you switch it off, then 30 seconds later switch it back on, then 15 seconds later switch it off, after 7.5 seconds on again, and so on after half the preceding interval. After two minutes, is the lamp on or off?

8.5.2 Reading and Links

Physics. Aristotle. Translated by R.P. Hardie and R.K. Gaye. 2006. http://classics.mit. edu/Aristotle/physics.html

BBC Radio 4. 2016. In Our Time. Podcast. 22 September. Zeno's Paradoxes. https:// www.bbc.co.uk/sounds/play/b07vs3v1

8.6 Cool Down Exercise

To conclude, a quick exercise to help you cool down.

8.6.1 Exercise

1. Does 0.999 .... = 1?
2. Is zero an even or an odd number?

8.6.2 Reading and Links

BBC Radio 4. 2004. In our time. Podcast. 13 May. Zero. https://www.bbc.co.uk/ sounds/play/p004y254

Reading and References

Aczel, A.D. 2016. *Chance: A Guide to Gambling, Love, the Stock Market and Just about Everything Else*, New York: Thunder's Mouth Press.

Akerlof, G. 1970. The market for lemons: Quality, uncertainty and the market mechanism, *Quarterly Journal of Economics*, 84, 488–500.

Albert, D. 2012. On the origin of everything. Sunday Book Review. *The New York Times*, 23 March.

Aristotle. *The Politics. Book 1*. Translated by T. A. Sinclair. Revised and Re-presented by T.J. Saunders. 1981. London: Penguin Books Ltd.

Axelrod, R. 1984. *The Evolution of Cooperation*, New York: Basic Books.

Axelrod, R. 2006. *The Evolution of Cooperation* (Revised ed.), New York: Perseus

Axelrod, R. and Hamilton, W.D. 1981. Evolution of cooperation, *Science*, 211, 1390–1396.

Baker, R.D. 2002. Probability paradoxes: An improbable journey to the end of the world, *Mathematics Today*, December, 185–189.

Banz, R.W. 1981. The relationship between return and market value of common stocks, *Journal of Financial Economics*, 9(1), 3–18.

Barberis, N.C. 2013. Thirty years of prospect theory in economics: A review and assessment. *Journal of Economic Perspectives*, 27 (1), 173–196.

Barberis, N. and Huang, M. 2008. Stocks as lotteries: The implications of probability theory for security prices, *American Economic Review*, 98(5), 2066–2100.

Barberis, N., Huang, M. and Thaler, R. 2006. Individual preferences, monetary gambles, and stock market participation: A case for narrow framing, *American Economic Review*, 96(4), 1069–1090.

Barrow, J.D. 2012. *100 Essential Things You Didn't Know about Sport*, London: The Bodley Head.

Barrow, J.D. 2008. *100 Essential Things You Didn't Know You Didn't Know*, London: The Bodley Head.

Bayes, T. and Price, R. 1763. An essay towards solving a problem in the doctrine of chances. By the late Rev. Mr. Bayes, communicated by Mr. Price, in a letter to John Canton, M.A. and F.R.S, *Philosophical Transactions of the Royal Society of London*, 53, 370–418.

Bedwell, M. 2015. Slow thinking and deep learning: Tversky and Kahneman's Cabs, *Global Journal of Human-Social Science*, 15, 12.

Benartzi, S. and Thaler, R.H. 1995. Myopic loss aversion and the equity premium puzzle, *Quarterly Journal of Economics*, 110(1), 73–92.

Benford, F. 1938. The law of Anomalous numbers, *Proceedings of the American Philosophical Society*, 78(4), 551–572.

Bernoulli, J. 2005. [1713]. *The Art of Conjecturing, Together with Letter to a Friend on Sets in Court Tennis (English translation)*, translated by Edith Sylla, Baltimore: Johns Hopkins University Press.

Black, F. 1992. Beating the market: Yes, it can be done, *Economist*, 5 December, 23–26. Republished in Economist Online. 16 October, 2013. https://www.economist.com/free-exchange/2013/10/16/beating-the-market-yes-it-can-be-done

Bostrom, N. 2006. Do we live in a computer simulation? Nick Bostrom, *New Scientist*, 8–9.

Bostrom, N. 2003. Are you living in a computer simulation? *Philosophical Quarterly*, 53(211), 243–255.

Boumen, S. and Jacobsen, B. 2002. The Halloween indicator: "Sell in May and go away": Another puzzle, *American Economic Review*, 92(5), 1618–1635.

Brierley, J. 2017. *Unbelievable?* London: SPCK.

Brown, A., Reade, J. and Vaughan Williams, L. 2019. When are prediction market prices most informative? *International Journal of Forecasting*, 35(1), 420–428.

Brumfiel, G. 2008. The testosterone of trading, *Nature*, 14 April.

Carrazedo, T., Curto, J.D. and Oliveira, L. 2016. The Halloween effect in European sectors, *Research in International Business and Finance*, 37, 489–500.

Chiappori, P., Levitt, S. and Groseclose, T. 2002. Testing mixed-strategy equilibria when players are heterogeneous: The case of penalty kicks in soccer, *American Economic Review*, 92, 1138–1151.

Clotfelter, C. and Cook, P.J. 1993. The 'Gambler's Fallacy' in lottery play, *Management Science*, 39(12), 1521–1525.

Coates, J.M. and Herbert, J. 2008. Endogenous steroids and financial risk taking on a London trading floor, Proceedings of the National Academy of Sciences of the United States of America, 15(16), 6167–6172.

Collins, A., McKenzie, J. and Vaughan Williams, L. 2019. When is a talent contest not a talent contest? Sequential performance bias in expert evaluation. *Economics Letters*, 177, April, 94–98.

Conrad, J., Dittmar, R.F. and Ghysels, E. 2013. Ex ante skewness and expected stock returns. *Journal of Finance*, 68(1), 85–124.

Cross, F. 1973. The behavior of stock prices on Fridays and Mondays. *Financial Analysts Journal*, 29(6), 67–69.

Danziger, S., Levav, J. and Avnaim-Pesso, L. 2011. Extraneous factors in judicial decisions, *PNAS*, 108(17), 6889–6892.

DeGiorgi, E.G. and Legg, S. 2012. Dynamic portfolio choice and asset pricing with narrow framing and probability weighting, *Journal of Economic Dynamics and Control*, 36(7), 951–972.

Dohmen, T.J. 2008. The influence of social forces: Evidence from the behavior of soccer referees, *Economic Inquiry*, 46(3), 411–424.

Dowie, J. 1976. On the efficiency and equity of betting markets, *Economica*, 43(170), May, 139–150.

Eckhardt, W. 1997. A Shooting-room view of Doomsday, *The Journal of Philosophy*, 94(5), 244–259.

Edwards, A. 2013. Ars conjectandi three hundred years on, *Significance*, June, 39–41.

Ellenberg, J. 2015. *How Not To Be Wrong. The Hidden Maths of Everyday Life*, London: Penguin Books.

Ellerton, P. 2014. Why facts alone don't change minds in our public debates, *The Conversation*. May 13. https://theconversation.com/why-facts-alone-dont-change-minds-in-our-big-public-debates-25094

Fama, E.F. 1970. Efficient capital markets : A review of theory and empirical work, *Journal of Finance*, 25(2), May, 383–417.

Feinstein, A. et al. 1985. The Will Rogers Phenomenon – Stage migration and new diagnostic techniques as a source of misleading statistics for survival in cancer, *New England Journal of Medicine*, 312(25), 1604–1608.

Fenton, N., Neil, M. and Berger, D. 2016. Bayes and the law, *Annual Review of Statistics and its Applications*, 3, 51–77.

Fortune, P. 1991. Stock market efficiency: An autopsy? *New England Economic Review*, March, 17–40.

Galton, F. 1907. Vox Populi, *Nature*, 75, 7 March.

Garrett, T., Paton, D. and Vaughan Williams, L. 2020. Taxing gambling machines to enhance public and private revenue, *Kyklos*, 73(4), 500–523.

Gigerenzer, G. 2003. *Reckoning with Risk. Learning To Live With Uncertainty*, London: Penguin Books.

Golec, J. and Tamarkin, M. 1998. Bettors love skewness, not risk, at the horse track. *Journal of Political Economy*, 106(1), 205–225.

Good, I.J. 1995. When batterer turns murderer, *Nature*, 375, 541.

Gordon, K.H. 1921. Group judgements in the field of lifted weights, *Psychological Review*, 28(6), November, 398–424.

Green, T.C. and Hwang, B.-H. 2012. Initial public offerings as lotteries: Skewness preferences and first-day returns, *Management Science*, 58(2), 432–444.

Griffith R.M. 1949. Odds adjustments by American horse-race bettors, *American Journal of Psychology*, 62(2), 290–294.

Grossman, S.J. and Stiglitz, J. 1980. The impossibility of informationally efficient markets, *American Economic Review*, 70(3), June, 393–408.

Haggard, K.S. and Witte, H.D. 2010. The Halloween effect: Trick or treat, *International Review of Financial Analysis*, 19(5), 379–387.

Haigh, J. and Vaughan Williams, L. 2008. Index betting for sports and stock indices. In *Handbook of Sports and Lottery Markets*, Eds. D. Hausch and W. Ziemba, pp. 357–383, North Holland.

Hanson, R. and Oprea, R. 2009. A manipulator can aid prediction market accuracy, *Economica*, 76(302), 304–314.

Henery, R.J. 1985. On the average probability of losing bets on horses with given starting price odds, *Journal of the Royal Statistical Society. Series A (General)*, 148(4), 342–349.

Hirshleifer, D. and Shumway, T. 2003. Good day sunshine: Stock returns and the weather, *Journal of Finance*, 58(3), June, 1009–1062.

Hooper, M. 2013. Richard Price, Bayes' theorem and god, *Significance*, February, 36–39.

Horgan, J. 2012. Science will never explain why there's something rather than nothing, *Scientific American*, 23 April.

Hu, W.-Y. and Scott, J.S. 2007. Behavioral obstacles in the annuity market, *Financial Analysts Journal*, 63(6), 71–82.

Jacobsen, B. and Zhang, C.Y. 2012. The Halloween indicator: Everywhere and all the time, Working Paper. Massey University, University of New Zealand.

Jervis, R. 2002. Signaling and perception. In *Political Psychology*, Ed., K. Monroe, Earlbaum.

Kachelmeier, S.J. and Shehata, M. 1992. Examining risk preferences under high monetary incentives: Experimental evidence from the People's Republic of China, *American Economic Review*, 82(5), 1120–1141.

Kahneman, D. and Tversky, A. 1984. Choices, values, and frames, *American Psychologist*, 39(4), April, 341–350.

Kahneman, D., Sibony, O. and Sunstein, C.R. *Noise: A Flaw in Human Judgment*. London: William Collins.

Keane, S.M. 1987. *Efficient Markets and Financial Reporting*, Edinburgh: Institute of Chartered Accountants of Scotland.

Kelly, J.L. 1956. A new interpretation of information rate. *Bell System Technical Journal*, 35(4), 917–926.

Koszegi, B. and Rabin, M. 2006. A model of reference-dependent preferences, *Quarterly Journal of Economics*, 121(4), 1133–1165.

Koszegi, B. and Rabin, M. 2007. Reference-dependent risk attitudes, *American Economic Review*, 97(4), 1047–1073.

Koszegi, B. and Rabin, M. 2009. Reference-dependent consumption plans, *American Economic Review*, 99(3), 909–936.

Krueger, T.M. and Kennedy, W.F. 1990. An examination of the super bowl stock market predictor, *Journal of Finance*, 45(2), 691–697.

Kucharski, A. 2016. *The Perfect Bet. How Science and Maths are Taking the Luck Out of Gambling*, London: Profile Books Ltd.

Lamont, O.A. and Thaler, R.A. 2003. Anomalies: The law of one price in financial markets, *Journal of Economic Perspectives*, 17(4), 191–202.

Lee, M. and King, B. 2017. Bayes' Theorem: The maths tool we probably use every day. But what is it? *The Conversation*. 23 April.

Leslie, J. 1998. *The End of the World: The Science and Ethics of Human Extinction*, Abingdon: Routledge.

Levitt, S.D. and Dubner, S.J. 2015. *When to Rob a Bank*, London: Penguin Books.

Lewis, M.A. 2020. Bayes' theorem and Covid-19 testing, *Significance*, 22 April.

Malkiel, B. 2003. The Efficient Market Hypothesis and Its Critics. CEPE Working paper 91. Princeton University.

McGraw, A.P., Mellers, B.A. and Tetlock, P.E. 2005. Expectations and emotions of Olympic athletes, *Journal of Experimental Social Psychology*, 41(4), 438–446.

Mankiw, N.G. and Zeldes, S.P. 1991. The consumption of stockholders and nonstockholders, *Journal of Financial Economics*, 29(1), 97–112.

Mantonakis, A., Rodero, P., Lesschaeve, I. and Hastie, R. 2009. Order in choice: Effects of serial position on preferences, *Psychological Science*, 20(1), 1309–1312.

Mlodinow, L. 2009. *The Drunkard's Walk. How Randomness Rules Our Lives*, London: Penguin Books.

Moskowitz, C. 2016. Are we living in a computer simulation? Clare Moskowitz, *Scientific American*, 7 April.

Moskowitz, T.J. and Wertheim, L.J. 2011. *Scorecasting*, New York: Random House.

Nevill, A.M., Balmer, N.J. and Williams, A.M. 2002. The influence of crowd noise and experience upon refereeing decisions in football, *Psychology of Sport and Exercise*, 3(4), 261–272.

New Scientist. 2015. *Chance: The Science and Secret of Luck, Randomness and Probability*, Ed. M. Brooks, London: Profile Books Ltd.

Nozick, R. 1969. Newcomb's problem and two principles of choice. In Essays in Honor of Carl G. Hempel, Ed. N. Rescher et al., New York: Springer.

Orkin, M. 1991. *Can You Win? The Real Odds for Casino Gambling, Sports Betting, and Lotteries. With a chapter on Prisoner's Dilemma*, New York: W.H. Freeman and Company.

Parfit, D. 1998. Why anything? Why this? Part 1, *London Review of Books*, 20(2), 22 January, pp. 24–27.

Parfit, D. 1998. Why anything? Why this? Part 2, *London Review of Books*, 20(3), 5 February, pp. 22–25.

Paton, D., Siegel, D. and Vaughan Williams, L. 2010. Gambling, prediction markets and public policy, *Southern Economic Journal*, 76(4), 878–883.

Paton, D., Siegel, D. and Vaughan Williams, L. 2009. The growth of gambling and prediction markets: Economic and financial implications, *Economica*, 76(302), 219–224.

Paton, D. and Vaughan Williams, L. 2005. Forecasting outcomes in spread betting markets: Can bettors use 'quarbs' to beat the book? *Journal of Forecasting*, 24(2), 139–154.

Paton, D. and Vaughan Williams, L. 1998. Do betting costs explain betting biases? *Applied Economics Letters*, 5(5), 333–335.

Paton, D., Siegel, D. and Vaughan Williams, L. 2002. A Policy Response to the E-Commerce revolution: The case of betting taxation in the UK, *Economic Journal*, 12(480), F296–F214.

Paton, D., Siegel, D. and Vaughan Williams, L. 2004. Taxation and the demand for gambling: New evidence from the United Kingdom, *National Tax Journal*, 57(4), 847–861.

Paton, D., Siegel, D. and Vaughan Williams, L. 2002. Gambling taxation: A comment, *Australian Economic Review*, 34(4), 437–440.

Pope, D.G. and Schweitzer, M.E. 2011. Is Tiger Woods loss-averse? Persistent bias in the face of experience, competition and high stakes, *American Economic Review*, 101(1), 129–157.

Post, T., Van den Assem, M.J., Baltussen, G. and Thaler, R.H. 2008. Deal or no deal? Decision making under risk in a large-payoff game show, *American Economic Review*, 98(1), 38–71.

Poundstone, W. 2019. *How to Predict Everything*. London: Oneworld Publications.

Poundstone, W. 2015. *How to Predict the Unpredictable: The Art of Outsmarting Almost Everyone*. London: Oneworld Publications.

Poundstone, W. 2005. *Fortune's Formula. The Untold Story of the Scientific System that beat the casinos and Wall Street*, New York: Hill and Wang.

Price, J. and Wolfers, J. 2014. Right-oriented bias: A comment on Roskes, Sligte, Shalvi, and De Dreu (2011), *Psychological Science*, 25(11), 2109–2111.

Puga, J., Krzywinski, N. and Altman, N. 2015. Points of significance: Bayes' Theorem, *Nature Methods*, 12(4), April, 277–278.

Reade, J. and Vaughan Williams, L. 2019. Polls to probabilities: Comparing prediction markets and opinion polls, *International Journal of Forecasting*, 35(1), 336–350.

Reade, J., Singleton, C. and Vaughan Williams, L. 2020. Betting markets for English Premier League results and scorelines: Evaluating a forecasting model, *Economic Issues*, 25(1), 87–106.

Robertson, B., Vignaux, G.A. and Berger, C.E.H. 2016. *Interpreting evidence: Evaluating Forensic Science in the Courtroom*, 2nd Edn, Chichester: Wiley.

Roskes, M., Sligte, D. Shalvi, S., Carsten, K. and De Dreu, W. 2011. The right side? Under time pressure, approach motivation leads to right-oriented bias. *Psychological Science*, 22(11), 1403–1407.

Rozeff, M.S. and Kinney, W.R. 1976. Capital market seasonality: The case of stock returns, *Journal of Financial Economics*, 3, October, 379–402.

Salop, S.C. 1987. Evaluating uncertain evidence with Sir Thomas Bayes: A note for teachers, *Economic Perspectives*, 1(1, Summer), 155–160.

Saunders, E. 1993. Stock prices and wall street weather, *American Economic Review*, 83(5), 1337–1345.

Saville, B., Stekler, H. and Vaughan Williams, L. 2011. Do polls or markets forecast better? Evidence from the 2010 US Senate Elections, *Journal of Prediction Markets*, 5(3), 64–74.

Schmidt, B. and Clayton, R. 2017. Super bowl indicator and equity markets: Correlation not Causation, *Journal of Business Inquiry*, 17(2), 97–103.

Science Buddies. 2012. Probability and the birthday paradox, *Scientific American*, 29 March.

Significance. 2021. A school named Bayes. June, 18, 3.

Silver, N. 2012. *The Signal and the Noise: The Art and Science of Prediction*. London: Allen Lane.

Skorupski, W.P. and Wainer, H. 2015. The Bayesian flip. Correcting the prosecutor's fallacy. *Significance*. August, 16–20.

Smith, M.A. and Vaughan Williams, L. 2010. Forecasting horse race outcomes: New evidence on odds bas in UK betting markets, *International Journal of Forecasting*, 26(3), 543–550.

Smith, M., Paton, D. and Vaughan Williams, L. 2009. Do bookmakers possess superior skills to bettors in predicting outcomes? *Journal of Economic Behavior and Organization*, 71(2), 539–549.

Smith, M.A., Paton, D. and Vaughan Williams, L. 2006. Market efficiency in person-to-person betting, *Economica*, 73(292), 673–689.

Smith, M.A., Paton, D. and Vaughan Williams, L. 2005. An assessment of quasi-arbitrage opportunities in two fixed-odds horse-race betting markets. In *Information Efficiency in Financial and Betting Markets*, Ed. L. Vaughan Williams, Chapter 4, Cambridge: Cambridge University Press.

Spence, M. 1973. Job market signaling, *Quarterly Journal of Economics*, 87(3), 355–374.

Stangroom, J. 2009. *Einstein's Riddle. Riddles, Paradoxes and Conundrums to Stretch Your Mind*, London: Bloomsbury Publishing Plc.

Stewart, I. 2019. *Do Dice Play God? The Mathematics of Uncertainty*, London: Profile Books Ltd.

Stiglitz, J.E. 1975. The theory of 'Screening,' education, and the distribution of income, *American Economic Review*, 65(3), 283–300.

Stiglitz, J. E. 1981. *Information and the Change in the Paradigm of Economics*, Nobel Prize Lecture, December 8.

Stovall, R. 1989. The super bowl predictor, *Financial World*, 158(72), 24 January.

Summers, L.H. 1985. On economics and finance, *Journal of Finance*, 40(3), 633–635.

Surowiecki, J. 2004. *The Wisdom of Crowds: Why the Many are Smarter Than the Few and How Collective Wisdom Shapes Business, Economies, Societies, and Nations*, Doubleday.

Sydnor, J. 2010. (Over)insuring modest risks, *American Economic Journal: Applied Economics*, 2(4), 177–199.

Talwalkar, P. 2016. *The Irrationality Illusion. How to Make Smart Decisions and Overcome Bias*, CreateSpace Independent Publishing Platform.

Talwalkar, P. 2015a. *Math Puzzles Volume 1. Classic Riddles and Brain Teasers in Counting, Geometry, Probability, and Game Theory*, Scotts Valley: CreateSpace Independent Publishing Platform.

Talwalkar, P. 2015b. *Math Puzzles Volume 2. More Riddles and Brain Teasers in Counting, Geometry, Probability, and Game Theory*, Scotts Valley: CreateSpace Independent Publishing Platform.

Talwalkar, P. 2015c. *Math Puzzles Volume 1. Even More Riddles and Brain Teasers in Counting, Geometry, Probability, and Game Theory*, Scotts Valley: CreateSpace Independent Publishing Platform.

Talwalkar, P. 2015d. *40 Paradoxes in Logic, Probability, and Game Theory*, Scotts Valley: CreateSpace Independent Publishing Platform.

Talwalkar, P. 2014. *The Joy of Game Theory. An Introduction to Strategic Thinking*, Scotts Vally: CreateSpace Independent Publishing Platform.

Terrell, D. 1994. A test of the gambler's fallacy: Evidence from pari-mutuel games, *Journal of Risk and Uncertainty*, 8, 309–317.

Tetlock, P. and Gardner, D. 2016. *Superforecasting: The Art and Science of Prediction*, London: Random House.

Thalheimer, R. and Ali, M.M. 1995. The demand for parimutuel horse race wagering and attendance, *Management Science*, 41(1), 129–43.

Thorp, E.O. 1966. *Beat the Dealer: A Winning Strategy for the Game of Twenty-One*, New York: Random House.

Tijms, H. 2019. *Surprises in Probability – Seven Short Stories*, Boca Raton: CRC Press. Taylor & Francis Group.

Treynor, J. 1987. Market efficiency and the bean jar experiment, *Financial Analysts Journal*, 43, 50–53.

Trotta, R. 2014. Is the All-There-Is all there is? *Significance*, June, 69–71.

Tversky, A. and Kahneman, D. 1982. Evidential impact of base rates. In *Judgment under Uncertainty: Heuristics and Biases*, Eds. Kahneman, D., Slovic, P. and Tversky, A., Abingdon: Routledge.

Vaughan Williams, L. 2020. Joe Biden: How betting markets foresaw the result of the 2020 US election. *The Conversation*. 13 November. https://theconversation.com/joe-biden-how-betting-markets-foresaw-the-result-of-the-2020-us-election-150095

Vaughan Williams, L. 2019. Strictly come dancing: Research shows that the luck of the draw matters in talent shows. *The Conversation*. 5 September. https://theconversation.com/strictly-come-dancing-research-shows-that-the-luck-of-the-draw-matters-in-talent-shows-122896

Vaughan Williams, L., Sung, M. and Johnson, J.E.V. 2019. Prediction markets: Theory, evidence and applications, *International Journal of Forecasting*, 35(1), 266–270.

Vaughan Williams, L. 2018. Written evidence (PPD0024). House of Lords Political Polling and Digital Media Committee. 16 January. http://data.parliament.uk/writtenevidence/committeeevidence.svc/evidencedocument/political-polling-and-digital-media-committee/political-polling-and-digital-media/written/72373.pdf

Vaughan Williams, L. 2016. How the wisdom of crowds could solve the mystery of Shakespeare's 'lost plays.' *The Conversation*. 14 April. https://theconversation.com/how-the-wisdom-of-crowds-could-solve-the-mystery-of-shakespeares-lost-plays-57705

Vaughan Williams, L. 2015. What happened to MH370? Prediction markets might give us the answer. *The Conversation*. 4 August. https://theconversation.com/what-happened-to-mh370-prediction-markets-might-give-us-the-answer-45528

Vaughan Williams, L. 2015. The Nobel Prize prediction industry: far from perfect, but pretty impressive. *The Conversation*. 12 October. https://theconversation.com/the-nobel-prize-prediction-industry-far-from-perfect-but-pretty-impressive-48916

Vaughan Williams, L. 2015. Forecasting the decisions of the US Supreme Court: Lessons from the 'Affordable Care Act' judgment, *Journal of Prediction Markets*, 9(2), 64–78.

Vaughan Williams, L. 2014. The Churchill Betting Tax, 1926–30: A historical and economic perspective, *Economic Issues*, 19(2), 21–38.

Vaughan Williams, L. 2012. *The Economics of Gambling and National Lotteries. The International Library of Critical Writings in Economics*, Cheltenham: Edward Elgar.

Vaughan Williams, L. Ed. 2011. *Prediction Markets: Theory and Applications*, Routledge International Studies in Money and Banking, Abingdon: Routledge.

Vaughan Williams, L. Ed. 2005. *Information Efficiency in Financial and Betting Markets*, Cambridge: Cambridge University Press.

Vaughan Williams, L. 2003a. *The Economics of Gambling*, London: Routledge.

Vaughan Williams, L. 2003b. *Betting to Win: A Professional Guide to Profitable Betting*, Harpenden: Oldcastle Books Ltd.

Vaughan Williams, L. 2000. Can forecasters forecast successfully? Evidence from UK betting markets, *Journal of Forecasting*, 19(6), 505–513.

Vaughan Williams, L. 1999. Information efficiency in betting markets: A survey, *Bulletin of Economic Research*, 51(1), 1–39.

Vaughan Williams, L., Garrett, T. and Paton, D. 2020. Taxing gambling machines to enhance tourism, *Journal of Gambling Business and Economics*, 13(2), 83–90.

Vaughan Williams, L. and Reade, J. 2016. Prediction markets, social media and information efficiency, *Kyklos*, 69(3), 518–556.

Vaughan Williams, L., Sung, M., Fraser-Mackenzie, P., Peirson, J. and Johnson, J.E.V. 2016. Towards an understanding of the origins of the favourite-longshot Bias: Evidence from online poker, a real-world natural laboratory. *Economica*, 85(338), 360–382.

Vaughan Williams, L. and Paton, D. 2015. Forecasting the outcome of closed-door decisions: Evidence from 500 years of betting on papal conclaves, *Journal of Forecasting*, 34(5), 391–404.

Vaughan Williams, L. and Reade, J. 2015. Forecasting elections, *Journal of Forecasting*, 35(4), 308–328.

Vaughan Williams, L. and Siegel, D. 2013. *The Oxford Handbook of the Economics of Gambling*, New York: Oxford University Press.

Vaughan Williams, L. and Paton, D. 2013. The taxation of gambling machines: A theoretical perspective. In *The Oxford Handbook of the Economics of Gambling*, Ed. L. Vaughan Williams and D. Siegel, New York: Oxford University Press.

Vaughan Williams, L. and Stekler, H. 2010. Sports forecasting, *International Journal of Forecasting*, 26(3), 445–447.

Vaughan Williams, L. and Vaughan Williams, J. 2009. The cleverness of crowds. *Journal of Prediction Markets*, 3(3), 45–47.

Vaughan Williams, L. and Paton, D. 1998. Why are some favourite-longshot biases positive and others negative? *Applied Economics*, 30(11), 1505–1510

Vaughan Williams, L. and Paton, D. 1997. Why is there a favourite-longshot bias in British racetrack betting markets? *Economic Journal*, 107, 150–158.

Vaughan Williams, L. and Paton, D. 1997. Does information efficiency require a perception of information inefficiency? *Applied Economics Letters*, 4(10), 615–617.

Vaughan Williams, L. and Paton, D. 1996. Risk, return and adverse selection: A study of optimal behaviour under asymmetric information. *Rivista di Politica Economica*, 11–12, 63–81.

Walker, M. and Wooders J. 2001. Minimax play at Wimbledon, *American Economic Review*, 91, 1521–1538.

Waniek, M., Niescieruk, A., Michalak, T. and Rahwan, T. 2015. Spiteful Bidding in the Dollar Auction. *Proceedings of the Twenty-Fourth International Joint Conference on Artificial Intelligence*.

Zhang, C.Y. and Jacobsen, B. 2021. The Halloween indicator, "Sell in May and go away": Everywhere and all the time, *Journal of International Money and Finance*, 110, February, 1–49.

Solutions to Exercises

1.1.2 Solutions

Question a. The formula for Bayes' Theorem can be represented as:
Updated (posterior) probability given new evidence = ab / [ab + c (1 − a)]

a is the probability that the hypothesis is true before the arrival of new evidence. This is called the prior probability.

b is the probability of the evidence arising given that the hypothesis is true.

c is the probability of the evidence arising given that the hypothesis is not true.

Question b. P (HIE) = P (H) . P (EIH) / [P (H) . P (EIH) + P (EIH′) . P (H′)]

Question c. P (H) is the probability that the hypothesis is true before the new evidence. This is the prior probability (also represented as **a**). P (EIH) is the probability of the evidence given that the hypothesis is true (also represented as **b**). P (EIH′) is the probability of the evidence given that the hypothesis is not true (also represented as **c**).

Question d.
No.
P (HIE) = P (EIH) P (H) / P (E) ... Bayes' Theorem
This can be expanded to: P (HIE) = P (EIH) P (H) / [P (EIH) P (H) + P (EIH′) P (H′)]
From the first equation, P (HIE) only equals P (EIH) when P (H) = P (E), i.e. P (H) / P (E) = 1

a. P (HIE) = P (H) . P (EIH) / [P (H) . P (EIH) + P (EIH′) . P (1 − P (H)]

b. No. There is a very good chance that you might feel warm being out in the sun, but there are many other reasons why you might feel warm other than being out in the sun.

Question e. 7 + 5 / (7 + 10) = 12/17 = 70.6%.

Smaller numbers update the probability more quickly. Larger numbers anchor more to the baseline probability.

Question f. Hypothesis (H) = you chose the biased die.
Evidence (E) = the die lands on 6.
 What is the chance that you have chosen the biased die given that you rolled 6, i.e. what is P (HIE)?

> P (HIE) = P (EIH) . P (H) / [P (H) . P (EIH) + P (EIH′) . P (1 − P (H)]
>
> P (EIH) = P (6 I Biased) = 1/2 (it's weighted such that half the time it rolls a 6)
>
> P (H) = P (Biased) = 1/2 as there are two dice, each with an equal probability of being the biased die.
>
> P (EIH′) = P (6 I Not Biased) = (chance of rolling a six if the die is not biased) = 1/6
>
> So, P (HIE) = P (EIH) . P (H) / [P (H) . P (EIH) + P (EIH′) . P (1 − P (H)]
> = 1/2 × 1/2 / [(1/2 × 1/2 + 1/6 (1/2)) = 1/4 / (1/4 + 1/12) = 1/4 / 1/3
> = 3/4.

Alternative solution
 Probability the die is biased given that a 6 is rolled = ab / [ab + c (1 − a)].
 So, the hypothesis is that the die is biased. The evidence is that it lands on 6.

> a is the prior probability that the chosen die is biased (the hypothesis is true) = 1/2
>
> b is the probability of rolling 6 (the evidence) if the die is biased (the hypothesis is true) = 1/2
>
> c is the probability of rolling 6 if the die is not biased (the hypothesis is not true) = 1/6

Probability the die is biased = 1/2 × 1/2 / [(1/2 × 1/2 + 1/6 (1/2)] = 1/4 / (1/4 + 1/12) = 1/4 / 1/3 = 3/4

Question g.
Hypothesis (H) = you chose the weighted coin.
 Event B = the coin landed on Heads.
 What is the chance that you have chosen the weighted coin given that it landed on Heads, i.e. what is P (HIE)?

> P (HIE) = P (EIH) . P (H) / [P (H) . P (EIH) + P (EIH′) . P (1 − P (H)]
>
> P (EIH) = P (Heads I Weighted) = 3/4 (it's weighted such that 75% of the time it lands on Heads)

P (H) = 1/2, as there are two coins, each with an equal probability of being the weighted coin.

P (EIH') = P (Heads I Not Weighted) = 1/2, as a fair coin lands Heads half the time.

So, P (HIE) = 1/2 × 3/4 / [1/2 × 3/4 + 1/2 (1/2)] = 3/8 / (3/8 + 1/4) = 3/8 / 5/8 = 3/5.

Alternative solution

Probability the coin is weighted given that it lands on Heads = ab / [ab + c (1 − a)].

a is the prior probability that the chosen coin is weighted = 1/2

b is the probability of the coin landing Heads if it is weighted = 3/4

c is the probability of the coin landing Heads if the coin is not weighted = 1/2

Probability the coin is weighted = 1/2 × 3/4 / [1/2 × 3/4 + 1/2 (1/2)] = 3/8 / (3/8 + 1/4) = 3/8 / 5/8 = 3/5

Question h.
In terms of the probability that your colleague is a genuine perfect Dice Predictor (H1) compared to a guesser (H2), what is the Bayes Factor?

$$\text{Bayes Factor} = P (E \text{ I } H1) / P (E \text{ I } H2)$$

P (E I H1) = 1
P (E I H2) = 1/6 × 1/6 = 1/36
Bayes Factor = 1 / 1/36 = 36

What are the posterior odds if you were originally perfectly split between believing in his powers and believing he was a guesser?

Posterior odds = O (H1) . P (E I H1) / P (E I H2)
Posterior odds = 1/1 × 36 = 36

What are the posterior odds if you originally believed that he was 100 times more likely to be a guesser than a genuine perfect Dice Predictor?

Posterior odds = 1/100 × 36 = 0.36

Question i.
P (Evidence I H1) = 1, where H1 is the hypothesis that the coin toss is rigged to always land heads.

P (Evidence I H2) = $(1/2)^{12}$ = 1/4096, where H2 is the hypothesis that the coin is fair.

Bayes Factor = P (E I H1) / P (E I H2) = 1 / (1/4096) = 4096

Posterior odds if prior probability of cheating is 1/1,000 = 1/1,000 × 4,096 = 4.096

The posterior odds represent how many times better the hypothesis of rigged coin tosses explains the data than the hypothesis of a fair series of coin tosses.

1.2.2 Solutions

Question a.

a is the prior probability, i.e. the probability that a hypothesis is true before the new evidence arises. This is 0.2 (20%) because 20% of the taxis in New Amsterdam are white.

b is the probability the new evidence would arise if the hypothesis is true. This is 0.8 (80%). There is an 80% chance that the witness would say the taxi was white if it was indeed white.

c is the probability the new evidence would arise if the hypothesis is false. This is 0.2 (20%). There is a 20% chance that the witness would be wrong and identify the taxi as white if it was yellow.

Inserting these numbers into the formula, ab / [ab + c (1 − a)], gives:
  Posterior probability = 0.2 × 0.8 / [0.2 × 0.8 + 0.2 (1 − 0.2)] = 0.16 / [0.16 + 0.16] = 0.5 = 50%

Question b.

a is the prior probability, i.e. the probability that a hypothesis is correct before the new evidence arises. This is 0.5 (50%) based on the previous calculation.

b is the probability the new evidence would occur if the hypothesis is correct. This is 0.7 (70%). There is a 70% chance that the witness would say the taxi was white if it was indeed white.

c is the probability the new evidence would arise if the hypothesis is false. This is 0.3 (30%). There is a 30% chance that the witness would be wrong and identify the taxi as white if it was yellow.

Inserting these numbers into the formula, ab / [ab + c (1 − a)], gives:

Posterior probability = 0.5 × 0.7 / [0.5 × 0.7 + 0.3 (1 − 0.5)] = 0.35 / [0.35 + 0.15] = 0.7 = 70%

Question c.

a is the prior probability, i.e. the probability that a hypothesis is true before the new evidence arises. This is 0.7 (70%) based on the previous calculation.

b is the probability the new evidence would arise if the hypothesis is true. This is 0.5 (50%). There is a 50% chance that the witness would say the taxi was white if it was indeed white.

c is the probability the new evidence would arise if the hypothesis is false. This is 0.5 (50%). There is a 50% chance that the witness would be wrong and identify the taxi as white if it was yellow.

Inserting these numbers into the formula, ab / [ab + c (1 − a)], gives:

Posterior probability = 0.7 × 0.5/ [0.7 × 0.5 + 0.5 (1 − 0.7)] = 0.35 / [0.35 + 0.15] = 0.7 = 70%. Same as before. The new witness is right exactly 50% of the time and wrong 50% of the time, so adds no new information. The posterior (updated) probability stays the same as the prior probability.

Question d.

Mr. Smith is correct 50% of the time and so adds no new information. He may as well toss a coin to choose the colour.

Mr. Jones is correct 100% of the time and so adds vital information. If he says the taxi was white, it was white.

Mr. Evans gets it wrong 100% of the time and so adds vital information by reversing the information. If he says the taxi was white, it was yellow.

So, Mr. Smith is not at all useful to investigators. Mr. Jones and Mr. Evans are equally very useful to investigators, for different reasons.

1.3.2 Solutions

To calculate just how likely the beetle is to be rare given that we see the pattern on its back, we apply Bayes' Theorem.

Posterior probability = ab / [ab + c (1 − a)]

a is the prior probability of the hypothesis (beetle is rare) being true. b is the probability we observe the pattern, and the beetle is rare (hypothesis is true). c is the probability we observe the pattern, and the beetle is not rare (hypothesis is false).

In this case, a = 0.01 (1%); b = 0.95 (95%); c = 0.02 (2%).

So, updated probability = ab / [ab + c (1 − a)] = 0.0095 / [0.0095 + 0.02 (1 − 0.01)] = 0.0095 / (0.0095 + 0.0198) = 0.0095 / 0.0293 = 0.324.

So there is a 32.4% chance that the beetle is rare when the nature lover spots the distinctive pattern on its back.

Alternative solution

P (H) = 0.01

P (EIH) = 0.95

P (EIH′} = 0.05

P (HIE) = P (EIH) . P (H) / [P (EIH) . P (H) + P (EIH′) . P (H′)] = 0.95 × 0.01 / (0.95 × 0.01 + 0.05 × 0.99) = 0.0095 / (0.0095 + 0.0495) = 0.0095 / 0.059 = 0.161

What are a, b, c? What are P (H), P (EIH), P (EIH′), and P (HIE)?

a = 0.01

b = 0.95

c = 0.02

P (H) = 0.01

P (EIH) = 0.95

P (EIH′) = 0.05

P (HIE) = 0.161

1.4.3 Solutions

1. Using the formula, ab / [ab + c (1 − a)]

   a is the prior probability, i.e. the probability that a hypothesis is true before you see the new evidence. Before the new evidence (the test), this chance is put at 1 in 100 (0.01), as 1% of the people who visit the surgery have the virus. So, a = 0.01.

   b is the probability of the new evidence if the hypothesis is true. The probability of the new evidence (the positive result on the test) if the hypothesis is true (the patient has the virus) is 95%, since the test is 95% accurate. So, b = 0.95.

   c is the probability of the new evidence if the hypothesis is false. The probability of the new evidence if the hypothesis is false (does not have the virus) is 5%, because the test is 95% accurate, and we can only expect a false positive 5 times in 100. So, c = 0.05.

Using Bayes' Theorem, the updated (posterior) probability = 0.01 × 0.95 / [(0.01 × 0.95) + 0.05 (1 − 0.01)] = 0.0095 / (0.0095 + 0.0495) = 0.161.

So the chance the doctor should give to the patient having the virus, if tested positive, is 16.1%.

2. What is the probability that the player is guilty if tested positive?

$$ab / \left[ab + c\left(1-a\right) \right]$$

a is the prior probability, i.e. the probability that a hypothesis (the player is using banned substances) is true before you see the new evidence. Before the new evidence (the test), this chance is estimated at 1 in 10 (0.1), as 10% of the players are using the banned substances. So, a = 0.1.

b is the probability of the new evidence if the hypothesis is true. The probability of the new evidence (the positive result on the test) if the hypothesis is true (the player is guilty) is 90%, since a positive test is 90% accurate. So, b = 0.9.

c is the probability of the new evidence if the hypothesis is false. The probability of the new evidence (negative result on the test) if the hypothesis is false (the player is not taking the banned substances) is 85%, so we can expect a false positive 15 times in 100 if the player is not guilty. So, c = 0.15.

Using Bayes' Theorem, the updated (posterior) probability = 0.1 × 0.9 / [(0.1 × 0.9) + 0.15 (1 − 0.1)] = 0.09 / (0.09 + 0.135) = 0.09 / 0.225 = 0.4.

So, the probability that an entrant to the tennis tournament tests positive and is taking the banned substances is 40%.

3. a. Sensitivity = TP / (TP + FN) = 66 / (66 + 4)
   b. Specificity = TN / (TN + FP) = 827 / (827 + 3)
   c. PPV = TP / (TP + FP) = 66 / (66 + 3)
   d. NPV = TN / (TN + FN) = 827 / (827 + 4)
   e. LR+ = Sensitivity / (1 − Specificity)
   f. LR− = (1−sensitivity)/specificity

4. a. Sensitivity = 90 / (90 + 10) = 90%
   b. Specificity = 750 / (750 + 150) = 83%

5. a. Sensitivity is the proportion of those with the disease who test positive.
   b. Specificity is the proportion of those without the disease who test negative.

c. The larger the Positive Likelihood Ratio, the greater the likelihood of disease when testing positive; the smaller the Negative Likelihood Ratio, the less the likelihood of disease when testing negative.

6. Percentage of infected in the vaccinated group = 52 / 56,230 = 0.00092

Percentage of infected in the placebo group = 88 / 8,428 = 0.01044

Calculate the Vaccine Efficacy.

The formula to use is:

Vaccine Efficacy = 1 – risk in vaccinated arm of the trial / risk in unvaccinated armof the trial = 1 – 0.00092 / 0.01044

Vaccine Efficacy = 1 – 0.08812 = 91.19%

Vaccine efficiacy refers to performance in controlled clinical trials, while vaccine efficiency refers to real-world performance in the wider population.

1.5.2 Solutions

Question 1: We can solve the Lucy Jones problem using Bayes' Theorem. The updated probability that a hypothesis is true after obtaining new evidence, according to the a, b, c formula of Bayes' Theorem, is equal to:

$$ab / \left[ab + c\left(1-a\right) \right]$$

a is the prior probability, i.e. the probability that a hypothesis is true before the new evidence.

b is the probability of the new evidence if the hypothesis is true.

c is the probability of the new evidence if the hypothesis is false.

Before the new evidence (the test), this chance is 1 in 1,000 (0.001)
So a = 0.001.

The probability of the new evidence (entry to the academy) if the hypothesis is true (Lucy will become a professional chess player) is 100% since all professional players must be graduates of the academy.
So b = 1.

The probability we would see the new evidence (admission to the academy) if the hypothesis is false (Lucy will not become a professional player) is 20/999 since 999 of those who took the test will not become professional players, and of these 20 will gain entry to the academy.
So c = 20/999.

Substituting into Bayes' equation gives:

Posterior probability = ab/ [ab + c (1 − a)] = 0.001 × 1 / [0.001 × 1 + 20/999 (1 − 0.001)] = 0.001 / (0.001 + 0.02) = 0.001 / 0.021 = 1 / 21 = 0.0476 = 4.76%

So, using Bayes' Theorem, the chance that Lucy Jones, who gained admission to the chess academy, will become a professional player, is not 98% as intuition might suggest, but just 4.76%.

Looked at another way, of a thousand children in total, in terms of probability 20 will be admitted to the academy and fail to become professional chess players. Only 1 in 1,000 will enter the academy and actually become a professional player. Therefore, the probability that Lucy will become a professional chess player is now 1 in 21, i.e. 4.76%.

Alternative solution

We can also solve the problem using the traditional notation version of Bayes' Theorem.

$$P (HIE) = P (EIH) . P (H) / [P (EIH) . P (H) + P (EIH') . P (H')]$$

Before the new evidence (the test), this chance is 1 in 1,000 (0.001).

$$\text{So } P (H) = 0.001.$$

The probability of the new evidence (entry to the chess academy) if the hypothesis is true (Lucy will become a professional chess player) is 100% since all professional players must be graduates of the academy.

$$\text{So } P (EIH) = 1.$$

The probability we would see the new evidence (entry to the academy) if the hypothesis is false (Lucy will not become a professional player) is 20/999 since 999 of those who took the test will not become professional players, and of these 20 will have been admitted to the academy.

$$\text{So } P (EIH') = 20/999.$$

Substituting into Bayes' equation gives:

$$P (HIE) = 0.001 × 1 / [0.001 × 1 + 20/999 (1 − 0.001)] = 0.001 / (0.001 + 0.02) = 0.001 / 0.021 = 0.0476 = 4.76\%$$

Alternative calculation: 20 aspiring chess players are admitted to the academy and will fail to become professional chess players. One aspiring chess player enters the academy and will become a professional chess player. Probability of becoming a professional chess player if admitted to the academy = 1/21 = 4.76%.

Question 2:
Without any other information, the prior probability of securing one of the five positions is down to the number of graduate applications made. There are 1,000 so she has a 5 in 1000, or 1 in 200 chance of being offered a position.

$$a = P(H) = 5/1000 = 0.005$$

We now receive new information that she has been offered an interview.
The probability of being offered an interview, given that she will be offered a position, P(EIH), is 1, since we are told all the employees have to attend an interview.

$$b = P(E|H) = 1$$

995 applicants (1,000 minus the five successful applicants) will not be offered the entry level position, and of these 50 will be interviewed.

$$c = P(E|H') = 50/995$$
$$P(H|E) = 0.005 \times 1 / [(0.005 \times 1 + 50/995 (1-0.005)] = 0.005 / 0.055 = 0.91$$
$$= 9.1\%$$

Alternative calculation: 5 applicants are interviewed and offered positions. 50 applicants are interviewed and not offered positions. Probability of being offered a position if interviewed = 5/55 = 0.91 = 9.1%.

Without any other information, the prior probability that she secures one of the five positions at Apple, P (H), is a 5 in 1,000 or 1 in 200 chance. This is because there are 1,000 graduate applications, out of which 5 will be successful.

$$a = P(H) = 5 / 1,000 = 0.005$$

We now receive new information that she has been offered an interview.
The probability of being offered an interview, given that she will be offered a position, P (EIH), is 1, since we are told all the employees have to attend an interview.

$$b = P(EIH) = 1$$

995 applicants (1,000 minus the 5 successful applicants) will not be offered the entry level position, and of these 50 will be interviewed.

$$c = P(EIH') = 50/995$$

$$P(HIE) = 0.005 \times 1 / [(0.005 \times 1 + 50/995 (1 - 0.005)] = 0.005 / 0.055 = 0.91 = 9.1\%$$

Alternative calculation: 5 applicants are interviewed and offered positions at Apple. Fifty applicants are interviewed and not offered positions. Probability of being offered a position if interviewed $= 5/55 = 0.91 = 9.1\%$.

1.6.2 Solution

There are three things to estimate. The first is the Bayesian prior probability ("**a**"). This is the prior robability of the hypothesis (he is guilty) being true. The second is the probability that the new evidence would have arisen if the hypothesis was true ("**b**"). It is the probability that the neighbour would have identified him if he did break into the warehouse. The third is the probability that the new evidence would have arisen if the hypothesis was false ("**c**"). In this case, you need to estimate the probability of the neighbour identifying him if he did *not* break into the warehouse.

You can now use Bayes' Theorem to estimate the (posterior) probability that your friend did commit the crime based on all available information.

$$(\text{Posterior}) \text{ probability} = \mathbf{ab} / [\mathbf{ab} + \mathbf{c} (1 - \mathbf{a})]$$

Consider what numbers you think reasonable for a, b, and c, with reasons, insert those numbers, and calculate the result.

For illustrative purposes, let us estimate a, b, and c to be 0.1, 0.8, and 0.2, respectively.

The calculation and the simple algebraic expression that we have identified in this setting is:

$$ab / \left[ab + c(1-a) \right]$$

a is the prior probability of the hypothesis (he is guilty) being true. This can also be represented by the notation P (H). In the example, $a = 0.1$.

b is the probability the neighbour identifies him conditional on the hypothesis being true, i.e. he is guilty. This can also be represented by the notation P (EIH), i.e. probability of E (the evidence) given the hypothesis is true, P (H). In the example, $b = 0.8$.

c is the probability the neighbour identifies him conditional on the hypothesis not being true, i.e. he is not guilty. This can also be represented by the notation P (EIH'), i.e. probability of E (the evidence) given the hypothesis is false, P (H'). In the example, $c = 0.2$.

In our example, $a = 0.1, b = 0.8, c = 0.2$.

Using Bayes' Theorem, the updated (posterior) probability that your friend is guilty is:

$$ab \ / \ [ab + c \ (1 - a)] = 0.08 \ / \ (0.08 + 0.18) = 0.08 \ / \ 0.26$$

Posterior probability = 0.308 = 30.8%

1.7.2 Solutions

1. There are four suspects, each of whom is equally likely to be guilty. So the prior probability of guilt for each is 12.5% (0.125).

2. There are now four suspects left, each of whom is equally likely to be guilty. So the new probability that each of the remaining suspects is guilty is 25% (0.25).

3. The new clue means that the probability of guilt we assign to Daisy rises to 40%. The probability we should assign to each of the other three suspects is 20% each (60% divided by 3).

4. The 20% probability of guilt we assigned to Tony should now be distributed to the remaining suspects. It should be in proportion to the prior probability of guilt before his elimination, in accordance with the principles of Bayesian updating. So the 20% should be distributed in a 4-2-2 proportion, i.e. 10% to Daisy, 5% to Delilah, and 5% to Joe. The posterior probability becomes 50% (Daisy) and 25% each to Delilah and Joe.

5. None of them, since no remaining suspect is more than 50% likely to have committed the crime.

1.8.1 Solution

The prior probability that the bus will arrive between 8 am and 8.10 am is 60% (100% minus 10% minus 30%).

When the bus does not arrive in that time slot, we should distribute the 60% to the other two possibilities in proportion to their prior probability of guilt before the elimination of the 8 am–8.10 am time slot, in accordance with the principles of Bayesian updating. This is in the ratio of 1:3, i.e. 15% to pre-8 am and 45% to post-8.10 am.

So, the updated probabilities are 25% that the bus arrived before 8 am (10% + 15%) and 75% that the bus will still arrive (30% + 45%).

1.9.2 Solution

An increase in **a** or **b**, or both, would increase the subjective probability of Desdemona's guilt in Othello's eyes, while an increase in **c** would reduce it.

A reduction in **a** or **b**, or both, would reduce the subjective probability of Desdemona's guilt in Othello's eyes, while a reduction in **c** would increase it.

1.10.1 Solution

Bayes' Theorem shows that the probability that a hypothesis is true given the evidence arising is not equal to the probability of the evidence arising given that the hypothesis is true. Put another way, P (HIE) only equals P (EIH) when P (H) = P (E), i.e. P (H) / P (E) = 1.

To conflate these two things is known as the Prosecutor's Fallacy.

In fact, P (HIE) = P (H) . P (EIH) / P (E)

$$P (E) = P (H) . P (EIH) + P (EIH') . P (H')$$

Thus, P (HIE) = P (H) . P (EIH) / [P (H) . P (EIH) + P (EIH') . P (H')].

where, P ((HIE) is the probability that the hypothesis is true, given the new evidence. This is the posterior (or updated) probability.

P (H) is the probability that the hypothesis is true before the new evidence. This is the prior probability (**a**).

P (EIH) is the probability of the evidence arising given that the hypothesis is true (**b**).

P (H') is the probability that the hypothesis is not true (**1 − a**).

P (EIH') is the probability of the evidence arising given that the hypothesis is not true (**c**).

Now, let's say that P (H) is the probability that a defendant in court is guilty of the crime as charged.

In this case, by representing P (HIE) as identical to P (EIH), when the former may be significantly smaller than the latter, jurors can be confused into convicting an innocent defendant.

2.1.1 Solutions

Intuitively, the answer is 1/2. It cannot be the card that is yellow on both faces, so it must either be the blue–blue card or the blue–yellow card.

In fact, this solution is wrong. The visible blue face could be one of three blue faces, not two. It could be the blue face of the blue–yellow card, or one of the other two blue faces on the blue–blue card. Of these three possibilities, in two the hidden face is blue; in one it is yellow. You can demonstrate this to yourself by labelling the blue faces on the two cards as Blue 1, Blue 2, and Blue 3. Only one of these three blue faces has a yellow face on the other side. So, the probability that the other face is blue (the blue–blue card) is 2/3 not 1/2.

The reason for the failure of common intuition may be that the number of possible cards (two) is being considered instead of the number of possible blue faces (three).

2.2.2 Solutions

Question 1. You should switch to the pyramid.
There was a 1 in 3 chance at the outset that your original choice contained the cheque. This does not change when the host opens the sphere, which he knows to be empty. There was a 2 in 3 chance that it was either the sphere or the pyramid before the host opened the sphere and by opening the sphere, which the host knows to be empty, that can be eliminated. So the chance it is the pyramid is now 2 in 3, compared to 1 in 3 for your original choice, the cube.

Question 2. It makes no difference whether you switch or not.
There was a 1 in 3 chance at the outset that your original choice contained the prize. There was a 2 in 3 chance that it was a different shape. The host does not know where the prize was secretly placed and is therefore giving you no new information. It is the same as asking you as the contestant to choose a container to open. If you randomly opened the cylinder, which might have contained the prize, this means there are now two shapes left (cube and cone). Each of these started with a 1 in 3 chance of containing the prize. The chance of each remaining shape being the winning one rises to a half in each case. So it makes no difference whether you switch or not.

Question 3. If the host knows where the car is located, you should let the host open the boxes, then switch. At the start, you have 1 chance in 10 of having selected the winning box. If you open five more boxes, the chance you will find the prize is 6/10, since you have opened six of the ten boxes. There was at the start a 9 in 10 chance at the outset that one of the other boxes contains the prize. If the host now opens eight of these boxes, all of which he knows do not contain the prize, the remaining box will have a 9/10 chance, therefore, of being the winning box. If the host doesn't know where the car is, however, no new information is being introduced when he (or she) opens a box. In this case, you should open the additional five boxes, as this will give you a 6/10 chance of finding the car. If the host doesn't know where the car is, the chance it is in either of the remaining two boxes is in both cases 1 in 2, which is a smaller chance. If you did choose this option, however, you would now be indifferent between sticking with your original box and switching.

2.3.1 Solution

When the prisoner expressly asked the warden to name one of the *other* men who will be executed, he is asking the warden not to name him, whether he is to be pardoned or not. So he obtains no new information about his fate when the warden names which one of the other two prisoners will be executed. New information is a requirement for changing the probability that something will happen or not. So the probability that he will be pardoned remains at 1/5.

2.4.1 Solutions

1. You should switch to either box 3 or box 4.

There was a 1 in 4 chance at the outset that your original choice, box 1, contained the prize. This does not change when the host opens the box which she knows to be empty. There was a 3 in 4 chance that it was either box 2 or box 3 or box 4. By opening box 2, which she knows to be empty, that can be eliminated. So the chance it is either box 3 or box 4 is now 3 in 4 in total (or 3/8 each), compared to 1 in 4 for your original choice, box 1.

2. It makes no difference whether you switch or not.

There was a 1 in 4 chance at the outset that your original choice, the black box, contained the prize. There was a 3 in 4 chance that it was either the white box or the grey box or the brown box, i.e. 1/4 for each individual box.

By randomly opening a box (the host doesn't know which box contains the prize), she is giving you no new information. It is the same as asking you to choose a box to open. So the chance of each of the remaining three boxes rises to 1/3 each. So it makes no difference whether you switch or not.

2.5.1 Solution

Bassanio should switch to the lead casket, unless he has additional information which could be usefully exploited, such as any clues he might be offered. His original choice of the gold casket has a 1 in 3 chance of containing the portrait. The silver and lead caskets combined have a 2 in 3 chance of containing the portrait. That remains the case after Portia opens the silver casket because she was forced to open that casket, given that she knew that the lead casket was the correct choice. So the lead casket has a 2 in 3 chance of being the correct casket. If Portia had no idea which casket contained the portrait, and randomly opened the silver casket, the remaining choices (the lead casket and the gold casket) would have an identical chance of being the correct casket, other things equal.

2.6.2 Solutions

1. The answer is 2/3. With two children, there are four possibilities of equal likelihood, i.e. daughter–daughter, son–son, daughter–son, and son–daughter. Each element of the binary pairs can be distinguished by any discriminating factor, such as age or height or location. We know that she has a daughter, so the son–son binary pair can be eliminated. This leaves the following possibilities: daughter–daughter (say older daughter and younger daughter); daughter–son (say older daughter and younger son); son–daughter (say older son and younger daughter). In two of these binary pairs, the other element is a son, whereas one binary pair contains a daughter. So the probability that her other child is a son is 2/3.

2. The answer is 1/2. With two children, there are four possibilities of equal likelihood, i.e. Girl–Girl, Girl–Boy, Boy–Girl, and Boy–Boy. Each element of the binary pairs can be distinguished by any discriminating factor, such as age or height or in this case location. We know that her daughter is with her in the park, so the Boy–Boy

binary pair can be eliminated. This leaves the following possibilities: Girl–Girl (girl in the park and girl at home) and Girl–Boy (girl in the park and boy at home). We can eliminate Boy–Girl (boy in the park and girl at home). That leaves two binary pairs, Girl–Girl and Girl–Boy. In one of these binary pairs, the other element is a girl, whereas the other binary pair contains a boy. So the probability that her other child is a boy is 1/2.

3. Both are equally likely. The possibilities are HH, TT, HT, and TH. All are equally likely, so HH (both Heads) and HT (Heads followed by Tails) are equally likely.

4. One Head and one Tail are twice as likely as two Heads. Note that HH, TT, HT, and TH are all equally likely, so HT and TH (one lands Heads and the other Tails) are twice as likely as HH (both heads).

5. A brother and a sister are twice as likely as two brothers: BB, SS, BS, and SB. All are equally likely, so BS and SB (older brother and younger sister OR older sister and younger brother) are twice as likely as BB (both brothers).

2.7.2 Solutions

1. Does it matter that the son has this unusual name? It does. If you know that there are two children, one in each house, and one is a boy, the chance the other child is a girl is 2 in 3. This is because there are three remaining options: Boy–Boy (one in each house); Boy–Girl (boy in the house on your left, girl in the house on your right); Girl–Boy (girl in the house on your left, boy in the house on your right). Two of these three options include a girl.

If you find out further that the neighbours have a child living in each of their houses, and that one of them (you don't know which) is a son called by the rare name of Vermont, you are left with (to all intents and purposes) just two options. The options are a son in one of the houses named Vermont and a son in the other house not named Vermont *or* a son named Vermont in one of the houses and a daughter in the other. You can almost certainly (but not entirely) eliminate the option that both neighbours have a son living at home called Vermont. So, there are now two realistic possibilities, and they are equally likely – that a boy called Vermont lives in one of the houses and a girl lives in the other house, or a boy not named Vermont lives in the other house. So the chance that the child living on the other house is a girl is very close to 1 in 2.

2. With the additional information that at least one of the children is a girl born in the first week of January, the sample space is now composed of Girl 1 born in the first week of January and a boy born in any week of the year (52 possibilities), Girl 1 born in the first week of January and Girl 2 born in any week of the year (52 possibilities), Girl 2 born in the first week of January and a boy born in any week of the year (52 possibilities), and Girl 2 born in the first week of January and Girl 1 born in any week of the year (52 possibilities). This makes a sample space of 208 possibilities. This is nearly correct, but note that Girl 1 born in the first week of the January and Girl 2 born in the first week of January is the same event as Girl 2 born in the first week of January and Girl 1 born in the first week of January. We shouldn't count the same event twice, so we remove this to give a sample space of 207. Of these 207 possibilities, 103 (104 minus the case where both girls are born in the first week of January) involve a girl born in the first week of January. So, the conditional probability that both children are girls is equal to 103/207.

3. With the additional information that at least one of the children is a boy born on a Saturday, the sample space is now composed of Boy 1 born on a Saturday and a girl born on any day of the week (seven possibilities), Boy 1 born on a Saturday and Boy 2 born on any day of the week (seven possibilities), Boy 2 born on a Saturday and a girl born on any day of the week (seven possibilities), and Boy 2 born on a Saturday and Boy 1 born on any day of the week (seven possibilities). This makes a sample space of 28 possibilities. This is nearly correct, but note that Boy 1 born on a Saturday and Boy 2 born on a Saturday is the same event as Boy 2 born on a Saturday and Boy 1 born on a Saturday. We shouldn't count the same event twice, so we remove this to give a sample space of 27. Of these 27 possibilities, 13 (14 minus the case where both boys are born on a Saturday) involve a boy born on a Saturday. So, the conditional probability that both children are boys, given that at least one is a boy born on a Saturday, is equal to 13/27.

Alternative Solution:
Using Bayes' Theorem, let A be the event that both children are boys. Let B be the event that at least one child is born on a Saturday.

Now, P (B I A) = 13/49. This is because there are 49 possible combinations of days of the week on which the two boys could be born on, of which 13 (see above) include a boy born on a Saturday.

P (A) = 1/4. The possibilities are boy–boy, girl–girl, boy–girl, and girl–boy. Of these four possibilities, one is boy–boy.

P (B) = 27/196. This is because there are 196 (14 × 14) possible combinations of family unit by day of the week and gender, and of these 169 (13 × 13) combinations that do not include a boy born on a Saturday: $196 - 169 = 27$. So the probability that at least one child is a boy born on a Saturday= $27/196$.

Now, $P(A \text{ I } B) = P(B \text{ I } A) \times P(A) / P(B)$

Substituting into Bayes' Theorem,

$$P(A \text{ I } B) = 13/49 \times \tfrac{1}{4} / (27/196) = 13/27$$

2.8.1 Solution

It doesn't matter. What you stand to gain by switching equals what you stand to lose.

If you could open the box first, it would only make a difference to your decision if you had additional information, such as the maximum or minimum amount that might be in a box. If, for example, the maximum amount allowed in a box is £150, and the first box you open has £100 inside it, there cannot be £200 in the other box, so there must be £50. So sticking to the original choice in this example would be the optimal strategy.

2.9.1 Solutions

1. Square root of $365 \times 1.2 = 22.93$. So the answer is 23 people.

2. Square root of $100 \times 1.6 = 16$.

3. $1 - (364/365)^{99} = 1 - 0.76 = 0.24$.

4. In a group of 24 people, there are, in fact, 276 pairs of people from which to choose. Therefore, a group of 24 people generates 276 chances, each of size 1/365, of having at least two people in the group sharing the same birthday.

This is derived as follows:

In a group of 24 people, there are, according to the standard formula, $^{24}C_2$ pairs of people (called 24 Choose 2) pairs of people.

Generally, the number of ways k things can be chosen from n is:

$$^n C_k = n!/(n-k)!\,k!$$

Here n! (n factorial) is $n \times n - 1 \times n - 2 \dots$ down to 1. Similarly, for k! Thus, $^{24}C_2 = 24! / 22!\,2! = 24 \times 23 / 2 = 276$.

These chances have some overlap: if A and B have a common birthday and A and C have a common birthday, then inevitably so do B and C. The probability that at least two people in the group of 24 do not share a birthday is:

$$(364/365)^{276} = 0.469$$

So the odds that at least two of the 24 people share the same birthday = 1 − 0.469 = 0.531 = 53.1%

5. Each person in the room has the same probability (1/365) of being born on any given day of the year. So the probability that the second person in the room has the same birthday as this is 1/365.

2.10.1 Solutions

1. The mean interval between trains is 30 minutes, so the average expected wait would seem to be 15 minutes if you arrive at a random time.

But it is three times as likely that you will arrive during the 45 minutes interval as during the 15 minutes interval. There is, therefore, three times the chance of waiting 22.5 minutes (halfway along the 45 minutes interval) as 7.5 minutes (halfway along the 15 minutes interval).

So your expected wait is 3 × 22.5 minutes plus 1 × 7.5 minutes, divided by 4. This equals 75 divided by 4 or 18.75 minutes (18 minutes, 45 seconds).

2. The answer will almost certainly exceed the average of all those who attend the health club that day. The reason is that it is more likely that when you turn up you will come across those who stay there for a long time than a short time.

3. Your best estimate, based on the single observation, is 40 seconds for the first car to arrive at the light. There is already a car in front of you that you estimate would expect to wait 40 seconds, and has been there for at least some time before you. So your expected wait time is less than 40 seconds.

2.11.2 Solutions

Hospital patients are not a random sample of the population but consist disproportionately of those who are older, who are frail, who are smokers, as well as suffering from pneumonia. For the sake of simplicity and illustration, assume that patients are admitted to hospital for one of two reasons, either for smoking-related illnesses or pneumonia. In this case, tests for pneumonia on those in hospital are likely to show its existence less among smokers than non-smokers, because the smokers are already hospitalised for smoking-related illnesses unrelated to pneumonia. The non-smokers would be there exclusively because they are suffering from pneumonia. More generally, if two factors influence being selected into a sample ("collide on selection", hence "collider bias"), they may seem associated when they are in fact independent.

2.12.1 Solutions

1. In three successive years, Justice recorded a better average than Jeter. Over the whole period, though, the batting average for Derek Jeter was 0.300 (385/1284), superior to David Justice, on 0.298 (312/1046).

So who is the better baseball player with the bat? There is no definitive answer to that.

2. X is more successful in the first trial: 80% compared to 78%.

   X is more successful in the second trial: 50% compared to 40%.

   Y is more successful overall: 80/105 = 76.2% compared to 100/140 (71.4%).

2.13.1 Solutions

1. Transfer 50 from B to A.

The arithmetic mean of B now rises from 70 to 75. The arithmetic mean of A rises from 25 to 30.
   Or, transfer 60 from B to A.
   The arithmetic mean of B now rises from 70 to 72.5. The arithmetic mean of A rises from 25 to 32.

2. Move the data point 100 from B to A.

The arithmetic mean of A now rises from 70 to 73.75. The arithmetic mean of B rises from 120 to 123.33.

3. The element which is moved has to lie between the arithmetic means of the two sets.

3.4.1 Solutions

1. According to the Doomsday Argument, what is our best estimate, in 2021, of how much longer the University of Bologna will continue to exist?

   Solution: 933 years. It was founded in 1088.

   A caveat is that the Doomsday Argument might indicate that humans will not survive as long as another 933 years.

2. What spread of years of existence is left with a 50% probability? What about with a 60% probability?

   50% probability: $1/3 \times 933$ years to 3×933 years = 311 years to 2,799 years.

   60% probability = $1/4 \times 933$ years to 4×933 years = 233 years to 3,732 years.

   Same caveat as above.

3. If a randomly selected giant tortoise is 20 years old, and we know nothing more about it, what is our best estimate of its future life expectancy?

   The Doomsday Argument does not apply to biological life expectancy. The average life expectancy of a giant tortoise is about 100 years, so our best estimate of its future life expectancy is not another 20 years, but closer to 80 years. If the 100 years is expected life expectancy at birth, the tortoise's remaining life expectancy will in fact be a bit more than 80 years, as it has already survived 20 years in which it might have already died.

3.5.1 Solutions

1. There are 64 available luxury apartments in a one-day sell-off, one of which you need to accept. You are offered details of each in turn and must choose to accept or decline within three minutes. If

you decline, you are offered the next in the list. If you accept, it is yours.

How many should you assess and reject before starting to consider which of the others accept?

Solution: $0.37 \times 64 = 23.68$. You should assess and reject 24 before considering which of the rest to accept.

What is the probability that you have selected the best apartment?

Solution: About 37%.

2. What if there was a 50% chance that your selection would turn out to be unavailable? How many should you now assess and reject before starting to consider which of the others to accept?

Solution: $0.25 \times 64 = 16$. You should assess and reject 16 before considering which of the rest to accept.

What is the probability now that you have selected the best apartment?
Solution: About 25%.

3. Using an established rule of thumb, how many apartments should you assess and reject before starting to consider which of the rest to accept, in order to maximise the chance of selecting a good option if not the very best?

Solution: Square root of $64 = 8$.

4. What is the probability that you would get the best apartment if you could choose randomly which to choose and obtain that?

Solution: $1/64 = 1.56\%$.

3.10.1 Solution

The top sequence is more likely to be a fake sequence. It contains no runs of at least four Heads or Tails in a row, which we would expect in a real experiment.

4.1.1 Solutions

1. The probability of throwing a double-6 in one throw is 1/36. The probability of *not* throwing a double-6 in one throw is 1 – 1/36 = 35/36. The probability of *not* throwing a double-6 in 21 throws is 35/36 to the power of 21 = 0.5534. So, the probability of throwing a double-6 in 21 throws of the dice = 1 – 0.5534 = 0.4466.

The edge against the Butch = 0.4466 – 0.5534 = 0.1068 = 10.68%.

2. P (you win) = 100 / (100 + 400) = 1/5 = 20%.

   P (opponent wins) = 400 / (100 + 400) = 4/5 = 80%

3. This is not correct. It is like saying that the chance of a Heads on one toss of a coin is 50% so the chance of two Heads is 100%. The actual probability of being shot down if the pilot flies 50 missions equals $1 - (0.98)^{50} = 0.64$, or 64%.

4. P (Tail ∪ even number) = P (Tail) + P (even number) – P (Tail ∩ even number) = 1/2 + 1/3 – (1/2 × 1/3) = 5/6 – 1/6 = 4/6 = 2/3.

5. The answer is 5/10. Although there are only nine marbles left in the bag, you have been given no information about the colour of the selected marble, so there remains a sample space of ten possibilities, of which five are yellow.

6. The probabilities are identical. Note that the odds of heads–heads and of tails–tails are irrelevant. Either way, the game continues. The probability of heads–tails is 99/100 × 1/100 = 99/10,000. The probability of tails–heads = 1/100 × 99/100 = 99/10,000. These are the relevant probabilities, and they are identical.

7. The simple intuitive solution is that, of the three plain chocolates, two were in the cube and one in the cone, so the conditional probability that the plain chocolate came from the cube is 2/3.

Using Bayes' Theorem, H is the hypothesis that the chocolate came from the cube and E is the event that the chocolate is plain.
So, P (H I E) = P (E I H) . P (H) / P (E) = 1 × 1/2 / 3/4 = 2/3.
Note that:

P (E I H) = 1, because all chocolates in the cube are plain.

P (H) = 1/2, because there are two containers, equally likely to be chosen.

P (E) = 3/4, because there are four chocolates in total, of which three are plain.

8. The planes being examined are those that went on a mission and survived to return to base. The bullet holes on these planes did not prevent them surviving. What about those that failed to return to base? Since few planes returned with bullet holes in the engine, a reasonable conclusion is that the engine should be prioritised for reinforcement. This is an example of what is known as survivorship bias.

4.2.2 Solutions

1. At 2-2, the prize should be shared evenly by the players.
2. At 3-2, the expected chance of player 1 winning when leading 3-2 = 50% × 1 + 50% × 50% = 50% + 25% = 75%. Expected chance of player 2 winning = 25%.

4.3.1 Solutions

Probability of winning exactly 5 games out of 8 =

$$^8C_5 = n!/(n-r)!\, r! \times 0.5^8 = 8!/3!\,5! \times 1/256 = 56/256 = 7/32 = 0.218$$

4.4.1 Solutions

Question 1.

a. 18/37 = 48.6%.
b. No.

Question 2. Since the edge is against you (60% to 40%), the fewer games you play the better. If you play one game, you have a 40% chance of winning the match. The probability of winning the match declines from there. The more games you play the less likely you are to win the match. It is the same principle as a recreational player facing a professional tennis player over one point or over a match. It is much more likely that the recreational player might win the single point than a full game, and the chance of winning an entire set is much less than that and so on.

Solution: If you play one game, your probability of winning is 0.4.

If you play three games, you win the match if you win two games or three games.

There is one way you could win three games. The probability of this equals $0.4 \times 0.4 \times 0.4 = 0.064$.

There are three ways you could win exactly two games out of three.

a. Win game 1, win game 2, lose game 3 = $0.4 \times 0.4 \times 0.6 = 0.096$.

b. Win game 1, lose game 2, win game 3 = $0.4 \times 0.6 \times 0.4 = 0.096$.

c. Lose game 1, win game 2, win game 3 = $0.6 \times 0.4 \times 0.4 = 0.096$.

These chances add up to: $3 \times 0.096 = 0.288$.

Total probability = $0.064 + 0.288 = 0.352$, i.e. 35.2%

We can also derive this using the binomial theorem.

Generally, in a fixed sequence of n Bernoulli trials, the probability of exactly r successes is:

$$^{n}C_r \times p^r (1-p)^{n-r}$$

This is the binomial distribution. Note that it requires that the probability of success on each trial be constant. It also requires only two possible outcomes.

So, what is the chance of winning exactly two games out of three played?

Probability of exactly two wins out of three = $^{3}C_2 \times (0.4)^2 \times (0.6)^1 = 3! / 2! (3-2)! \times 0.16 \times 0.6 = 3 \times 0.096 = 0.288$.

Probability of three wins out of three = $^{3}C_3 \times (0.4)^3 = 0.064$.

Probability of at least two wins out of three = $0.288 + 0.064 = 0.352$, i.e. 35.2%.

4.5.2 Solutions

Calculate the subjective odds (against) in this table assuming that f, the fixed fraction of losses undiscounted by the bettor, is a half.

1. a. $Q / (1 - Q) = 1/4 / 3/4 = 1/3$ (as $f = 1/2$ and $q = 1/2$)

   b. $Q / (1 - Q) = fq/(1 - fq) = 2/5 / 3/5 = 2/3$ (as $f = 1/2$ and $q = 4/5$)

2. A series of bets at shorter odds.

3. In principle, we might expect both to be the same or similar. It is the size of the odds, not whether it is a favourite or not, which is at the heart of the idea of the favourite-longshot bias. In practice, this is an empirical issue.
4. At least an above-average return.
5. Tennis, because there are only two outcomes. In football there are three (including the draw) and in horse racing usually rather more. This means that the odds about the favourite are likely to be shorter in tennis than in football and shorter still than in a typical horse race. In practice, this is an empirical issue.
6. Can knowledge of the Gambler's Fallacy be used in principle to improve your chances of winning the UK National Lottery main draw? No, because the draw is random.
7. Can knowledge of the Gambler's Fallacy be used in principle to improve the expected return to a £2 ticket on the UK National Lottery main draw? Yes, if some numbers are more popular than others. If more popular numbers come up, the expected payout to an individual winner will be less, because the prize pool will be shared among more winners.

4.6.1 Solutions

$P(r;\mu) = (e^{-\mu})\,(\mu^r)\,/\,r!$

$P(4;2) = (2.71828^{-2})\,(2^4)\,/\,4!$

$P(4;2) = (0.13534)\,(16)\,/\,24$

$P(4;2) = 0.09$

4.7.1 Solutions

When playing blackjack at the tables, with cards not shuffled back into the deck, will basic strategy generate a profit? Solution: Only if you are lucky.

Will card counting generate a profit? Solution: Yes, if you do it right and you are not unlucky or thrown out of the casino.

4.8.2 Solution

Probability of losing n fair coin tosses $= 1/2^n$.

Total loss with starting stake of x, with three losses of coin toss $= x + 2x + 4x = 7x$.

So martingale strategy suggests a bet of $7x + x = 8x$.

Loss after n losing rounds $= x + x^2 + \ldots + x^n$.

So martingale bet $= (x + x^2 + \ldots + x^n) + x = x^{n+1}$.

This strategy always wins a net x.

This holds so long as there is no finite stopping point at which the next martingale bet is not available (such as a maximum bet limit) or can't be afforded.

So, let us assume that everyone has some number of losses, which means that they don't have enough money to pay a stake large enough for the next round that it would cover the sum of the losses to that point. Call this run of losses n.

Probability of losing n times $= 1/x^n$

So, the player wins x with a probability of $(1 - 1/x^n)$.

Total losses after n losing bets $= (x + x^2 + \ldots + x^n) = (x^{n+1} - x)$

Expected gain is equal to the probability of not folding multiplied by the gain plus the probability of folding multiplied by the loss.

$$\text{Expectation} = (1 - 1/x^n) \cdot x - 1/x^n (x^{n+1} - x)$$
$$= x - x/x^n - x + x/x^n = 0$$

The intuitive explanation for the zero expectation is that the player wins a gain (x) with a very good probability $(1 - 1/x^n)$ but with a small probability $(1/x^n)$ makes a much greater loss $(x^{n+1} - x)$.

More generally, for an increment of x,

$$\text{Expectation} = (1 - 1/x^n) \cdot x - 1/x^n (x^{n+1} - x)$$
$$= x - x/x^n - x + x/x^n = 0$$

If the odds are tilted against the bettor, the expected gain in a finite series of coin tosses is less than zero, but the same principle applies.

4.9.1 Solutions

$$F = Pw - (Pl / W)$$

where

F = Kelly criterion fraction of capital to bet

W = Amount won per amount wagered (i.e. win size divided by loss size)

Pw = Probability of winning

Pl = Probability of losing

1. x = 70% − 30% = 40%
2. 60% − 40% = 20%
3. W = 2

So, F = 50% − 50% /2 = 50% − 25% = 25%.

4. W = 2

So, F = 60% − 40% / 2 = 60% − 20% = 40%.

5. When win probability = 1, as F = 1 − (0/W) = 1.

4.10.1 Solutions

In the case of ensemble averages, what eventually removes the randomness from the sample?
Solution: The size of the sample.

In the case of time averages, what removes randomness?
Solution: The time devoted to the process.

4.11.2 Solutions

A. Buy Call Option

1. Outcome A
Total Goals = 200; your gross profit = 200 − 175 = 25 × £10 = £250.
Your premium = 5.5 × £10 = £55; your net profit = £250 − £55 = £195.

2. Outcome B

Total Goals = 140; if you exercise the option, your gross profit = 140 − 175 = −35 × £10 = −£350.

In practice, you do not exercise the option, so your gross profit (gross loss) is zero.

Your loss = your premium = 5.5 × £10 = £55; your net profit = −£55.

3. Outcome C

Total Goals = 180. Your gross profit = 180 − 175 = 5 × £10 = £50.

Your premium = 5.5 × £10 = £55; therefore, your net profit = £50 − £55 = −£5.

B. Buy Put Option

1. Outcome A

Total Goals = 200; if you exercise the option, your gross profit = 175 − 200 = −25 × £10 = −£250.

In practice, you do not exercise the option, so gross profit (gross loss) is zero.

Your premium = 11.5 × £10 = £115.

Therefore, your net profit = −£115.

2. Outcome B

Total Goals = 140; your gross profit = 175 − 140 = 35 × £10 = £350.

Your premium = 11.5 × £10 = £115; your net profit = £350 − £115=£235.

3. Outcome C

Total Goals = 170; your gross profit = 175 − 170 = 5 × £10 = £50.

Your premium = 11.5 × £10 = £115; therefore, your net profit = £50 − £115 = −£65.

C. Sell Call Option

1. Outcome A

Total Goals = 200; your gross profit = 175 − 200 = −25 × £10 = −£250.

Your premium = 3.0 × £10 = £30; your net profit = £30 − £250 = −£220.

2. Outcome B

Total Goals = 140.

If the buyer exercises the option, your gross profit = 140 − 175 = 35 × £10 = £350. In practice, they do not exercise the option, so gross profit (gross loss) is zero.

Your premium = 3.0 × £10 = £30; therefore, your net profit = £30.

3. Outcome C

Total Goals = 180; your gross profit = 175 − 180 = −5 × £10 = −£50.

Your premium = 3.0 × £10 = £30; your net profit = £30 − £50 = −£20.

D. Sell Put Option

1. Outcome A

Total Goals = 200; if the buyer exercises this option, your gross profit = 200 – 175 = 25 × £10 = £250.

In practice, the buyer does not exercise the option, so gross profit (gross loss) is zero. Your premium = 8.5 × £10 = £85; net profit = £85.

2. Outcome B

Total Goals = 140; your gross profit = 140 – 175 = 35 × £10 = –£350.
Your premium = 8.5 × £10 = £85; therefore, your net profit = £85 – £350= –£265.

3. Outcome C

Total Goals = 170; your gross profit = 170 – 175 = –5 × £10 = –£50.
Your premium = 8.5 × £10 = £85; net profit = £85 – £50 = £35.

Summary Puzzle

World Cup – Total Goals
Based on a market of 166-170.

Strike	Option	Price Bid	Offer	Option	Price Bid	Offer
165	Call	6.5	9.5	Put	3.0	6.0

Stake = £10 per goal.

1. Assuming the market makes up at 190:

   What is your net profit if you BUY a CALL OPTION? Solution = 25 × £10 – 9.5 × £10 = £155.

   What is your net profit if you BUY a PUT OPTION? The option is not exercised. You pay the premium = £60. Net profit = –£60.

   What is your net profit if you SELL a CALL OPTION? Solution = 6.5 × £10 – 25 × £10 = –£185.

   What is your net profit if you SELL a PUT OPTION? Solution = 3 × £10 = £30. The option is not exercised.

2. Repeat if the market makes up at 130.

   What is your net profit if you BUY a CALL OPTION? The option is not exercised. Net loss = premium. Net profit = –£95.

   What is your net profit if you BUY a PUT OPTION? Solution = 35 × £10 – 6 × £10 = £290.

What is your net profit if you SELL a CALL OPTION? The option is not exercised. Net profit = 6.5 × £10 = £65.

What is your net profit if you SELL a PUT OPTION? Solution = 3 × £10 − 35 × £10 = −£320.

5.1.1 Solutions

Statement 1: All flamingos are pink.

Statement 2: If something is not pink, then it is not a flamingo.

Are statements 1 and 2 logically distinct or equivalent?
Solution: They are logically equivalent.
Is either statement true?
Solution: Both statements are untrue.

6.5.1 Solutions

Pope and Schweitzer demonstrate, using a database of millions of putts, that professional golfers are significantly more likely to make a putt for par than a putt for birdie, even when all other factors, such as distance to the pin and break, are allowed for. They find it is because golfers see par as the "reference" score, and so a missed par is viewed (subconsciously or otherwise) by golfers as a greater loss than a missed birdie. They perform differently in consequence. Equivalent birdie putts tend to come up slightly too short relative to par putts. They are also significantly more likely to miss the hole to the left or right.

6.6.1 Solutions

Is a tax on betting stakes a commodity tax or an ad valorem tax?
Solution: A commodity (or unit or specific) tax.

Is a tax on losses by bettors (operator gross profits) a commodity tax or an ad valorem tax?
Solution: Ad valorem tax.

7.1.1 Solutions

1. In the "Live or Die" scenario, there are two Nash equilibria (both drive on right side of the road or both drive on left side of the road) but no dominant strategy equilibrium. This is an example of the more general rule that not every Nash equilibrium is a dominant strategy equilibrium. Every dominant strategy equilibrium is, however, a Nash equilibrium, as there is no incentive for the parties to deviate from the equilibrium, e.g. the Prisoner's Dilemma.

2. Yes. Steal is the dominant strategy for each player. Steal–Steal is the dominant strategy equilibrium and also the Nash equilibrium.

7.2.1 Solution

The strategies are identical.

7.3.2 Solution

A probability of 2/3 call and 1/3 raise might be implemented by examining the second hand on a watch. If it is in the first 40 seconds of the current minute, call, otherwise raise.

8.1.1 Solution

Two cards. The card showing an even number (4) and the yellow card.

8.2.1 Solution

There is no missing pound. Starting with the original £30 paid at check-in, £3 was returned to the guests, decreasing their total payment to £27. The bell boy also pocketed £2, lowering the original £30 by a further £2, to £25, which was the final bill.

The error in the story is to add the £2 made by the bell boy to the £27 paid by the guests. Either the £3 refunded to the guests should be added to the £27 to total the original £30, or the £2 pocketed by the bell boy should be subtracted from the £27 to arrive at £25.

In summary, the £2 should have been subtracted from the £27 paid by the guests, not added, to give £25, which is the amount earned by the hotel from the room.

8.3.1 Solution

Draw a right-angled triangle, vertical length (a) and horizontal length (b) equal to 1.

Then, the length of the hypotenuse of the triangle, c, can be derived from the length of the adjacent (a) and opposite (b) sides, using Pythagoras' Theorem.

$a^2 + b^2 = c^2$

So, $1^2 + 1^2 = c^2$.

So $c^2 = 2$.

$c = \sqrt{2}$

This is a line of finite length, representing a number of infinite length.

8.5.1 Solutions

There is no straightforward solution to the question. There are three ways, however, of summing the series.

Method 1:

$1 - 1 + 1 - 1 + 1 - 1 + \ldots = (1 - 1) + (1 - 1) + (1 - 1) + \ldots = 0 + 0 + 0 + 0 + \ldots = 0$. The lamp is off?

Method 2:

$1 - 1 + 1 - 1 + 1 - 1 + \ldots = 1 - (1 + 1) - (1 + 1) - (1 + 1) - \ldots 1 - 0 - 0 - 0 - \ldots = 1$. The lamp is on?

Method 3:

Let S = 1 − 1 + 1 − 1 + 1 − 1.......

So, 1 − S = 1 − (1 − 1 + 1 − 1 + 1 − 1 ...) = 1 − 1 + 1 − 1 − 1 − 1... = S.

So, 1 − S = S.

So, 2S=1.

Therefore, S = 1/2.

So the series: 1 − 1 + 1 − 1 + 1 − 1 − ... equals 0.5. The lamp is neither *on* nor *off* but half on, half off? Or is it perhaps shining half as brightly as it was at the start?

You decide!

8.6.1 Solutions

1. Let S = 0.999......

   So, 10S = 9.99999......

   10S − S = 9S = 9

   Therefore, S = 1.

   But S = 0.999 ...

   Therefore, 0.999..... = 1.

   Link here: https://www.youtube.com/watch?p=PL492A573042CCB9A5&feature=plpp&v=G_gUE74YVos&app=desktop. YouTube presentation. By singingbanana, 28 April, 2011.

2. a. If you divide an even number by two, it produces a whole number without fractions. For example, 12 gives 6, 10 gives 5, and 8 gives 4. In contrast, odds numbers do not. For example, 11 divided by 2 gives 5.5. Divide 0 by 2, and you get 0, which is a whole number.

   b. An even number lies between two odds numbers. For example, 10 lies between 9 and 11. Zero lies between two odds numbers, −1 and 1.

   c. Adding an even number to an odd number produces an odd number. For example, adding 12 to 7 gives 19, an odd number. Adding zero to an odd number produces an odd number. For example, 0 plus 5 equals 5.

Conclusion: Zero is an even number.

Link here: https://spark.adobe.com/v/BtydmpUIwxQ Adobe Voice presentation. By Leighton Vaughan Williams.

Index

A

Abnormal returns, 140, 192–193, 198
Ad hoc hypothesis, 188
Ad valorem taxation, 219, 220
Adverse selection, 193–195
Algorithm for truth, 187
Ancestor simulations, 174
Annuitisation puzzle, 207
Arbitrage (arb), 166–167
Aristotle, 162, 241–242
Asymmetric information, 194, 196
"At-the-money" option, 164

B

Base rate fallacy, 15
Basic strategy, 151
Bayes and beetle, 19–20
Bayes and Bobby Smith problem, 27–30
Bayes and broken window, 31–33
Bayes and false positives problem,
    20–22
    examples, 22–23
    sensitivity, 23–24
    specificity, 23–24
    vaccine efficacy, 24–25
Bayes at theatre, 37–40
Bayes factor, 4–6
Bayesian bus problems, 36–37
Bayesian detective problem, 34–35
Bayesian "prior," 6, 10
Bayesian reasoning, 2, 3, 42, 96
Bayesian Taxi Problem, 13–17
Bayesian updating, 3, 42
Bayes in courtroom, 40–44
Bayes' Theorem, 1–4, 7
    derivation, 7–9
    intuitive presentation, 9
    proportional form, 4
    variables, 7
"Bean Jar" experiment, 210
Beat the Dealer, 150
Bell boy paradox, 238–239

Benford's Law, 115–118
Berkson's Bias, 82–83
Berkson's Paradox, 82–84
Bernoulli trials, 134
Bertrand Duopoly model, 226
Bertrand's Box Paradox, 47–48
Binomial distribution, 133, 136, 278
Birthday Problem, 75–79
Blackjack, 151
Boy–Girl Paradox, 62–68

C

Calendar effects, 198
Call option, 162–164
Calls, 162
Card counting, 150–151
Chevalier's Dice Problem, 123–128
Coherent superposition, 177
Collider bias, 47, 82, 84
Commodity taxation, 219
Copenhagen Interpretation, 171, 177
Copernican principle, 101, 102
Cosmological Constant, 179
Curious and classic market anomalies,
    197–201

D

Deadly Doors Problem, 59–60
"Deal or No Deal" game show, 205
Devil's Shooting Room Paradox, 153
Divergent series, 239, 241
Dollar auction, 228
Dominant strategy, 227, 233
Doomsday Argument, 100–102
Dunning–Kruger effect, 216

E

Edge, 156–157
Efficiency and inefficiency of markets,
    191–196
Efficient, 193

Efficient Market Hypothesis (EMH), 191, 192
El Clasico game, 232
EMH, *see* Efficient Market Hypothesis
Ensemble averaging, 159
Ensemble probability, 160
Entropy, 189–190
Equitable, 193
Equity Premium Puzzle, 206
Ergodic process, 159
Euler's number, 107
Exchange Paradox, 73
Expected Value Paradox, 158–161
Expert forecasting, 191, 216

F

Faking Randomness, 119–120
False negative (FN), 23
False positive (FP), 23
Favourite-longshot bias, 140–146
Fermat's Last Theorem, 111, 130
Filter, 183
Financial puzzles, 203–208
Fine-tuned universe puzzle, 179–185
Four Card Problem, 237–238
Fractional Kelly strategy, 157

G

"Gains from trade," 194
"Galton's ox" experiment, 209
Gambler's Fallacy, 142, 143
Gambler's Ruin problem, 127, 128
Game theory
    mixed strategies, 232–235
    Nash equilibrium, 223–228
    repeated game strategies, 230–231
Girl named Florida Problem, 68–72
Goalkeeper's strategy, 235
God's Coin Toss Problem, 97–100
"Golden Balls" game show, 227–229
Golden Ratio (1.62), 118
Good Judgment Project, 215

H

Halfers, 94, 95
Halloween Effect, 198

Halloween Indicator, 198
Hempel's paradox, 171–173
Henery hypothesis, 141
Henery odds transformation, 143, 145
Holiday effect, 198

I

Indices, 195
Information efficiency, 191–193
Information production function, 193
Initial public offering (IPO), 206
Inspection Paradox, 80–82
"In-the-money" option, 164
Inverse/Prosecutor's Fallacy, 20, 21, 28, 41
IPO, *see* Initial public offering

J

January effect, 198

K

Kelly criterion, 156, 157
Kelly strategy, 156–157, 161
Ketchup anomalies, 203–208
Keynesian Beauty Contest, 113–115

L

Landscape of reality, 183
Laplace's Rule of Succession, 6–7
Last Universal Common Ancestor (LUCA), 184
"The Law of Anomalous Numbers," 116
Law of One Price (LOOP) principle, 204
Level 1 rationality, 113
Level 2 rationality, 114
Level 3 rationality, 114
Likelihood Ratio, 24
Lindy effect, 101, 102
"Live or Die game," 225, 227
LOOP, *see* Law of One Price principle
Loss aversion, 205–207
LUCA, *see* Last Universal Common Ancestor

M

"Many gods" objection, 112
"Many-worlds" interpretation (MWI), 171, 178
Marilyn Vos Savant, 50, 63
Market anomalies, 198, 201
Market for Lemons, 193
Martingale betting system, 152–155
Mathemagic exercise, 239–241
Monty Hall Problem, 49–53
    derivation, 54–55
Multiple Comparisons Fallacy, 77
MWI, *see* "Many-worlds" interpretation

N

Narrow framing, 141, 207
Nash equilibrium, 223–228
Necktie Paradox, 74
Negative Likelihood Ratio, 24
Negative Predictive Value (NPV), 24
Neo-classical synthesis, 194
Net revenue, 220
Newcomb–Benford Law, 116
Newcomb's Paradox, 91–92
Newcomb's Problem, 91–92
Newton–Pepys Problem, 132–138
Nick Bostrom, 174–176
Noah's Law, 112
Non-participation puzzle, 207
NPV, *see* Negative Predictive Value

O

Observer selection effect, 110
Occam's Leprechaun, 188
Occam's Razor, 186–190
One-boxers, 92
Optimal Stopping Problem, 91, 104
"Out-of-the-money" option, 164
Overfitting, 186
Over-round (OR), 145

P

Pari-mutuel markets, 142
Pascal–Fermat "Problem of Points," 130–132

Pascal's Mugging Problem, 111, 112
Pascal's Triangle, 111, 131–132
Pascal's Wager, 111–113
Penalty-taker's strategy, 234–235
Poisson distribution, 148–150
Portia's Challenge, 61–62
Positive Likelihood Ratio, 24
Positive Predictive Value (PPV), 23
Possibility Theorem, 173
Posterior odds, 5
PPV, *see* Positive Predictive Value
Prediction markets, 211–213
Presumptuous Philosopher Problem, 99
Prior Indifference Fallacy, 38
Prior odds, 5
Prior probability, 42, 95
Prisoner's Dilemma, 226–228, 230
Prosecutor's Fallacy, 6, 16, 20–21, 28, 35, 41–42, 44, 265
Prospect theory, 141, 203–208
Put option, 162, 164–165

Q

"Quantum Suicide" thought experiment, 177
Quantum vacuum, 184, 185
Quantum world thought experiments, 176–179
Quasi-arbitrages (Quarbs), 167

R

Random walk, 192
Raven paradox, 171; *see also* Hempel's paradox
Reference class, 97
Reference weighting, 205
Risk-love, 142
Running count, 151

S

Sally Clark case, 40–43
Saving hypotheses, 188
Scale invariance, 117
Schelling points, 228
Schrödinger's Cat, 176
Screening, 196

Second Law of Thermodynamics, 189
Secretary Problem, 104
Self-Indication Assumption (SIA), 96, 98
Self-Sampling Assumption (SSA), 96, 97
Semi-strong form information, 192, 193
Sensitivity, 23–24
Shakespeare, William, 37, 39, 61, 103
SIA, *see* Self-Indication Assumption
Signalling, 195–196
Simpson's Paradox, 85–87
Simulated world question, 174–175
Sleeping Beauty problem, 93–96
Slower lane, spending time in, 110–111
Solomonoff Induction, 186, 188
Specificity, 23–24
Spread betting, 165
SSA, *see* Self-Sampling Assumption
Stop looking and start choosing,
    103–108
St. Petersburg Paradox, 153
Strike price, 163
Strong form information, 192
Super Bowl Indicator, 198–200
Superforecasting, 215–218
Symmetry/asymmetry paradox, 180

T

Target Sum, staking strategies, 138–139
Taxation anomalies, 219–220
Technologically mature civilisation, 174
Thales of Miletus, 162
Thirders, 94, 95
Thomson's Lamp thought experiment, 242

Three Prisoners Problem, 56–58
Time averaging, 159
Time probability, 159
"Tit for Tat" strategy, 231
"Total Goals in the World Cup,"
    165–166
True Negative (TN), 23
True positive (TP), 23
Tulip bulb mania, 162
Two-boxers, 92
Two Envelopes Problem, 73–75

U

Uncle points, 160

V

Vaccine Efficacy, 24–25
Vacuum state, 185

W

Wason selection task, *see* Four Card
    Problem
Wave function, 177, 178
Weak form information, 192–193
Weekend effect, 198
Will Rogers Phenomenon, 88–89
Wisdom of crowds, 191, 209–213

Z

Zeno's paradox, 241–242

Printed in the United States
by Baker & Taylor Publisher Services